A TO Z
OF
CHEMISTS

NOTABLE SCIENTISTS

A to Z
OF
CHEMISTS

ELIZABETH H. OAKES

Facts On File, Inc.

A TO Z OF CHEMISTS

Notable Scientists

Copyright © 2002 by Facts On File

Facts On File, Inc.
132 West 31st Street
New York NY 10001

Library of Congress Cataloging-in-Publication Data

Oakes, Elizabeth, 1964-
    A to Z of chemists / Elizabeth Oakes.
      p. cm.
    Includes bibliographical reference and index.
    ISBN 0-8160-4579-8
    1. Chemists—Biography. I. Title.
    QD21.034 2002
    540′.92′2′—dc21
    [B]   2002068685

Text design by Joan M. Toro
Cover design by Cathy Rincon
Chronology by Dale Williams

Printed in the United States of America

VB Hermitage 10 9 8 7 6 5 4 3 2 1

This book is printed on acid-free paper.

CONTENTS

List of Entries   vii

Acknowledgments   ix

Introduction   xi

Entries A to Z   1

Entries by Country of Birth   245

Entries by Country of Major Scientific Activity   249

Entries by Year of Birth   253

Chronology   255

Bibliography   259

Index   263

LIST OF ENTRIES

Alder, Kurt
Ampère, André-Marie
Anfinsen, Christian Boehmer
Arrhenius, Svante August
Aston, Francis William
Avogadro, Lorenzo Romano
    Amedeo Carlo
Baekeland, Leo Hendrik
Baeyer, Adolf von
Barton, Derek H. R.
Becquerel, Antoine-Henri
Berg, Paul
Bergius, Friedrich
Berzelius, Jöns Jakob
Bishop, Hazel Gladys
Bosch, Carl
Brady, St. Elmo
Brønsted, J. N.
Buchner, Eduard
Bunsen, Robert Wilhelm
Butenandt, Adolf
Calvin, Melvin
Carothers, Wallace Hume
Carr, Emma Perry
Carver, George Washington
Cavendish, Henry
Clark, Josiah Latimer
Cori, Gerty Theresa Radnitz
Cornforth, Sir John Warcup
Crick, Francis Harry Compton
Crutzen, Paul J.

Curie, Marie
Dalton, John
Daniell, John Frederic
Davy, Sir Humphry
Debye, Peter
Diels, Otto
Eigen, Manfred
Elion, Gertrude Belle
Faraday, Michael
Fischer, Emil
Fischer, Ernst Otto
Fischer, Hans
Flory, Paul
Franklin, Rosalind
Friedel, Charles
Fukui, Kenichi
Gay-Lussac, Joseph-Louis
Gilbert, Walter
Good, Mary Lowe
Graham, Thomas
Grignard, François-Auguste-
    Victor
Haber, Fritz
Hahn, Otto
Harden, Arthur
Hassel, Odd
Haworth, Sir Walter
Herzberg, Gerhard
Heyrovský, Jaroslav
Hill, Henry Aaron
Hinshelwood, Sir Cyril

Hodgkin, Dorothy Crowfoot
Hoobler, Icie Gertrude Macy
Hückel, Erich
Ingold, Sir Christopher
Joliot-Curie, Frédéric
Joliot-Curie, Irène
Jones, Mary Ellen
Just, Ernest
Karrer, Paul
Kendrew, Sir John Cowdery
King, Reatha Clark
Klug, Sir Aaron
Kuhn, Richard
Langmuir, Irving
Lavoisier, Antoine-Laurent
Le Beau, Désirée
Lee, Yuan Tseh
Lehn, Jean-Marie
Leloir, Luis Federico
Lewis, Gilbert Newton
Libby, Willard Frank
Maria the Jewess
Martin, Archer John Porter
Massie, Samuel Proctor
McMillan, Edwin M.
Mendeleyev, Dmitri Ivanovich
Moissan, Ferdinand-
    Frédéric-Henri
Molina, Mario
Moore, Stanford
Mulliken, Robert S.

Natta, Giulio
Nernst, Walther Hermann
Newlands, John Alexander
    Reina
Nobel, Alfred
Norrish, Ronald G. W.
Northrop, John Howard
Nyholm, Sir Ronald Sydney
Ochoa, Severo
Onsager, Lars
Osborn, Mary J.
Ostwald, Wilhelm
Pasteur, Louis
Pauling, Linus Carl
Pennington, Mary Engle
Perutz, Max Ferdinand
Polanyi, Michael
Porter, Sir George
Pregl, Fritz
Prelog, Vladimir
Priestley, Joseph
Prigogine, Ilya

Proust, Joseph-Louis
Ramart-Lucas, Pauline
Ramsay, Sir William
Richards, Ellen Henrietta
    Swallow
Richards, Theodore William
Robinson, Sir Robert
Rowland, Frank Sherwood
Rutherford, Lord Ernest
Ružička, Leopold
Sabatier, Paul
Sanger, Frederick
Saruhashi, Katsuko
Seaborg, Glenn Theodore
Semenov, Nikolai
Simon, Dorothy Martin
Singer, Maxine
Soddy, Frederick
Solomon, Susan
Stanley, Wendell Meredith
Staudinger, Hermann
Stein, William H.

Sumner, James Batcheller
Svedberg, Theodor
Synge, Richard
Telkes, Maria
Tiselius, Arne Wilhelm
    Kaurin
Todd, Baron Alexander
Urey, Harold
Virtanen, Artturi Ilmari
Wallach, Otto
Werner, Alfred
Wieland, Heinrich Otto
Wilkinson, Sir Geoffrey
Willstätter, Richard Martin
Windaus, Adolf Otto
    Reinhold
Wittig, Georg
Wöhler, Friedrich
Woodward, Robert Burns
Wrinch, Dorothy Maud
Ziegler, Karl
Zsigmondy, Richard Adolf

ACKNOWLEDGMENTS

Bill Baue's work on this project, as the head writer and researcher, deserves recognition and praise. He is thorough and always interested in portraying the person behind the scientist and the everyday details of human experience that reflect meaningfully on the extraordinary scientific achievements represented in this book. For assistance with photographs, I thank those scientists who graciously responded to my requests and the staff at many libraries and archives who helped. I would also like to express my gratitude to the University of Montana Mansfield Library, where much of the research for this book was completed, and to the authors of the many science reference books I consulted.

Finally, my sincerest thanks go to Frank K. Darmstadt, my editor at Facts On File, for his lightning-fast e-mail responses and overall support throughout the project, and the rest of the staff for their invaluable support.

INTRODUCTION

Here are the stories of more than 150 chemists, men and women from different parts of the world, who have contributed significantly to the field since the first century. This includes Jöns Jakob Berzelius, who instituted the system of combining abbreviations of elements with numerical subscripts, leaving his stamp forever on the field, and Gertrude Belle Elion, who developed drugs to cure diseases as diverse as leukemia, gout, herpes, malaria, and arthritis. More famous names, such as Louis Pasteur and Marie Curie, are also among this book's entries. While this encyclopedia brings together an array of well-known and lesser-known chemists, providing the basic biographical details of their life, the focus is on their work as chemists, with their scientific achievements presented in everyday language that makes even the most complex concepts accessible.

THE SCIENTISTS

A to Z of Chemists comprises some well-known scientific greats of history, as well as contemporary scientists whose work is just verging on greatness. Among these are minority scientists who have often been excluded from books such as this. Here, a majority of the 150 scientists were or are, first and foremost, chemists, but there is a handful of physicists, biologists, and others who contributed or contribute significantly to chemistry as well.

To compile the entrant list, I relied largely on the judgment of other scientists, consulting established reference works, such as the *Dictionary of Scientific Biography*, chemistry periodicals, awards lists, and publications from chemistry organizations and associations. Despite this process, I cannot claim to present the "most important" historical and contemporary figures. Time constraints and space limitations prevent the inclusion of many deserving chemists.

THE ENTRIES

Entries are arranged alphabetically by surname, also taking into consideration the name by which the chemist is most commonly known. The typical entry provides the following information:

Entry head: Name, birth/death dates, nationality, and field of specialization.

Essay: Essays range in length from 750 to 1,200 words, with most averaging around 900. Each contains basic biographical information—date and place of birth, family information, educational background, positions held, prizes awarded, etc.—but the greatest amount of attention is given to the scientist's work. Names in small caps within the biography provide easy reference to other scientists represented in the book.

In addition to the alphabetical list of scientists, readers searching for names of individuals from specific countries can consult the list by nationality in the back. In addition, they are also listed by scientific activity and by year of birth. Finally, the Chronology lists entrants by their birth and death dates.

A

Alder, Kurt
(1902–1958)
German
Organic Chemist

Kurt Alder's name is inextricably linked with that of OTTO DIELS, his mentor and collaborator. With Diels, Alder discovered one of the most ubiquitous reactions in the natural world, what is now known as the Diels-Alder reaction. Throughout his short-lived career (he died at the age of 55), he applied this reaction to different combinations, yielding practical results (such as his invention of a synthetic rubber), and other scientists utilized the Diels-Alder reaction in subsequent discoveries, such as the synthesis of morphine. Alder shared the 1950 Nobel Prize in chemistry with his compatriot, Diels.

Kurt Alder was born on July 10, 1902, in Königshütte, Germany. His father, Joseph Alder, was a schoolteacher in Kattowitz, in the Upper Silesia region. Poland usurped this region after World War I, prompting the Alder family to move to Berlin to retain their German citizenship. Alder attended the Oberrealschule in Berlin, then moved on to the University of Berlin in 1922 to study chemistry. He transferred to the Christian Albrecht University (now the University of Kiel), where he conducted his doctoral research on azo carboxylic ester under Otto Diels. He submitted his dissertation, "On the Causes of the Azoester Reaction," in 1926 to earn his doctorate in chemistry.

After graduation, Alder stayed on at the university, working as an assistant in Diels's laboratory. Alder collaborated with his mentor on the famous experiments that discovered the Diels-Alder reaction. The process conjoins a dienophile, or double-bonded molecule, with a conjugated diene, which is a molecule containing two adjacent double bonds. Alder and Diels reported on the reaction between the dienophile of acrolein and the diene of butadiene in their 1928 publication of their results.

In 1930 the University of Kiel promoted Alder to a lectureship in organic chemistry and in 1934 promoted him again, this time to the title of extraordinary professor. Alder left academia for industry in 1936, when he took up the directorship of scientific research at the Bayer Werke laboratory at I. G. Farbenindustrie in Leverkusen. The following year he elucidated the Alder-Stein rules (in collaboration with G. Stein), predicting the stereochemical sensitivity of diene reactions.

Also during his tenure in the industry, he reversed the diene reaction to investigate the dissociation of components (as opposed to the adduction that he had been examining); he discovered in the process that five-carbon cyclic dienes are much less stable than six-carbon dienes. In the late 1930s, he elicited the Diels-Alder reaction between butadiene and styrene (a dienophile) in the presence of peroxides to form a synthetic rubber called Buna S, which became an important substitute for natural rubber during World War II when resources became scarce.

In 1940 Alder returned to the academy when Cologne University appointed him to its chair for experimental chemistry and chemical technology. While there, Alder discovered yet another form of diene synthesis, what he called a "substituting addition," to identify a new type of reaction—a concerted or *ene*, reaction. Alder remained at Cologne throughout the 1940s, serving as dean of its Department of Philosophy from 1949 through 1950. Alder and Diels shared the 1950 Nobel Prize in chemistry for their joint role in identifying the chemical reaction named after them. The award recognized the significance of their discovery as a process that has been the basis of ongoing scientific advancement in the understanding of the natural world. M. Gates, for example, utilized the reaction in his synthesis of morphine, the pharmacological foundation of poppy-based narcotics such as opium and heroin.

Besides the Nobel Prize, Alder also received the Emil Fischer Memorial Medal from the Association of German Chemists in 1938, as well as induction into the Deutsche Akademie der Naturforscher Leopoldina in Halle. In 1950 the University of Cologne conferred on him the status of honorary Doctor of Medicine, and four years later the University of Salamanca, in Spain, awarded him with an honorary doctorate. In 1955 Alder joined 17 other Nobel laureates in using their collective moral leverage to effect world peace by urging nations to denounce war. He died in Cologne on June 20, 1958, only 55 years old, after suffering from declining health.

Ampère, André-Marie
(1775–1836)
French
Chemist, Physicist, Mathematician, Philosopher

André-marie Ampère is considered one of the founders of electromagnetic theory. His name is most often evoked in measuring electrical current in the unit of amps, a shortening of his name introduced after his death by Lord Kelvin. Experimentally, Ampère is remembered for his demonstration that two parallel wires conducting electricity attract each other when the flow heads in the same direction but repel each other when the flow of electricity heads in opposite directions.

The *amp*, the unit for measuring electrical current, was named for André-Marie Ampère, one of the founders of electromagnetic theory. (*AIP Emilio Segré Visual Archives*)

Ampère was born on January 22, 1775, in Polémieux, near Lyons, France. His father was a wealthy merchant and civic official who had his son tutored privately, although Ampère was largely an autodidact: By the age of 13, he had written a treatise on conic sections. In 1793, in the midst of the French Revolution, the Republican army guillotined his father, eliciting a nervous breakdown in Ampère. He recovered well enough to start teaching mathematics at a Lyons school in 1796. He married in 1799 and the following year saw the birth of his son, Jean-Jacques-Antoine, who later became an important historian and philologist. His wife, however, died after only five years of marriage, further destabilizing Ampère emotionally.

In 1802 Ampère's monograph on probability earned him renown. Entitled *Considerations on the Mathematical Theory of Games*, it correlated the odds of winning to the bankroll that backs a gambler. Before this publication, the École Centrale in Bourg had appointed him professor of physics and chemistry; after its publication, the lycée in Lyons hired him away as a professor of mathematics. In 1805 he moved to Paris, where the École Polytechnique hired him as an assistant lecturer in mathematical analysis, then in 1808 Napoléon appointed him as the first inspector general of France's university system, a position he retained for the rest of his life. He continued to teach through the remainder of his career as well; by 1809 he had been promoted to a full professorship of mathematics.

Over the next decade, Ampère studied and conducted research in a wide range of fields: psychology, philosophy, physics, and chemistry. In 1814, for example, he hypothesized the same theory on the molecular makeup of gases as Amedeo Avogadro had three years earlier (although Avogadro's hypothesis did not receive validation in his own lifetime). Ampère's professional appointments also reflected his diversity of expertise: In 1819 he taught philosophy at the University of Paris, the following year the university appointed him to an assistant professorship in astronomy, and in 1824 the Collège de France appointed him to the chair in experimental physics.

On September 11, 1820, Ampère attended a séance at the Académie des Sciences during which Dominique Arago demonstrated Hans Christian Oersted's experiment whereby a current of electricity deflected a compass needle, suggesting a magnetic force in traveling electricity. At the next séance, Ampère demonstrated that electrical currents traveling in the same direction through parallel wires attracted each other but repelled each other when traveling in opposite directions. The history of science marks this event as the genesis of the field of electromagnetics (although Ampère called it "electrodynamics" at the time, to emphasize the motion of the electricity and to distinguish it from electrostatics, or electrical forces in stasis).

Ampère expounded his theories of electromagnetism in his 1822 text *Collection of Observations on Electrodynamics* and more completely in his 1827 text *Notes on the Mathematical Theory of Electrodynamic Phenomena Deduced Solely from Experiment*. In these he theorized and demonstrated that a solenoid (his name for a coiled wire with electricity flowing through it) acts as a bar magnet, proposed that the degree of deflection of a compass needle by an electrical current could measure the strength of that current (an idea that resulted later in the development of the galvanometer), and announced what came to be known as Ampère's law, a mathematical calculation of magnetic force based on the separation, orientation, and magnitude of current emanating from two parallel current-carrying conductors. Ampère equated electricity in motion with magnetism; he applied this idea at the molecular level, suggesting that molecules are surrounded by an electrical current because of the perpetual motion of what are now known as *electrons*. In this sense, he presaged the electron shell model.

Ampère died of pneumonia on June 10, 1836, while conducting his duties as inspector general. His headstone epitaph, "Happy at last," bears testament to the grievous life he endured with the premature deaths of his father and wife. About a half century after his death, Lord Kelvin honored him by dubbing the unit of electrical current the *ampere*, or *amp*.

⊠ **Anfinsen, Christian Boehmer**
(1916–1995)
American
Biochemist

Christian Anfinsen shared the 1972 Nobel Prize in chemistry with STANFORD MOORE and WILLIAM H. STEIN for their work linking the structure of proteins to their biological functioning. Anfinsen focused his research on the protein enzyme ribonuclease, discovering how it folds into a three-dimensional structure that determines its function.

Christian Boehmer Anfinsen was born on March 26, 1916, in Monessen, a town on the outskirts of Pittsburgh, Pennsylvania. His mother, Sophie Rasmussen, was of Norwegian descent, as was his father, Christian Anfinsen, who was an engineer by trade. Anfinsen attended Swarthmore College, where he earned his bachelor of arts degree in 1937. He then pursued graduate study at the University of Pennsylvania, where he received his master's of science in organic chemistry in 1939.

Anfinsen then traveled to Copenhagen, Denmark, where he served for a year as a visiting investigator through an American Scandinavian Foundation fellowship. He returned to the United States to enroll in the doctoral program in biochemistry at Harvard University, writing his dissertation on methods to identify eye retina enzymes and earning his doctorate in 1943.

Anfinsen remained in Cambridge, Massachusetts, working as an instructor in biochemistry at the Harvard Medical School until he was conscripted into the war effort, in which he

served in the United States Office of Scientific Research and Development from 1944 through 1946. For the next two years, 1947 to 1948, he traveled to Sweden as an American Cancer Society senior fellow working on flavoproteins under Hugo Theorellin at the biochemical division of the Medical Nobel Institute. In the meanwhile, Harvard University had promoted him to an assistant professorship in 1945, and upon his return in 1948, the university again promoted him to an associate professorship.

In 1950 the National Institutes of Health (NIH) appointed Anfinsen director of its Laboratory of Cellular Physiology at the National Heart Institute. Over the next dozen years he conducted the ribonuclease research that earned him renown. (He spent one of those years, 1954, in Carlsberg, Denmark, conducting research under Kaj Linderstrøm-Lang with the funds granted him by the Rockefeller Foundation as part of its Public Service Award.) He obtained ribonuclease, or Rnase, an enzyme containing 124 amino acids folded into one specific configuration to elicit the breakdown of ribonucleic acid, from bovine pancreas; in addition, he obtained an extracellular nuclease from *Staphyloccus aureus*. He exposed these enzymes to high temperatures and chemicals such as urea, which acted to denature the enzymes from their three-dimensional structure. Anfinsen observed that these denatured enzymes lost certain functions but regained those functions upon renaturation, or return to their natural three-dimensional configuration. In this way, Anfinsen confirmed the interrelationship between structure and function. In 1959 he reported some of his findings in *The Molecular Basis of Evolution*.

In 1962 Anfinsen returned to Harvard for a year as a professor of biological chemistry, but the next year the NIH enticed him back to head the Laboratory of Chemical Biology at the National Institutes of Arthritis, Metabolic and Digestive Diseases. That year he reported his thermodynamic hypothesis using the action of cysteine as his focal point, since this amino acid controlled the catalytic capabilities of the enzyme. Near the end of his two decades back at the NIH, his marriage to Florence Bernice Keneger, with whom he had had three children, ended in divorce in 1978. The following year he married Libby Esther Schulman Ely. From 1981 to 1982 he spent a year as visiting professor of biochemistry at the Weizmann Institute of Science in Israel. Upon his return, Johns Hopkins University hired him as a professor of biology, a position he retained for the remainder of his career.

Anfinsen continued to be active scientifically, conducting research on geological deposits of bacteria that could prove instrumental in deactivating toxins such as chemical weapons. As a Nobel laureate, he also exerted his influence politically, convincing the United States Congress in 1988 to deny the Department of Defense's request for $300 million in funding for biological weapons research. On May 14, 1995, at the age of 79, Anfinsen suffered a heart attack at his Pikesville home in the Baltimore suburbs. He was pronounced dead that day at Northwest Hospital Center in Randallstown, Maryland.

One of the founders of physical chemistry, Svante August Arrhenius *(E. F. Smith Collection, Rare Book & Manuscript Library, University of Pennsylvania)*

⊠ **Arrhenius, Svante August**
(1859–1927)
Swedish
Physical Chemist

Svante Arrhenius almost failed his 1884 doctoral examination on chemical dissociation in solution, the very same topic that earned him the Nobel Prize two decades later in 1903. As a founding member of the Ionists, proponents of the idea that water suspends negatively and positively charged ions in a chemical solution, Arrhenius walked the line between physics and chemistry and thereby helped establish the field of physical chemistry.

Svante August Arrhenius was born on February 19, 1859, in Vik, a town in the Kalmar

district of Sweden. His mother was Carolina Thunberg, and his father, Svante Gustav Arrhenius, surveyed land and oversaw the local castle, set on Lake Mälar near Uppsala, before holding a position at the University of Uppsala. His uncle, Johan Arrhenius, was a famous botanist who served as secretary of the Swedish Agricultural Academy.

Arrhenius graduated from the Cathedral School in Uppsala in 1876 and proceeded to the University of Uppsala, where he studied mathematics, physics, and chemistry, earning his bachelor's degree in 1878. He continued at the university for three years of doctoral study, but neither inorganic chemist Per Theodor Cleve or spectroscopist Tobias Thalén (the only potential thesis advisers at the university) expressed faith in his dissertation topic: the electrical conductivity of solutions. Rather than abandon his intuition, Arrhenius arranged to conduct his research in absentia at the Physical Institute of the Swedish Academy of Sciences under physicist Eric Edlund.

In his dissertation, Arrhenius sought to solve the chemical mystery of why salt solution conducts electricity when neither salt nor water alone does; he theorized that solids fracture into charged particles, or *ions*, when dissolved. This radical notion earned him the derision—or perhaps the envy—of Cleve and Thalén, who granted him the lowest possible passing grade (fourth-class pass for the dissertation, third-class pass for the defense). This blackballing disqualified him for a docentship at the university level, effectively prohibiting his utilizing his doctorate.

Undeterred, Arrhenius sent copies of his dissertation, which he had written in French but had translated into German, to prominent chemists who might understand his work; the old guard joined Cleve and Thalén in dismissing it, while younger chemists received it with enthusiasm. Thirty-one-year-old Wilhelm Ostwald traveled to Uppsala in August 1884 to offer Arrhenius a position at Riga Polytechnicum, a Latvian university where he was a chemistry professor. Although Arrhenius had to decline due to his father's illness, he used the offer to secure a docentship in physical chemistry at the University of Uppsala in November of that year and later visited Ostwald in Riga in February 1886.

Arrhenius's influence continued to spread: In 1886 Oliver Lodge published a summary and review of the dissertation in English, and over the next several years, thanks to an 1886 grant from the Swedish Academy of Sciences, Arrhenius conducted postdoctoral travels to visit the laboratories of Friedrich Kohlrausch and

Walther Nernst in Würzburg, Germany; of Ludwig Boltzmann in Graz, Austria; and of Jacobus van't Hoff in Amsterdam, the Netherlands. This group formed the nucleus of the Ionists, who believed in Arrhenius's radical notion of ions, a theory that is now accepted as chemical fact.

In 1891 Arrhenius turned down an offer from the University of Giessen (Germany) to accept a faculty appointment at the Technical University in Stockholm. Three years later he married Sofia Rudbeck, his student and assistant, and together the couple had one son, Olev Wilhelm, before divorcing in 1896. The previous year, in 1895, the university promoted him to a professorship in physics. After his tenure as rector of the Technical University from 1896 to 1902, Arrhenius received the 1903 Nobel Prize, though the committee was confounded whether to award him the prize in physics or chemistry, as his work trod the boundary between the two disciplines.

Arrhenius published a series of texts based on lectures: *Text-Book of Electrochemistry* (1902) was made up of his 1897 lectures at his home institution; *Theories of Chemistry* (1907) was composed of his 1904 lectures at the University of California at Berkeley; *Theories of Solutions* (1912) included his Silliman lectures at Yale University; and *Quantitative Laws of Biological Chemistry* (1915) consisted of his 1914 Tyndall lectures at the Royal Institution in London.

In 1905 the Swedish Academy of Sciences in Stockholm established the Nobel Institute of Physical Chemistry expressly to appoint Arrhenius as its director. That year he also married his second wife, Maria Johansson, who bore him three children—Ester, Anna-Lise, and Sven. Four years later, in 1909, the academy completed building laboratories for Arrhenius and the institute.

Arrhenius's interests ranged far and wide. In 1896 he published a paper ("On the Influence of Carbonic Acid in the Air upon the Temperature of the Ground") that foretold of the "greenhouse effect" that was first confirmed in the late 20th century. In 1908 he published *Worlds in the Making,* wherein he posited the notion that life travels from planet to planet by spores. He also theorized that the second law of thermodynamics, which predicts the end of the universe due to ever-increasing entropy, might be counterbalanced by galaxies that reverse the law (where entropy decreases).

In addition to the Nobel Prize, Arrhenius also won the 1902 Davy Medal of the Royal Society, the first Willard Gibbs Medal of the Chicago section of the American Chemical Society in 1911, and the 1914 Faraday Medal from the British Chemical Society. He remained director of the Nobel Institute of Physical Chemistry until his death on October 2, 1927, in Stockholm.

⊠   **Aston, Francis William**
(1877–1945)
English
Chemist, Physicist

Francis Aston was an exacting experimentalist who designed and built his own mass spectrometers to identify the precise masses of chemical elements. In the process he confirmed the existence of isotopes in stable elements; it was previously believed that only radioactive elements had isotopes, or atoms of the same element with a different number of neutrons in the nucleus. Aston received the Nobel Prize in chemistry in 1922 in recognition of the significance of this discovery.

Francis William Aston was born on September 1, 1877, in Harbone, near Birmingham, England. He was the third of seven children born to Fanny Charlotte Hollis, the daughter of a gun maker, and William Aston, a farmer and metal merchant. He attended the Harbone Vicarage School as a child, then graduated at the top of his class in mathematics and science from Malvern College in 1893.

Francis William Aston's discovery that isotopes exist in stable elements, as well as in radioactive elements, won him the Nobel Prize in chemistry in 1922. *(E. F. Smith Collection, Rare Book & Manuscript Library, University of Pennsylvania)*

That same year he matriculated at Mason College in Birmingham to study chemistry under W. A. Tilden and P. F. Frankland and physics under J. H. Poynting. A Foster scholarship supported his study in 1898, but by 1900 he could no longer finance his education, so he put his chemical knowledge of fermentation to use as a brewer in Wolverhampton in order to save for continued schooling. In 1903 he returned to his college, which had been renamed Birmingham University, to study under Poynting. In 1907, while passing electrical currents though gases at low pressure to observe W. C. Crookes's dark space, he discovered a related phenomenon: a new dark space directly adjacent to the cathode later known as the Aston dark space.

In 1909 Joseph John Thomson invited Aston to serve as his personal assistant at Cavendish Laboratory of Cambridge University. Aston accepted and later earned the Clerk Maxwell Scholarship. There he improved on Wilhelm Wien's 1902 studies of neon by creating sharp parabolic lines corresponding to the velocity of the particle, thereby proving that substances have specific molecular weights. However, Aston could not figure out why the two neon parabolas corresponded to values of 20 and 22, as opposed to the weight 20.2 units determined by neon's discoverers William Ramsay and Morris Travers.

Before Aston could continue experimentation to resolve this discrepancy, World War I intervened, calling him into the service of the Royal Aircraft Factory in Farnborough as a chemist. In 1914 an experimental aircraft he was testing crashed, almost killing him. When he returned to Cambridge after the war, in 1919, he became a fellow of Trinity College, where he spent the remainder of his career. That year he also designed and built the mass spectrograph, the device upon which he conducted his most famous experiments.

Aston repeated his neon experiment with his new spectrograph, which yielded the same results as his earlier experiment: two parabolic lines indicating masses of 20 and 22. However, this time Aston observed that the line in his experiment was 10 times darker than that of Ramsay and Travers's line. Taking into account this differentiation, Aston found that the weighted average was 20.2, neon's exact mass as calculated by Ramsay and Travers. Aston thus confirmed the existence of isotopes in nonradioactive substances. He proceeded to confirm the existence of isotopes in other stable elements: sulfur, which had three isotopes of mass 32, 33, and 34; chlorine, which had two isotopes of mass 35 and 37; and silicon, which had three

isotopes of mass 28, 29, and 30. In all, he measured the isotopic masses of more than 50 elements, thereby identifying 212 of the 287 stable isotopes with his mass spectrometer (it is now on display at the Science Museum in London).

In December 1919, Aston introduced his *whole-number rule*, proposing that all atoms have a mass corresponding to a whole-number integer. He attributed slight deviations from whole numbers to isotopic constituents. Later, in 1927, he invented a stronger mass spectrograph on which he discovered that the masses of elements all deviate slightly from whole numbers. He attributed this deviation to what he called a "packing fraction," or the fractional deviance from the whole number due to the packing of particles into the nucleus. He plotted these fractions along a curve for all the elements and theorized that the fraction represented the relative stability of the atom, or, in other words, the amount of energy that could be released in destabilizing the atom. Aston thus foretold the potential power of nuclear reactions.

In a 1922 speech in acceptance of the Nobel Prize in chemistry, Aston warned of the dangers of such an immense form of energy creation as nuclear reactivity. Two days after receiving the Nobel Prize, he received the Hughes Medal from the Royal Society, which had inducted him as a member the previous year. Sixteen years later, the Royal Society granted him its Royal Medal, three years after he had built his third and most powerful mass spectrometer. This version never lived up to its potential, though, as other scientists were building more efficient devices to perform the same experiments. In 1941 the Institute of Physics awarded him its Duddell Medal and Prize.

Aston died on November 20, 1945, in Cambridge. He had remained a bachelor throughout his life, preferring the company of nature: He was an avid skier, golfer, and amateur astronomer. He also entertained himself with music as an accomplished pianist, violinist, and cellist. As he did not have any heirs to bequeath his ample savings (he also excelled at investing), he donated his estate to Trinity College and other scientific institutions, leaving to the entire scientific world his greatest contributions—the innovation of the mass spectrograph and the confirmation of stable-element isotopes.

⊠ Avogadro, Lorenzo Romano Amedeo Carlo
(1776–1856)
Italian
Physical Chemist

Amedeo Avogadro is considered one of the founders of physical chemistry. He is best remembered for hypothesizing that equal volumes of all gases (at the same temperature and pressure) will contain the same number of molecules, a hypothesis that became known as Avogadro's law, though it wasn't until after his death that the validity and significance of this assertion were acknowledged. He also assigned a number, Avogadro's number, as the constant by which the molecular weight of any gas could be calculated.

Lorenzo Romano Amedeo Carlo Avogadro di Quaregua e di Cerreto was born on August 9, 1776, in Turin, Italy. He hailed from a family of ecclesiastical lawyers, and, in fact, he commenced his career set to follow in their footsteps, earning a doctorate in ecclesiastical law in 1796. He practiced law full time for three years and continued to practice law part time for much of the remainder of his career.

In 1800, however, Avogadro switched the focus of his career to the sciences by taking private lessons in mathematics and physics. By 1806 he had become a demonstrator at the Academy of Turin, and three years later, in 1809, the College of Vercelli appointed him a professor of natural philosophy. That year JOSEPH LOUIS GAY-LUSSAC observed that all gases expand at the same rate when heated at the same rate.

Lorenzo Romano Amedeo Carlo Avogadro is best known for his hypothesis that equal volumes of all gases at the same temperature and pressure will have the same number of molecules. *(AIP Emilio Segré Visual Archives)*

Using this as his stepping-off point, Avogadro made his famous hypothesis that under the same conditions of temperature and pressure, equal volumes of any gases will contain the same number of molecules.

Published in the 1811 edition of *Journal de Physique*, the hypothesis did not become famous immediately, partly due to the obscurity of the journal and to his confusing nomenclature (he used *atom* and *molecule* interchangeably). The main reason, however, was because he failed to support his hypothesis with concrete experimental proof; in fact, his reputation as a rather shoddy experimentalist cast further doubt upon such a bold (though elegantly simple) assertion. Ironically, in 1814 ANDRÉ-MARIE AMPÈRE came to a similar conclusion regarding this molecular constancy and credited Avogadro for his hypothesis, but the theory still fell on deaf ears.

Avogadro continued his scientific pursuits despite the lack of recognition of the significance of his hypothesis. In 1820 the University of Turin appointed him to its chair of mathematical physics, a post he retained for the next three decades, during which time he conducted research on the electrical properties of substances, thermal expansion, and specific heat. He also suggested that simple gases, such as hydrogen and oxygen, might consist of diatomic molecules, as opposed to JOHN DALTON's "indivisible" atomic theory, which held that these gases consisted of single-atom molecules, a belief that was widely held at the time. Later in his tenure, from 1837 to 1841, Avogadro published a four-volume physics textbook, *Fisica dei corpi ponderabili*, in which he reiterated his assertion of his law of molecular constancy between gases, but the scientific community continued to ignore his work.

Avogadro died on July 9, 1856, in his birthplace of Turin. Four years later, in 1860, his work was vindicated when Stanislao Cannizzaro used it as the basis for a set of atomic weights that withstood close scrutiny, thus proving the validity of Avogadro's number posthumously.

Since 1860, technological and scientific advances have allowed for increasingly accurate calculations of Avogadro's number, or N. For example, in the late 19th century, reasonably accurate figures were attained from sedimentation equilibrium of colloidal particles. After the turn of the century, Robert Andrews Millikan's oil drop experiment allowed for even greater accuracy, and textbooks as late as 1958 cited the figure arrived at by this approach, 6.02×10^{23}. The first accurate determination of Avogadro's number preceded this by almost two decades: In 1941 R. T. Birge calculated N as 6.02486×10^{23}. Further advances have elicited the most current value of N as 6.022137×10^{23}. This number, if translated into fine grains of sand ($1/100$ inch in diameter) piled 50 feet deep, would cover the entire 262,000 square miles of the state of Texas and would be equivalent to the number of molecules in 1 mole of water—$1/25$ of a pint, or 18 grams.

B

⊠ Baekeland, Leo Hendrik
(1863–1944)
Belgian/American
Chemist

Leo Baekeland discovered the first fully synthetic substance that could be molded into almost any shape imaginable. He named his innovation Bakelite, partly after his own name, but its generic nomenclature is plastic. Other scientists had already discovered the same polymer from the reaction of carbolic acid, or phenol, with formaldehyde, but Baekeland devised a means of transforming it from a gooey resin into an infinitely transmutable substance by heating it under pressure and thereby ushered in the plastic age.

Baekeland was born on November 14, 1863, in Ghent, Belgium. His mother worked as a maid, and his father was a shoe repairman. He taught himself photography, then attended night classes at the Municipal Technical School of Ghent to learn the chemistry of film developing. Unable to afford the chemicals necessary for the process, he recycled his own by melting down a silver watch given to him as a gift and figuring out how to extract and purify silver nitrate from it for use in exposing his pictures.

Baekeland entered the University of Ghent at the age of 17 to study chemistry under F. Swarts, whose daughter, Celine (known as "Bonbon"), he later married. The couple had two children. Upon his graduation with a bachelor of science degree in 1882, the university retained him as a professor of chemistry while he pursued graduate studies. He earned his doctorate with honors two years later and served as a professor of chemistry and physics at the Government Higher Normal School of Science in Bruges from 1885 to 1887. From there he returned to the University of Ghent until 1889, when he immigrated to the United States after using a traveling fellowship to study in France and Britain.

In the United States, Baekeland first worked as a chemist for the E. & H.T. Anthony Company, a photographic paper manufacturer, before setting out on his own as a consultant and inventor of what the field lacked. He invented the first contact developing paper, an unwashed silver chloride emulsion, that succeeded on the commercial market. He then invented Velox (the validity of this assertion is dubious), otherwise known as "gaslight" paper because it could be developed with artificial light (which was gas at the time), and thus did not rely on direct sunlight for development. While his scientific claim may have been hyperbolic, his business sense was very accurate: In partnership with Leonardo Jacobi to manufacture Velox, he established the

Bakelite, an early plastic, was discovered by Leo Hendrik Baekeland, a prominent chemist and businessman in the early 20th century. *(E. F. Smith Collection, Rare Book & Manuscript Library, University of Pennsylvania)*

phenol (a coal-tar derivative) and formaldehyde (an embalming fluid derived from wood alcohol). Baekeland, however, hit upon a means of controlling pressure and temperature perfectly with his "bakelizer," a kind of industrial-strength pressure cooker, allowing him to transform the viscous, shellac-like by-product of the phenol-formaldehyde reaction into a hard-but-moldable substance—the perfect malleable synthetic.

After applying for a patent on July 13, 1907, and securing patent number 942,699 soon thereafter, Baekeland introduced polyoxybenzylmethylenglycolanhydride to the scientific community at the 1909 meeting of the New York chapter of the American Chemical Society and to the world thereafter as Bakelite. He founded and served as president of the General Bakelite Corporation in 1910 to manufacture the plastic, but he again exercised his business sense by licensing the use of his process to other manufacturers, who made such imitations as Redmanol and Condensite, while he simultaneously pushed Bakelite as the genuine article. By 1911 Bakelite had set up plants in the United States as well as in Germany, and by 1922 Baekeland had orchestrated the consolidation of several plastics makers into the Bakelite Corporation, of which he continued to serve as president until 1939, when the company folded into Union Carbide and Carbon Company.

Baekeland also served as president of the three major chemical associations in the United States: the American Chemical Society in 1924, which had awarded him its Nichols Medal in 1909; the American Electrochemical Society; and the American Society of Chemical Engineering, which had granted him its Perkin Medal in 1916. The Franklin Institute also bestowed its Franklin Medal on him in 1940. Baekeland died on February 23, 1944, in Beacon, New York.

Nepera Chemical Company, and George Eastman bought out the company in 1899 for $1 million.

Baekeland moved his family into Snug Rock, an estate on the banks of the Hudson River in the north of Yonkers, New York, and converted the barn into a lab to continue his experimentation. It was here that he discovered his next big hit: a synthetic substitute for shellac, the natural resin that served as an electrical insulator, among other things. He had followed in the footsteps of such scientific predecessors as ADOLF VON BAEYER, who combined

Baeyer, Adolf von
(1835–1917)
German
Chemist

Adolf von Baeyer struggled to establish himself throughout his early career, but recognition slowly accrued for him after he succeeded in synthesizing indigo and other dyes, such as triphenylmethane. Most notably, the king of Bavaria bestowed on him the noble title of *von* (probably more for his role in launching the German dye industry than for his scientific contribution), and then the scientist gained international recognition as the recipient of the 1905 Nobel Prize in chemistry.

Johann Friedrich Wilhelm Adolf Baeyer was born on October 31, 1835, in Berlin, Germany. His mother, Eugenie Hitzig, grew up in a family who hosted Berlin's literary coterie; his father, Johann Jacob Baeyer, performed geodetic surveys as a captain in the Prussian army before his promotion to general's rank. Baeyer charted his own path into science early on, as a boy performing experiments on plant nutrition at his grandfather's Müggelsheim farm; back in the confines of Berlin, he took to the test tubes with chemical experiments beginning at the age of nine. Three years later, he synthesized a previously unknown chemical compound, a double carbonate of copper and sodium. On his 13th birthday he initiated his life work, buying a chunk of indigo worth two talers for his first dye experiments.

Baeyer distinguished himself academically as well: His chemistry teacher at the Friedrich Wilhelm Gymnasium appointed him as his assistant. After graduating from secondary school in 1853, he entered the University of Berlin to study physics and mathematics. A stint in the Prussian army interrupted his study until 1856, when he returned to academia at the University of Heidelberg, intending to study chemistry

under Robert Bunsen. After an argument with the renowned chemist, however, he changed his mentor to Friedrich Kekulé, one of the few organic chemists at the time. He continued to collaborate with Kekulé even after returning to Berlin to complete his doctorate on arsenic methyl chloride, or cacodylic, in 1858.

His doctoral committee's tepid reception of his dissertation (they did not really understand it) prevented him from securing an academic position, so Baeyer traveled to Ghent, Belgium, to work with Kekulé, experimenting on uric acid for the next two years. Baeyer submitted the paper reporting these studies (in lieu of his dissertation) to the Berlin Institute of Technology, which hired him in 1860 as a lecturer in organic chemistry; he retained this position for the next dozen years, despite its paltry salary, because he had access to a well-appointed laboratory. In 1863 he made one of his most important discoveries: barbituric acid, a derivative of uric acid, which he apparently named after his girlfriend at the time, Barbara.

In 1868 Baeyer married Adelheid Bendemann, the daughter of a family friend, and together the couple had three children—Eugenie, Hans, and Otto. The year previous to his marriage, he had helped found the German Chemical Society to facilitate the sharing of scientific thought by assembling the best minds in the country to discuss advances in chemistry. And meanwhile, the University of Berlin had finally given him a professorship, but it was unsalaried. Despite Baeyer's distinction as an excellent clinical scientist, he experienced difficulty in establishing himself academically until 1872, when the University of Strasbourg finally appointed him to a full professorship of chemistry.

In 1865 Baeyer had commenced the research on indigo that would define his career. By 1866 he had devised an approximate structural formula for the compound, and later, in 1870, he synthesized indigo. He continued his indigo studies when he moved to the University of Munich as a professor of organic chemistry in 1875, and by 1883, he had developed an even more precise formula for the dye. He remained at the university for the next four decades, during which time he also developed a theory for the formation of carbon rings around compounds that became known as the Baeyer strain theory.

In 1885 King Ludwig II of Bavaria inducted Baeyer into the German nobility by conferring him with the title of von. Two decades later, Baeyer's contributions to chemistry were honored with the Nobel Prize, awarded on December 20, 1905. He continued to work for another decade, retiring from the university when he turned 80, in 1915. Two years later, as World War I raged, he died in his own family home in the country near Lake Starnberg on August 20, 1917.

⊠   **Barton, Derek H. R.**
(1918–1998)
English
Chemist

Derek Barton transformed the landscape of physical chemistry in one fell swoop by establishing conformational analysis, or the notion that chemical geometry corresponds to molecular function. This advancement in understanding also transformed the chemist's imagination from representing chemical reactions and structures in two dimensions to visualizing molecules more accurately in their three-dimensional configurations. Barton shared the 1969 Nobel Prize in chemistry with ODD HASSEL, who had established the geometry of a common steroid.

Derek Harold Richard Barton was born on September 8, 1918, in Gravesend, in Kent, England. His father, William Thomas Barton, was a carpenter (like his father before him) and a lumberyard owner who died when Barton was 17. After helping his mother, Maude Lukes, run the

family business for two years, Barton entered Gillingham Technical College. He studied there only a year before transferring to Imperial College of the University of London, where he graduated with first-class honors in 1940. He conducted his doctoral research on the synthesis of vinyl chloride (the black plastic used to make phonographic records) under Ian Heilbron and E. R. H. Jones and earned his Ph.D. in 1942.

During World War II Barton remained at Imperial to conduct military intelligence research on developing secret inks. In 1944 the Albright and Wilson Company in Birmingham hired him to work on synthesizing organic phosphorous compounds; however, he only lasted a year in industry before returning to academia as a junior lecturer in inorganic chemistry at Imperial. Four years later, in 1949, he took a position in organic chemistry, his actual area of specialization, at Imperial College.

During those four years, Barton commenced the research that would lead to his breakthrough realization. He investigated triterpenoids and steroids, recording correlations between their chemical structures and their molecular properties. Word of Odd Hassel's work deciphering the geometry of cyclohexane (a foundational molecule for triterpenoids and steroids) reached him, and he extended this work to other, more complex molecules. In 1949 the prominent steroid chemist Louis Fieser invited Barton to Harvard University as a visiting lecturer to fill the position vacated by ROBERT BURNS WOODWARD, who was taking a one-year sabbatical.

While attending a lecture by Fieser on some unsolved steroid mysteries, Barton had his epiphany. He realized that the three-dimensional shapes of molecules must correspond to their characteristics. He immediately expounded his theory, citing experimental confirmation, in a four-page paper entitled "The Conformation of the Steroid Nucleus" and published in the Swiss journal *Experientia,* a relatively obscure publication that soared in circulation after word got out about the significance of Barton's paper. Through this piece, Barton not only established the field of conformational analysis by asserting a correlation between molecular structure and molecular functions but also transformed the chemical imagination from a flat plane into a three-dimensional space of complex configurations.

Upon his return to England in 1950, Barton was promoted to a readership at Birkbeck College of the University of London, where three years later he was promoted again, to a professorship. In 1955 the University of Glasgow named him its regius professor, but in 1957 Barton returned to London to take up a full professorship at Imperial College, where he remained for the next two decades. During the 1950s he worked with free-radical chemistry to elicit reactions for chemical synthesis. He uncovered the biosynthetic process governing the transformation of opium poppies into morphine. He also generated photochemical reactions by casting ultraviolet light on samples to rupture chemical bonds into free radicals that he then manipulated to synthesize compounds. In 1958 he employed what became known as the Barton reaction to synthesize the steroid aldosterone, and in one experiment increased the world's supply of this electrolytic hormone from a matter of milligrams to over 60 grams.

In 1969 the Royal Swedish Academy of Sciences honored Barton and Hassel with the Nobel Prize in chemistry in recognition of Hassel's experimental determination of the chemical geometry of steroids and Barton's "gap-jumping" theory that correlated this chemical geometry to corresponding chemical characteristics, thereby demonstrating how the two seemingly disparate areas of form and function actually work hand in hand. Three years later Queen Elizabeth II knighted Barton, though he insisted on leaving behind the title of Sir Derek when he traveled outside of England.

In 1978 Barton relocated to Gif-sur-Yvette, France, as the director of research for the Centre

National de la Recherche Scientifique (CNRS) of the Institut de Chimie des Substances Naturelles (ICSN). This move thrilled his second wife, Professor Christiane Cognet, who hailed from France. (Barton's earlier marriage to Jeanne Kate Wilkins, which yielded one son, William Godfrey Luke Barton, ended in divorce.) Barton continued to invent new reactions, naming the components Gif reagents after the site of his research.

In 1986 Texas A&M University named Barton a distinguished professor, a post he held for the last dozen years of his life. During this time he received further accolades: The American Chemical Society granted him its 1989 Creative Work in Synthetic Chemistry Award and its 1995 Priestley Award. His second wife died in 1994, and his third wife, Judy Cobb, survived him when he passed away on March 16, 1998, in College Station, Texas.

⊠   **Becquerel, Antoine-Henri**
(1852–1908)
French
Physicist

While confirming that a luminescent salt of uranium emits X rays, Antoine-Henri Becquerel serendipitously discovered what his graduate student MARIE CURIE dubbed *radioactivity*, thus radically altering the face of the 20th century and the course of human history. For this discovery he shared the 1903 Nobel Prize in physics with Marie Curie and her husband, Pierre.

Becquerel was born on December 15, 1852, in Paris, France. He hailed from a long line of prominent scientists: He inherited the physics chair at the Musée d'Histoire Naturelle (as well as his interest in luminescence, or the emission of light from unheated substances) from his father, Alexandre Edmond, who had inherited the chair from his father, Antoine César, a physicist who fought in the Battle of Waterloo in 1815. Bec-

Antoine-Henri Becquerel, whose discovery of radioactivity in 1896 earned him the 1903 Nobel Prize in physics *(AIP Emilio Segré Visual Archives, William G. Myers Collection)*

querel attended secondary school at the Lycée Louis-le-Grand, graduating in 1872 to the École Polytechnique. After two years there he transferred to the École des Ponts et Chaussées. Also in 1874 he married Lucie Zoé Marie Jamin, the daughter of J. C. Jamin, a physics professor at the University of Paris. Tragically, she died in March 1878, shortly after giving birth to their only son, Jean, who later inherited the family's physics chair at the Musée d'Histoire Naturelle.

In 1875 Becquerel published his first scientific paper, in which he offered a formula explaining the relationship between the Faraday effect (the phenomenon discovered by MICHAEL FARADAY in 1845 whereby polarized light experiences a rotation of the plane on which the light beams vibrate when passing through a magnetic field) and the light beam's refraction when passing through a substance. The next year the École Polytechnique appointed Becquerel as a *répétieur* (tutor). A year later, in 1877, he earned his engineering degree from the École des Ponts et Chaussées and became an engineer at the National Administration of Bridges and Highways. In January 1878 his grandfather died, passing along the physics chair at the Musée d'Histoire Naturelle to Becquerel's father and Becquerel becoming his father's assistant.

Becquerel continued to conduct research, publishing papers on the magnetic properties of nickel, cobalt, and ozone in 1879, including his discovery that nickel-plated iron becomes magnetic when red hot. In the 1880s he took up his father's specialty, luminescence, and wrote his doctoral dissertation on how luminescence reacts to polarization; the University of Paris granted him his doctorate in 1888. The next year the Académie des Sciences inducted him into its fellowship.

In 1890 Becquerel married Louise Désirée Lorieux, the daughter of a mine inspector. When his father died in 1891, Becquerel inherited the family's chair of physics at the Musée d'Histoire Naturelle; simultaneously he inherited Alfred Potier's chair at the École Polytechnique. In 1894 the École des Ponts et Chaussées named him chief engineer, rounding out his three appointments at the three institutions with which he retained affiliation throughout his career. That year he also published his discovery of the influence of Earth's magnetic field on its atmosphere.

In 1895 William Röntgen announced his discovery of X rays, using cathode-ray tubes that evinced some luminescence. Becquerel immediately conducted research following up on the relationship between X rays and luminescence. He sought to induce X rays from a known luminescent material, the double sulfate of uranium and potassium, by placing these salts on top of photographic plates covered in black lightproof paper, then placing this in sunlight. At a meeting of the Académie des Sciences on February 24, 1896, he announced his results: The photographic plates were exposed, not by sunlight, which could not penetrate the black paper, but by X rays emitted by the luminescent salts in reaction to the sunlight, Becquerel proposed.

The next week Becquerel prepared this same setup—photographic plates wrapped in lightproof paper with luminescent salts taped atop—however, clouds covered the sun on February 26 and 27, so he set these packets aside in a dark drawer. On March 1 he decided to develop these plates for some unknown reason because without exposure to the sun, the luminescent salts could not produce X rays, according to his hypothesis. Remarkably, the photographic plates were completely exposed! The next day he announced his discovery to the academy: The photographic plates had been exposed not by X rays but by some other form of radiation. This new source of radiation was called Becquerel's rays until his graduate student Marie Curie coined the term *radioactivity*.

Once he had found radioactivity, he searched for better sources; by May 1896 he had discovered that uranium metal emits much more radioactivity than the salts he had discovered the energy in. He also researched every aspect of radioactivity he could conceive; by 1900 he had ascertained that uranium's radiation consists of electrons. Three years later the Royal Swedish Academy of Sciences honored Becquerel's discovery, as well as the Curies' role in furthering the understanding of radioactivity, by granting the trio the 1903 Nobel Prize in physics.

Although the full significance of Becquerel's discovery was yet to be realized, its potential significance was foretold with many prestigious honors: the 1900 Rumford Medal of the Royal Society, the 1901 Helmholtz Medal of the Royal Academy of Sciences of Berlin, and the 1905 Barnard Medal of the U.S. National Academy of Sciences. Becquerel died on August 24, 1908, in Croisic, in Brittany, France.

⊠   **Berg, Paul**
(1926–   )
American
Molecular Chemist

Paul Berg won the 1980 Nobel Prize in chemistry for his discovery of rDNA, or recombinant deoxyribonucleic acid, eight years earlier. Immediately after his discovery, however, he voluntarily abandoned rDNA research for a year, until the scientific field had a chance to discuss and establish the ethics of recombinant DNA research, which tread on sensitive territory such as the risk of science "controlling creation." In the infamous Berg letter, composed by the so-called Berg committee that he headed, he helped establish the boundaries of rDNA research.

Berg was born on June 30, 1926, in Brooklyn, New York. His parents were Sarah Brodsky and Harry Berg, and his brothers were Jack and Irving. Two books, Sinclair Lewis's *Arrowsmith* and Paul de Kruif's *Microbe Hunters*, inspired his early interest in chemistry. He attended New York's Abraham Lincoln High School, where his laboratory demonstrator and science club leader, Sophie Wolfe, further inspired his interest in science. (Wolfe also mentored two other future Nobel laureates: Arthur Kornberg and Jerome Karl.) Berg graduated in early 1943, then matriculated at Pennsylvania State College.

World War II interrupted Berg's undergraduate studies; he served in the U.S. Navy from 1944 through 1946, after which he returned to Penn State to continue his biochemistry studies. On September 13, 1947, he married Mildred Levy, who gave birth to their only son a year later. In 1948 he graduated from Penn State with a bachelor of science degree in biochemistry.

Berg pursued graduate studies in biochemistry at Western Reserve University, obtaining his doctorate in 1952. He conducted postdoctoral research under Herman Kalckar at the Institute of Cytophysiology in Copenhagen, Denmark, for the following year and then under

Paul Berg received the 1980 Nobel Prize in chemistry for developing the technology to create recombinant deoxyribonucleic acid (rDNA). *(Department of Special Collections, Stanford University Libraries)*

Arthur Kornberg at Washington University School of Medicine in St. Louis, Missouri, the year after that. He stayed on at Washington University, serving as a scholar in cancer research from 1954 through 1957 and concurrently as an assistant professor of microbiology from 1955 to 1959.

Berg established his reputation almost immediately upon joining Kornberg, who allowed the young scientist to pursue an unpromising hypothesis that ran against the theories established by two Nobel laureates. Berg disproved these theories when he discovered a new method for transforming fatty acids into their activated form (acyl coenzyme As, or coAs). This discovery led in turn to his codiscovery of aminoacyl transfer RNA (tRNA) synthetases and tRNA. When Kornberg abandoned Washington University for Stanford University School of Medicine, he imported his brain trust, including Berg and five other members of the microbiology department. Stanford appointed Berg an associate professor, later promoting him to a full professorship. In 1970 Stanford named him its Willson Professor of Biochemistry; the year before the university had appointed him

chairman of the biochemistry department, a position he retained until 1974. In 1973 the Salk Institute inducted him as a nonresident fellow, a position he has retained throughout his career.

In 1972 Berg discovered recombinant DNA when he fused DNA from the carcinogenic monkey virus SV40 with the lambda virus. Extrapolating the implications of his discovery, however, he immediately ceased this line of investigation until the scientific community could arrive at appropriate ethical guidelines for such research. He helped orchestrate the discussion by gathering together a group of preeminent scientists (the Berg committee), which composed the so-called Berg letter that called for a one-year moratorium on rDNA research. This led to the February 1975 Asilomar Conference, which culminated in the drafting of National Institutes of Health guidelines governing rDNA research in 1976. (The NIH revised these guidelines in 1994.)

In recognition of his discovery of rDNA and his tireless work in promoting the ethical self-governing of the scientific community in regards to rDNA research, Berg received the 1980 Nobel Prize in chemistry, together with his peers WALTER GILBERT and FREDERICK SANGER. Besides the Nobel Prize, Berg also won numerous other prestigious honors, including the 1959 Eli Lilly Prize in Biochemistry, the 1963 California Scientist of the Year, the 1974 V. D. Mattia Award of the Roche Institute for Molecular Biology, the 1980 Albert Lasker Medical Research Award, and the 1983 National Medal of Science. He was also inducted into the National Academy of Sciences and the American Academy of Sciences in 1966 and served as president of the American Society of Biological Chemists in 1975.

In 1985 Berg became the director of the Beckman Center for Molecular and Genetic Medicine. In 1991 he was named head of the NIH Human Genome Project Scientific Advisory Committee. Throughout the 1990s he collaborated with MAXINE SINGER in authoring three books: *Genes and Genomes: A Changing Perspec-*

tive, a graduate-level textbook published in 1990; *Dealing with Genes: The Language of Heredity*, a general-audience book published in 1992; and *Exploring Genetic Mechanisms*, published in 1997.

⊠ **Bergius, Friedrich**
(1884–1949)
German
Organic Chemist

Friedrich Bergius developed hydrogenation processes for converting coal into oil and gasoline and for converting wood into carbohydrates for human nutrition. Neither of these processes is particularly practical for general purposes, but in wartime, both proved vital. In fact, for the former process, Bergius shared the 1931 Nobel Prize in chemistry with CARL BOSCH, who continued to advance coal hydrogenation techniques after Bergius sold the patent.

Friedrich Karl Rudolf Bergius was born on October 11, 1884, in Goldschmieden, near Breslau, Germany (now Poland). His mother, Marie Haase, was the daughter of a classics and economics professor, and his father, Heinrich Bergius, headed a chemical factory. Young Friedrich took to chemistry early on, visiting his father's laboratories and conducting elementary experiments there. He "graduated" from his father's labs to those of a foundry in the Ruhr industrial district, where his father sent him for six months of metallurgical study.

In 1903 Bergius commenced his formal education at the University of Breslau, where he studied under Albert Ladenburg and Richard Abegg. After fulfilling his one-year military duty, he traveled to Leipzig University in 1905 to conduct doctoral research under Arthur Hantzsch. He later completed this research back in Abegg's lab and submitted his dissertation on the solvent properties of concentrated sulfuric acid to the University of Breslau, which granted him a Ph.D. in 1907.

Bergius spent the next two years conducting research as an assistant in WALTHER HERMANN NERNST's lab in Berlin and in FRITZ HABER's lab in Karlsruhe, where he attempted to synthesize ammonia with hydrogen and atmospheric nitrogen. In 1909 the Technische Hochschule of Hanover hired him as a university lecturer. He conducted research on the dissociation equilibrium, or breakdown, of calcium peroxide in Ernest Bodenstein's chemistry lab before setting up his own private laboratory to investigate chemical reactions under high pressure. He built a special apparatus to withstand stress under as much as 300 atmospheres of pressure without leaking.

While conducting hydrogenation experiments, Bergius succeeded in producing purer hydrogen cheaper than previously possible. After one of these experiments, he noticed that the by-products of peat, which he was using as his source of carbon, correlated to coal, prompting him to investigate the potential of hydrogenating coal as an energy source. Spurred on by gasoline and oil shortages in World War I, he applied his high-pressure methods at low temperatures and succeeded in converting coal into a gas substitute. He and collaborators Hugo Spract and John Billwiller received a patent for the process.

In 1914 Karl Goldschmidt of Goldschmidt Corporation financed Bergius's continuing experiments on coal hydrogenation, and soon thereafter he promoted the chemist to head the corporate research laboratories. After World War I, war reparations sucked the German economy dry, forcing Bergius to appeal for corporate sponsorship outside Germany; he turned to Britain and the United States. With this backing Bergius oversaw several thousand tests on 200 different types of coal in search of the optimal combination of materials and techniques for maximizing the yield of coal hydrogenation. He innovated several methods for increasing the efficiency of the coal hydrogenation process; first by pulverizing the coal, then suspending it in oil to create a slurry substance that not only burned evenly but also could be fed into the reactor continuously. Having perfected this coal liquefaction process as far as seemed practical in 1925, Bergius sold his patent rights.

Carl Bosch continued to advance coal hydrogenation and liquefaction techniques after Bergius sold the patent, and the Royal Swedish Academy of Sciences recognized the contributions of both men by awarding the 1931 Nobel Prize to them jointly. Bergius had since returned to investigations he had commenced in 1915, utilizing the same general principle as coal hydrogenation to convert wood into carbohydrates for human consumption. Specifically, he employed hydrolysis, or irrigating the wood with water in order to break the cellulose down into alcohol, yeast, and dextrose. In 1943 Bergius oversaw the establishment of a plant in Rheinau dedicated to producing foodstuffs from wood by this process, which proved instrumental in feeding the German populace during food shortages in World War II.

After the war Bergius emigrated from Germany, first to Austria, then to Madrid, Spain, where the government had contracted him to found a company. In 1946 he moved to Argentina, where he served as a technical consultant to the ministry of industries. He died in Buenos Aires on March 30, 1949, and was survived by his second wife, Ottilie Krazert, his two sons, and his one daughter. Besides winning the Nobel Prize, he also won the Liebig Medal of the German Chemical Society.

⊠   Berzelius, Jöns Jakob
(1779–1848)
Swedish
Chemist

Jöns Jakob Berzelius helped structure and advance chemistry into the modern era. He left his indelible mark on the field by instituting the symbology that combined abbreviations of elements with numerical subscripts, a system that revealed his synthesis of electrochemistry with the atomic theory. He identified the atomic weights of all but four of the elements known at the time, and he and his assistants added to the list of known elements by discovering cerium, selenium, lithium, thorium, silicon, vanadium, and several lanthanides, as well as pyruvic acid. Much of his work has been essential up through modern chemical studies, such as the use of selenium in television transmission and silicon in computer hardware and software.

Berzelius was born on August 20 or 29, 1779, near Linköping, Sweden. His father was a clergyman and schoolmaster who died, along with his mother, when Berzelius was young; rela-

During his lifetime Jöns Jakob Berzelius identified the atomic weights of all but four of the elements known *(E. F. Smith Collection, Rare Book & Manuscript Library, University of Pennsylvania)*

tives raised the boy thereafter. In 1796 he entered the University of Uppsala to study medicine, but he discovered chemistry while researching the dubious therapeutic effects of galvanism for his dissertation. He earned his medical degree in 1802 but distinguished himself the following year with his first of innumerable contributions to the field of chemistry, when he discovered a new element—cerium—while working in an unpaid position at the Caroline Medico-Chirurgical Institute in Stockholm. He did not abandon medicine, donating his expertise to care for the poor over the next several years.

In 1807 the surgical institute promoted Berzelius to a professorship of medicine, botany, and pharmacy. A year later the Royal Swedish Academy of Sciences inducted him into its fellowship, and he split his time between the two institutions for the majority of his career. Also in 1808 he collaborated with M. M. Pontin to discover ammonium amalgam and to introduce the mercury cathode. In 1811 he published a paper introducing his Latinate nomenclature to the field of chemistry, combining one- to two-letter symbols for the elements and numerical subscripts corresponding to the number of atoms in the molecule. This system remains in use to this day.

In 1815 the surgical institute recognized Berzelius's contributions to the field of chemistry by appointing him a professor of chemistry. He continued his comprehensive investigation of the elements and discovered another one, the photoconductive chemical selenium, in 1817. By the next year he had personally determined the atomic weight of 39 of the 49 elements known at the time, with his assistants ascertaining another six of the elements. One of his assistants, J. A. Arfwedson, discovered yet another element, lithium, in 1818. That year the Royal Swedish Academy of Sciences appointed Berzelius as its permanent secretary.

Throughout his researches Berzelius was developing a new theory synthesizing electrochemistry, or the use of electrolysis in chemical investigations, with atomic theory. He expounded this theory in his 1819 *Essay on Chemical Proportions*, a work that also introduced to chemistry the notion of dualism, or the division of all compounds into positive and negative components, though this belief later proved inaccurate. In 1821 he published his first annual survey of significant advances in chemistry, a tradition he maintained until 1848.

Berzelius's contributions to the field of chemistry were diverse and prodigious: Among other things, he introduced the term *protein* to the scientific vocabulary, discovered the existence of isomers, identified the catalysis phenomenon, and proposed the theory of radicals (the notion that stable atoms combine into molecules by binding radicals, or unconnected atoms). He continued to identify new elements, discovering silicon in 1824 and thorium in 1829; the next year his assistant N. G. Sefström discovered vanadium.

In 1832 Berzelius retired from his post at the surgical institute to devote himself to the academy. Upon his marriage in 1835, Charles XIV John, king of Sweden and Norway, named him a baron. Although some of his beliefs proved untenable, such as the application of his dualistic theory to organic substances and his adamant claim that chlorine was not an element, his contributions to the field of chemistry far outweigh his misjudgments. His masterful *Textbook of Chemistry* was considered the definitive work on the field in his time and was translated into six different languages (not including English, unfortunately). He died on August 7, 1848, in Stockholm.

Bishop, Hazel Gladys
(1906–1998)
American
Chemist

Hazel Bishop entered her profession as did so many of her female contemporaries and compatriots: They filled jobs left empty by all the male workers sent to war in the first half of the 1940s. Bishop, however, was not content with merely equaling her counterparts; she surpassed them by applying her knowledge of chemistry to capitalize on a market that few men comprehended—cosmetics. In a makeshift kitchen laboratory, she invented a new kind of lipstick that stayed on lips longer than any other product. The lipstick was, of course, a huge success, and Bishop had enough business acumen to reap the benefits of her product's popularity. She set up her own company to make and distribute the lipstick, then went on to establish a laboratory for developing new consumer products, and later left the scientific field altogether to work in finance on Wall Street.

Hazel Gladys Bishop was born on August 17, 1906, in Hoboken, New Jersey. She attended Barnard College, graduating in 1929. Thereafter, she attended night classes at Columbia University but did not continue long enough to earn a graduate degree. Instead, she worked for seven years, from 1935 to 1942, as an assistant in a dermatological laboratory.

As with hosts of other women, Bishop got her foothold in the chemical industry when World War II drained the American workforce of its pool of male labor, forcing companies to hire women. And as was the case of her sisters, she proved herself equal to her absent male counterparts, if not superior to most. From 1942 through the end of the war, Bishop worked as an organic chemist at the Standard Oil Development Company. Then, in 1945, she shifted from Standard to the Socony Vacuum Oil Company, where she filled a similar position in organic chemistry.

While conducting chemical research at Standard and Socony, Bishop simultaneously conducted her own research in a makeshift laboratory she had set up in her kitchen, applying her chemical expertise to the realm of cosmetics. By 1949 she had perfected a formula for a lipstick that stayed on lips longer.

In 1950 Bishop quit her post at Socony to strike out on her own. She founded Hazel Bishop, Inc., to produce and distribute her Lasting Lipstick. She marketed the kiss-proof lipstick with the line, "It stays on you, not on him." Testament to its success, rival companies appeared almost immediately, though they could not unseat Bishop's chemically superior product.

Bishop served as the president of her own company, steering it through its initial burst of popularity. However, she retained her presidency only until November 1951, when a disagreement with the majority shareholder of her stock forced her resignation. She turned around and sued for mismanagement of the company, regaining control of Hazel Bishop, Inc., by 1954, when it was bringing in $10 million in annual sales.

Bishop then proceeded to establish the Hazel Bishop Laboratories and broadened her research from cosmetics to consumer products in general. By 1955 the labs had developed a leather cleaner; subsequent developments included cosmetics and other personal-care products. Bishop organized new companies (under her corporate wing) to distribute many of these products individually. She continued to grow her corporation until 1962, when she decided to let it run itself while she pursued other endeavors.

That year she joined the prestigious brokerage firm Bache and Company as a registered agent. She enjoyed success on Wall Street similar to her industrial successes and later became a financial analyst for Evans and Company. Bishop also retained her association with the cosmetics industry by lending her cosmetics marketing expertise to the Fashion Institute of Technology in Manhattan.

Bishop lived throughout almost an entire century, and her career signaled a new era in women's lives, as Bishop ushered the way not only into a male-dominated industry but also into the male domain of business. Her success allowed her the freedom to live as she wished, not as was dictated to her. When she tired of heading the business she had founded, Bishop made a major career transition into finance, where she created another personal success story. Bishop, who never married, died on December 5, 1998, in Rye, New York.

⊠ **Bosch, Carl**
(1874–1940)
German
Chemist

Carl Bosch shared the 1931 Nobel Prize in chemistry with FRIEDRICH BERGIUS for their development of high-pressure methods of inducing chemical reactions; this represents the first time the Royal Swedish Academy of Sciences recognized industrial chemistry over pure science in awarding the prize. Specifically, Bosch and Bergius worked on the same high-pressure process of generating gasoline from coal and hydrogen; Bosch also industrialized FRITZ HABER's process for synthesizing ammonia, a crucial base ingredient of agricultural fertilizers and explosives.

Bosch was born on August 27, 1874, in Cologne, Germany. He was the eldest of six children born to Karl Bosch, a gas and plumbing supplier, and his wife, Paula. He graduated from gymnasium (the German equivalent of high school) in 1893 and, at his father's recommendation, delayed his entrance into university by apprenticing for a year in a metallurgy factory, after which he was certified as a journeyman. In 1894 he matriculated at the Technische Hochschule in Berlin to study metallurgy and mechanical engineering and two years later transferred to the University of Leipzig to study chemistry under Johannes Wislicenus. In 1898 he earned his doctorate, with a dissertation on carbon compounds.

Upon graduation, Bosch's father again advised him to favor industry over the academy,

so in 1899 Bosch left his job as a laboratory assistant at the university to join Badische Anilin und Soda Fabrik (BASF) in Ludwigshafen am Rhein. BASF, which specialized in coal-tar dyes, assigned him to investigate means of synthesizing indigo dye artificially with azo dyes and phthalic acid anhydride. Bosch might have remained in obscurity longer if he hadn't been assigned the routine job of verifying the experimental results of the prominent physical chemist WILHELM OSTWALD. In the process, however, Bosch not only failed to verify Ostwald's results in synthesizing ammonia catalytically from its elements but also demonstrated the veracity of his own results, thereby making a name for himself early on.

BASF next assigned Bosch to the problem of devising another method for synthesizing ammonia to replace diminishing supplies of Chilean saltpeter, or sodium nitrate, the key component in fertilizers and explosives. As a result, Bosch spent the first decade of the 20th century testing and recalibrating all the existing techniques for synthesizing ammonia, only to find each of them inefficient for industrial-scale production. In 1902 Bosch married Else Schilbach, and the couple had two children, a son and a daughter.

In 1904 Fritz Haber discovered a means for synthesizing copious amounts of ammonia by means of a high-temperature, high-pressure fusion of hydrogen with nitrogen, and in 1909 Bosch obtained the patent for this process from its inventor. Then came the hard work of transforming the process for industrial-scale production. Bosch oversaw more than 20,000 experiments to find a replacement for Haber's catalyst, the rare and expensive osmium. Alwin Mittasch, Bosch's handpicked head of chemical research, identified aluminum oxide as a feasible catalyst, a mere half year later. Bosch and Mittasch also identified a means of supplying enough hydrogen for large-scale ammonia production from a mixture of hydrogen and carbon monoxide.

The next challenge consisted in designing and constructing a vessel that could withstand the 200 atmospheres of pressure and the 600° Celsius temperature necessary to induce the reaction. Although this was technically an impossibility at the time, Bosch appointed Franz Lappe as chief engineer, and they proceeded to devise a strategy for surmounting this impossibility. Hydrogen robs metal of its carbon at high temperatures and pressures, making it brittle and volatile, so Bosch and Lappe designed a double-walled chamber, with the first barrier made of soft, hydrogen-porous iron (which is airtight with the other elements). The outer layer consisted of extra strength carbon steel, with thousands of holes bored in it to allow for hydrogen leakage. Bosch and his colleagues tested the prototype successfully on March 5, 1911, and industrial production of 20 metric tons of ammonia per day commenced in September 1913—just in time for the start of World War I, when it proved vital for food and weapons production.

In 1919 BASF named Bosch its managing director. Six years later, after saving the German chemical industry from dismantlement by the Versailles Treaty, Bosch orchestrated the consolidation of eight German dyestuffs companies into one, I. G. Farben, with himself as its president. That same year, 1925, Bosch bought Bergius's patent for producing gasoline by hydrogenating coal, a process that never proved commercially feasible but won the pair the 1931 Nobel Prize in chemistry. Essentially, the Royal Swedish Academy of Sciences recognized Bosch's and Bergius's applications of high pressure to chemical interactions, processes that continue in use to this day.

In 1935 I. G. Farben appointed Bosch chairman of its board of directors, and two years later the Kaiser Wilhelm Institute (later renamed the Max Planck Society) elected him as its president. During these years, however, the Nazi regime wrested control of the corporation from Bosch. His health deteriorated, and he died in Heidelberg on April 26, 1940, after suffering for years from intestinal diseases, compounded by depression.

⊠ Brady, St. Elmo
(1884–1966)
American
Chemist

St. Elmo Brady was the first African American to earn a doctorate in chemistry in the United States, from the University of Illinois. He then went on to lead a distinguished career at three of the prominent black colleges in the country: Tuskegee Institute in Tuskegee, Alabama; Howard University in Washington, D.C.; and Fisk University in Nashville, Tennessee. He headed the chemistry departments at the latter two, while he headed the entire science division at the former. In this way he strongly influenced the education of the first generations of African Americans to go to college in significant numbers.

Brady was born on December 22, 1884, in Louisville, Kentucky. He graduated from high school in Louisville with honors and matriculated at Fisk University, a black college in Tennessee, in 1904. There he studied under Thomas W. Talley, earning his bachelor of science degree in 1908. Brady landed a position at Tuskegee Institute (now Tuskegee University), another prominent black college, where his colleagues included Booker T. Washington and GEORGE WASHINGTON CARVER. Brady followed in the latter's footsteps by conducting scientific research on plants native to the region.

In 1913 Brady took a leave of absence from Tuskegee to pursue graduate studies at the University of Illinois. The next year he received his master of science in chemistry. He then received a fellowship in chemistry, allowing him to commence doctoral research under Clarence G. Derick on the interrelationship between structure and strength in organic acids. Brady wrote his dissertation on "The Divalent Oxygen Atom," for which he earned his Ph.D. in chemistry in 1916, thereby becoming the first African American to earn a doctorate in chemistry. He was also the first African American admitted into the chemistry honor society Phi Lamda Upsilon (in 1914) and one of the first African Americans admitted into the science honor society Sigma Xi (in 1915).

After receiving his doctorate, Brady returned to Tuskegee Institute, more out of a sense of obligation to return from his leave of absence than out of a real desire to return to the hostile environment of the deep South. Tuskegee appointed him to head its division of science. In 1920, after four years in this position, Howard University in Washington, D.C., offered him a chair as well as the chairmanship of its chemistry department. He accepted, partly glad just to escape the infernal heat of the South, where he couldn't even conduct the most basic chemistry experiments with a Bunsen burner because it was so hot on the Alabama campus. Brady organized the undergraduate program during his seven-year tenure at Howard, while also following the lead of his colleague, Carver, in conducting research on plants native to the region and native to the country.

In 1927 Fisk University offered Brady the chairmanship of its chemistry department, and he jumped at the chance to return to his alma mater. Unfortunately, the combined racism and sexism in Louisville prevented his wife, Myrtle, from securing a good job, so she remained in Washington, D.C., with their one son, who later became a physician. At Fisk, Brady conducted research on the chemical makeup of magnolia seeds and castor beans, publishing his results in academic journals.

During Brady's 25 years at the helm of the Fisk chemistry department, he oversaw the building of the first modern chemistry laboratory facilities at a black college, for which Fisk later honored him by naming the building after him

and his mentor, Thomas Talley. Talley's name also graced the lecture series he established to draw prominent scientists to visit the graduate program he established at Fisk. Brady also established a summer program in infrared spectroscopy at Fisk, in collaboration with the faculty of the University of Illinois.

Brady retired from Fisk in 1952 and returned to Washington, D.C.; however, instead of relaxing after his long career, Brady continued to be active in his field. Tougaloo College, near Jackson, Mississippi, sought his consultation on how best to develop a chemistry department. He also prompted them to build a new science building and recruit new faculty. He died three days after his 82nd birthday, on December 25, 1966.

⊠ **Brønsted, J. N.**
(1879–1947)
Danish
Physical Chemist

Johannes Nicolaus Brønsted advanced the field of physical chemistry by applying thermodynamic theory to electrochemical reactions, thus discovering important aspects of chemical interactions. He also contributed to the understanding of chemical affinities and ionic principles, but most important, he refined the definitions of acids and bases in accordance with their atomic status as proton rich or proton deficient, a theory he published simultaneously with Thomas Lowry based on independent research. For this work he received the 1928 Ørsted Medal.

Brønsted was born on February 22, 1879, in Varde, in the Jutland region of Denmark. His mother died soon after his birth, and his father, a civil engineer with the water dynamics corporation Hedeselskabet, remarried shortly thereafter. In 1891, when Brønsted was 12, the family moved from a Society for Cultivation of Heaths farm to the city of Aarhus. Two years later his father passed away, and his stepmother decided to move Brønsted and his sister to Copenhagen, where they would have access to the best education within the financial constraints of a fatherless household.

Brønsted attended secondary school at the Metropolitan School, graduating in 1897 to the Polytechnic Institute, where he honored his father's legacy by studying engineering. After earning an engineering degree in 1899, he continued with graduate studies in chemistry. He earned a *magister scientiarum* degree in 1902, and the next year he married his fellow chemical engineering student Charlotte Louise Warburg. The couple moved to Birkerød, where Brønsted worked as an electrical engineer for several years.

In 1905 Brønsted returned to academia as an assistant in chemistry at the University of Copenhagen, while simultaneously conducting doctoral studies. As a dissertation, he submitted a paper (the last of a trio on chemical affinities in reactions) reporting measurements of reactions between water and sulfuric acid. In 1908 the university awarded him a doctorate and appointed him a professor of chemistry (a post he retained throughout the remainder of his career). He taught elementary inorganic chemistry at his alma mater, the Polytechnic Institute, for a decade, until 1919, when the university exempted him from classroom instruction to free his time for research.

In 1912 Brønsted published *Outlines of Physical Chemistry*, a concise textbook that he revised in the 1930s to reflect advances in the application of thermodynamics to the field of physical chemistry. He occupied himself with this line of investigation, publishing 13 important papers on the thermodynamics of chemical affinities between 1906 and 1918. Throughout the last five of these years, he concentrated on measuring specific heat, counting the amount of calories burned in raising 1 gram of a substance 1° Celsius, as well as on examining the ability of solutions to dissolve substances.

In the 1920s Brønsted examined the intersection of thermodynamics and electrochemistry by investigating the mechanics of kinetic reactions, including the rates and intermediary steps of reactions. In 1921 he published "The Principle of the Specific Interaction of Ions," a paper proposing that the interaction between ions in solutions depends on the attraction of opposite charges. His collaborative experimentation with Volume K. La Mer yielded solubility measurements that confirmed the Debye-Hückel theory proposed by PETER DEBYE and ERICH HÜCKEL, as reported in "The Activity Coefficients of Ions in Very Dilute Solutions" in the *Journal of the American Chemical Society*. He also conducted experiments using cobaltammines that his lab inherited from S. M. Jørgensen.

In 1923 Brønsted and the British chemist Thomas Lowry published papers simultaneously reporting on research conducted independently of each other that reached the same conclusion. Both investigators proposed a new theory of acids and bases, refining the definition of acids as substances that release protons, while bases accept protons. Thus the status of a substance as an acid or base depends on the atomic configuration of its protons.

International recognition of the significance of this theory brought Brønsted increasing influence in the field of physical chemistry. Yale University in New Haven, Connecticut, invited him to the United States as a visiting professor from 1926 through 1927. While stateside, he applied to the International Education Board for funding to finance the construction of modernized laboratory facilities in Copenhagen. The board granted his request, allocating sufficient funds to establish the new University Physicochemical Institute under the umbrella of the University of Copenhagen. In 1930 Brønsted moved not only his laboratory but also his family's residence into the new facilities.

Brønsted turned his attention to politics after World War II, running for the Danish parliament on the issue of Schleswig, as both Denmark and Germany claimed sovereignty over this region in peninsular Jutland. On October 28, 1947, the electorate voted him into the position, but he died on December 17, 1947, before he could occupy the office. Two years later the British Chemical Society, which had initiated him as an honorary fellow in 1935, organized a memorial lecture in his name.

⊠ **Buchner, Eduard**
(1860–1917)
German
Biochemist

Eduard Buchner discovered the secret behind the fermentation of yeast into alcohol, revealing that the process required neither living yeast, as "vitalists" such as Louis Pasteur believed, nor decomposing yeast cells, as the "mechanists" believed. For this discovery he won the 1907 Nobel Prize in chemistry and became one of the founders of modern enzyme chemistry.

Buchner was born on May 20, 1860, in Munich, Germany. His mother was Friederike Martin, and his father, Ernst Buchner, edited a medical journal, practiced obstetrics, and taught forensic medicine as a professor at the University of Munich. His older brother, Hans, also became a scientist, specializing in bacteriology as a professor of hygiene at the University of Munich.

Eduard Buchner served in the field artillery after graduating from gymnasium, the German equivalent of high school. He then matriculated at the Technische Hochschule in Munich to study chemistry under Emil Erlenmeyer Sr. However, financial constraints prevented him from continuing there, so he spent the next four years working in canning factories in Munich and Mombach, until 1884 when his brother Hans helped finance his return to the University of Munich. There he studied chemistry under

ADOLF VON BAEYER at the Bavarian Academy of Sciences and botany under Karl von Nägeli at the Institute for Plant Physiology. Fermentation represented the intersection of chemistry with botany, so he focused his research on this process.

In 1886 Buchner published his first paper, on the ability of yeast cells to elicit fermentation aerobically, contradicting Pasteur's belief that it occurred only anaerobically. He then conducted research on pyrazole to earn his doctorate in 1888. By 1890 he had become one of Baeyer's teaching assistants, and within a year he was delivering his own lectures as a privatdozent (lecturer) at the Academy of Sciences. Baeyer granted him the funds necessary to continue his fermentation studies in a small laboratory, where he endeavored to extract the fluid of yeast by rupturing its cells. Before he could finish the experiments that later won him the Nobel Prize, however, the university's Board of Laboratory revoked Baeyer's grant, considering Buchner's work inconsequential.

Buchner left Munich for Kiel, whose university beckoned him with the directorship of its analytical chemistry department in 1893 and later an assistant professorship. In his three years there he conducted his famous fermentation experiments. Martin Hahn, the assistant of his brother Hans, showed him the simple method for cracking yeast cells by grinding them in black sand and diatomaceous earth with a pestle and mortar. Buchner then wrapped the doughy mass in a cloth and pressed the juices out hydraulically. To preserve the extract for experimentation, he added sucrose (as canners add sugar to fruit to create preservatives), not expecting to initiate the very process he was seeking: the heady bubbling of fermentation.

Buchner announced his discovery of fermentation in the absence of cells altogether—neither "vital" live ones nor decomposing ones—in his 1897 paper "Alcoholic Fermentation without Yeast Cells." The previous year the

University of Tübingen hired him as a professor of analytical pharmaceutical chemistry; the subsequent year Berlin's College of Agriculture appointed him director of its Institute of Fermentation Studies and full professor. By 1902, a mere four years later, he had investigated most of what there was to know about the active ingredient of fermenting yeast, what he termed *zymase* (from the Greek root *zyme*, which means "yeast" or "ferment.") The next year he published the comprehensive text *The Zymase: Fermentation* with his brother and Hahn.

At the turn of the century Buchner married Lotte Stahl, the daughter of a former colleague in the mathematics department at the University of Tübingen, and together the couple had three children: two sons and a daughter. In 1907 the Swedish Academy of Sciences bestowed its Nobel Prize in chemistry on Buchner for his discovering what later became known as *glycolysis*, but what he called as "yeast fermentation." Two years later, the University of Breslau appointed him chairman of its department of physiological chemistry, but he only lasted two years until the University of Würzburg wooed him there.

Buchner volunteered as a captain in the German army when World War I erupted. He fought at the head of an ammunition supply unit before his promotion to major in 1916. After a brief respite from battle, he returned to action in Foscani, Romania, where he died of shrapnel injuries on August 13, 1917. His legacy, however, was the establishment of the study of enzymes, which he considered to be the "foremen" in the production of plant and animal cells.

⊠   Bunsen, Robert Wilhelm
(1811–1899)
German
Chemist

Although Robert Bunsen is best remembered for the ubiquitous piece of laboratory equipment

Best known for his design improvements of the Bunsen burner, Robert Wilhelm Bunsen also introduced the spectroscopic method of chemical analysis. *(E. F. Smith Collection, Rare Book & Manuscript Library, University of Pennsylvania)*

named after him, the Bunsen burner, he didn't actually invent it but rather introduced a crucial design improvement. In addition, this was far from his most important contribution to science. More significant was his introduction, in collaboration with Gustav Kirchoff, of the spectroscopic method of chemical analysis. He also discovered the radical cacodyl, as well as the elements cesium and rubidium.

Robert Wilhelm Eberhard von Bunsen was born on March 31, 1811, in Göttingen, Germany, the youngest of four sons. His father was a librarian and professor of linguistics at the University of Göttingen, thus exposing his son to an academic environment from an early age. Bunsen attended school in Holzminden, graduating to his father's university to study chemistry. His keen intelligence was quickly apparent; he earned his doctorate at the age of 19.

A long period of travel throughout his homeland and Europe (financed in part by the German government) to conduct scientific studies ensued, during which he visited laboratories in Berlin, Giessen, Bonn, Paris, and Vienna (where he remained from 1830 through 1833). The University of Göttingen appointed him a lecturer upon his return. While studying arsenic's insolubility there, he discovered an antidote to the poison, which remains the best known anodyne: the use of iron oxide as a precipitating agent.

In 1836 Bunsen went to work at the University of Kassel, then two years later the University of Marsburg hired him. There he distilled arsenic with potassium acetate to arrive at a cacodyl (also known as alkarsine or Cadet's liquid), a malodorous compound that the named after the Greek term for "stinky" (kakōdlēs). However, his arsenic studies blinded him in one eye when the compound exploded, rocketing glass shrapnel into his eye.

After recovering, Bunsen collaborated with Lyon Layfair to devise a means to improve the efficiency of blast-furnace exhaust (which lost 50 to 80 percent of the heat produced) by recycling gases and other by-products, such as ammonia. In 1841 he invented the Bunsen battery, replacing the costly platinum electrode of Grove's battery with a carbon electrode. The next year he was elected a member of the Chemical Society of London. In 1846 he devised Bunsen's theory of geyser action by heating water at the bottom and middle of a tube to the boiling point, thereby generating an ejaculatory explosion.

In 1852 the University of Heidelberg appointed Bunsen to the chair vacated by Leopold Gmelin. There he performed electrolytic studies using chromic acid instead of nitric acid to generate pure metals of chromium, magnesium, aluminum, manganese, sodium, barium, calcium, and lithium. He innovated a means of measuring the specific heat of metals in order to determine their precise atomic weights by developing an ice calorimeter that gauged not the mass but rather the volume of the melted ice. Also in 1852 he commenced a collaboration with Sir Henry Roscoe that lasted the remainder of the decade. During this first year they determined the exponential equivalency between light radiated by the sun per minute and the chemical energy generated by combining equal volumes of hydrogen and chlorine gases to form hydrochloric acid. In 1853 Bunsen was inducted into the Académie des Sciences, and in 1858 he was named a foreign fellow of the Royal Society of London.

In 1859 Bunsen commenced collaboration with Gustav Kirchoff, a Prussian physicist, and together they innovated spectroscopy, or the study of light emitted by burning chemicals, by passing it through a prism, thereby facilitating the chemical analysis of its constituent parts. This process required a nonluminous flame burning at a high temperature, so Bunsen redesigned a burner (by premixing the gas and air *before* combustion) originally designed by the University of Heidelberg technician Peter Desaga. Thereafter the altered lab burner came to be known as the Bunsen burner. Using the Bunsen-Kirchoff spectroscope, which they had pieced together out of a prism, a cigar box, and two discarded telescopes, the collaborating scientists discovered the elements cesium (from the Latin for "blue gray," *caesius*, after its blue spectral lines) and rubidium (from the Latin for "red," *rubidus*, after the violet spectral lines) in 1861.

Bunsen received numerous honors recognizing his accomplishments: the 1860 Copley Medal of the Royal Society, the 1877 Davy Medal (with Kirchoff), and the 1898 Albert Medal. He retired in 1889, and a decade later, on August 16, 1899, in Heidelberg, he died after three days of peaceful sleep.

⊠   Butenandt, Adolf
(1903–1995)
German
Biochemist

The five years between 1929 and 1934 were fertile ones for Adolf Butenandt, who identified and isolated three sex hormones—female estrone and progesterone and male androsterone—during that time period. His discoveries proved instrumental in later developments of cortisone, a hormone-related substance used to counter arthritis, and birth control pills. Butenandt shared the 1939 Nobel Prize in chemistry with LEOPOLD RUŽIČKA, who also conducted sex hormone research, but they didn't receive the award until after World War II, in 1949, because the Nazi government did not permit them to accept the prize.

Adolf Friedrich Johann Butenandt was born on March 24, 1903, in Bremerhaven-Lehe (now Wesermünde), Germany. His mother was Wilhelmina Thomfohrde. His father, Otto Louis Max Butenandt, was a businessman. After graduating from the Bremerhaven Oberrealschule, he matriculated into the chemistry and biology programs at the University of Marburg in 1921. In 1924 he transferred to the University of Göttingen, where he studied biochemistry under Adolf Windaus, who went on to win the Nobel Prize in chemistry in 1928. Butenandt switched dissertation topics when the constitution of thyroxin, the hormone he had focused on up to that point, was described in 1926. In 1927 he completed his dissertation on rotenone, a compound contained in insecticides, and received his doctorate.

Butenandt continued his hormone research at Göttingen, refocusing on sex hormones, as not much was known about them at that point. In 1929 he identified a sex hormone that he found in pregnant women's urine, and he subsequently isolated it in a pure crystalline form. In his article announcing his discovery in the October issue of *Die Naturwissenschaften*, he dubbed this hormone *progynon* but subsequently renamed it *folliculine* (after its source, ovarian follicles). Edward Doisy, an American biochemist, announced his discovery of the same hormone at a lecture in August 1929, so the two clearly made the discovery simultaneously yet independently. In the end the hormone became known as *estrone*, as it originated in female estrogen.

In 1931 Butenandt collaborated with Kurt Tscherning to isolate androsterone, a male sex hormone. They distilled a mere 50 milligrams of the hormone in crystalline form from 4,000 gallons of male urine. From this sample they ascertained that the chemical composition of androsterone differed only slightly from that of estrone, and later that sex hormones are classified as steroids. That same year the University of Göttingen promoted Butenandt to head the organic and biochemical department of its chemistry laboratory. He also married his former research assistant, Erika von Ziegner, and together they had seven children, two boys and five girls.

In 1933 the Technische Hochschule in Danzig hired Butenandt as a professor of organic chemistry and director of its organic chemical institute. There he continued his sex hormone studies with assistance from the Rockefeller Foundation, a rare occurrence in the 1930s, as the American philanthropic organization was pulling its funding from German projects in opposition to the Nazi party. In 1934 he collaborated with Ulrich Westphal to isolate another female sex hormone, progesterone, from the corpora lutea of sows. In the process they determined the close relationship between progesterone and pregnanediol, another substance that Butenandt had discovered in the urine of pregnant women in 1931.

In 1935 Butenandt turned down an invitation from Harvard University for a professorship. A year later the Kaiser Wilhelm Institute for Biochemistry in Berlin-Dahlem appointed him its director, thereby overseeing all scientific research in Germany, while he held a simultaneous position as honorary professor at the University of Berlin. He maintained these positions until after World War II, when the Kaiser Wilhelm Institute was dismantled and reconstructed as the Max Planck Institute in Tübingen (where he had already moved it to avoid the risk of bombing in Berlin). The University of Tübingen also appointed him a professor of physiological chemistry in 1945.

The Royal Swedish Academy of Sciences awarded Butenandt and Leopold Ružička, who had discovered the molecular structure of male sex hormones the Nobel Prize in chemistry in 1939 for their work, but they didn't actually receive the medal until a decade later, after Europe had settled from the war. During the war Butenandt conducted research on the eye pigmentation of insects, and he extended this line of research at Tübingen. In 1953 he isolated the first insect hormone, ecdysone, a chrysalitic hormone that induces a caterpillar to transmogrify into a butterfly. His colleague Peter Karlson later demonstrated its relationship to mammalian sex hormones. Butenandt, in collaboration with Erich Hecker, also identified the first crystallized pheromone in silk spinners, bombykol, which is akin to sexual hormones.

In 1952 Butenandt moved to the University of Munich as a professor of physiological chemistry and director of its physiological chemistry institute. The Max Planck Institute followed him to Munich in 1956, and in 1960 he returned to its services as its president. He retired from this position in 1972, after having retired from his positions at the university the prior year.

Butenandt received many honors in his lifetime, including the Grand Cross for Federal Services of West Germany and the Adolf von Harnack Medal of the Max Planck Society. The French Legion of Honor inducted him as a commander in 1969. Butenandt suffered a long illness in the 1990s and died on January 18, 1995, at the age of 91.

C

⊠ Calvin, Melvin
(1911–1997)
American
Chemist

Melvin Calvin elucidated the photosynthetic process, whose steps are now known as the Calvin cycle, by tracking radioactive carbon dioxide as it is transformed into carbohydrates. He discovered that photosynthesis proceeds in the absence of light and later confirmed the primary elements that formed the atmosphere from which primitive life developed. For these achievements he received the 1961 Nobel Prize in chemistry.

Calvin was born on April 8, 1911, in St. Paul, Minnesota. His mother was Rose Irene Hervitz, a Russian émigré, and his father, Elias Calvin, was a factory worker who had been well educated before emigrating from Russia. A job in the Cadillac factory moved the family to Detroit, Michigan, where Calvin attended public school. He graduated from high school in 1927 and entered the Michigan College of Mining and Technology (now Michigan Technological University). The Great Depression forced him to leave college after two years to work in a brass factory, where he practiced hands-on chemistry. He returned to earn his bachelor's degree in 1931, then pursued his doctorate in chemistry under George Glockler at the University of Min-

nesota. He wrote his dissertation on the electron affinities of halogens, earning his Ph.D. in 1935.

A Rockefeller Fellowship funded Calvin's two years of postdoctoral work at the University of Manchester in England, where he studied catalytic reactions of metalloporphyrins (derivatives of chlorophyll and hemoglobin) under MICHAEL POLANYI, thus commencing his study of the process of photosynthesis. He returned in 1937 to instruct in chemistry at the University of California at Berkeley, which promoted him to assistant professor in 1941. The next year he married Marie Jemtegaard, a social worker of Norwegian descent, and together the couple had two daughters, Elin Bjorna and Karole Rowena, and one son, Noel Morgen.

During World War II Calvin contributed to the cause by serving as an investigator on the National Defense Research Council, then as a researcher obtaining pure oxygen, uranium, and plutonium for the Manhattan Project. After the war the university promoted Calvin to associate professorship in 1945 and to full professorship in 1947. In the intervening year Berkeley's Bio-Organic Chemistry Group of the Lawrence Radiation Laboratory (later renamed the Laboratory of Chemical Biodynamics) appointed him its director, a position he retained until 1980.

After World War II a radioactive isotope of carbon—carbon 14—became available; Calvin capitalized on this by fusing it with oxygen to form CO_2, the necessary component of photosynthesis. This carbon left a radioactive trace as it traveled through the photosynthetic process, making it feasible to track its progress. Calvin exposed chlorophyll and other by-products of the photosynthetic process to paper chromatography and photographic negatives simultaneously, to capture both the chemical composition of the compound on paper and the radioactive carbon on film. By comparing the progression of the carbon through various chemical compositions at different times (he took samples from the chlorella alga at intervals of less than a second), Calvin was able to establish the chemical transformations enacted by photosynthesis. The steps in the process are collectively called the Calvin cycle, in his honor.

Calvin discovered that photosynthesis, or the transformation of carbon dioxide into carbohydrates, does not depend on light, as was previously believed, since the process progresses even in light's absence. For elucidating such a complex process as photosynthesis, Calvin received the 1961 Nobel Prize in chemistry. The year before Berkeley had named him director of its Laboratory of Chemical Biodynamics. In his late career Calvin experimentally established hydrogen, carbon dioxide, and water as the chemical components from which life springs by bombarding them with radiation, which resulted in their transformation into organic molecules.

Calvin also garnered many garlands in his late career: He received the 1964 Davy Medal from the Royal Society, the 1978 Priestley Award and the 1981 Oesper Prize from the American Chemical Society, and the 1989 National Medal of Science. Calvin served his field in turn by acting as president of the American Society of Plant Physiologists from 1963 to 1964 and as president of the American Chemical Society in 1971. He retired as a university professor in 1980 but continued to conduct research on the generation of hydrocarbon fuel as a potential energy source, among other topics.

He died on January 8, 1997, at Alta Bates Hospital in Berkeley, California.

⊠ Carothers, Wallace Hume
(1896–1937)
American
Chemist

Wallace Hume Carothers helped to establish the study and synthesis of polymers, or sets of large molecules in long repetitive chains. He discovered a synthetic rubber that the DuPont company, where he worked for the last decade of his career, called "neoprene." His theoretical and laboratory work resulted in the posthumous discovery of nylon by his team of researchers. He was responsible for more than 50 patents, a testament to his scientific creativity.

Carothers was born on April 27, 1896, in Burlington, Iowa, the eldest of four children born to Mary Evalina McMullin and Ira Carothers. His father taught in a country school from the age of 19, working his way up eventually to become the vice president of the Capital City Commercial College in Des Moines, Iowa. His favorite sister, Isobel, went by the name "Lu" in the musical trio that became popular on radio, Clara, Lu, and Em.

Carothers graduated from North High School in Des Moines in 1914 and matriculated at his father's institution, the Capital City Commercial College, where he studied accounting. He completed the requirements of accounting in a mere year, graduating in 1915. Tarkio College in Missouri offered him a teaching assistantship in accounting while he studied science. After Carothers completed all the chemistry courses in the school's curriculum, the head of the chemistry department, Arthur M. Pardee, departed for World War I. Carothers, who failed his physical examination and could not serve in the military, took over the helm of the chemistry department in Pardee's absence. He graduated from Tarkio in 1920 with a bachelor of science degree.

Wallace Hume Carothers, shown here in the early 1930s, demonstrating the elastic properties of neoprene *(Hagley Museum and Library)*

Carothers pursued graduate study at the University of Illinois, completing his master's in chemistry in one year. He again secured a teaching position thanks to Pardee, who offered him a one-year appointment at the University of South Dakota. Carothers returned to the University of Illinois in 1922 to continue with doctoral study. In 1923 and 1924 the chemistry department granted him its highest award, the Carr fellowship. He wrote his dissertation under Roger Adams on aldehydes in reactions catalyzed by platinum. After earning his Ph.D. in 1924, the department of chemistry offered him a position, so he stayed on at the University of Illinois.

Also in 1924 Carothers published one of his first papers, "The Double Bond," in the *Journal of the American Chemical Society,* in which he applied physicist Irving Langmuir's notions of the double bonds in atoms to organic chemistry. Two years later Harvard University hired Carothers as an instructor in organic chemistry; in his second year there he lectured and served as a lab leader. In 1928 the chemical firm DuPont hired him as director of a new research program in charge of a team of chemists at its Experimental Station in Wilmington, Delaware.

In 1931, within three years of Carothers's arrival, DuPont marketed the team's first major discovery: a polymer that combined vinylacetylene with a chlorine compound to create a rubber analogue known as neoprene. Next he attempted to synthesize polymers through polycondensation, or the removal of water (or similar liquids) from a substance. He experimented with combinations of dibasic acids and dihydroxy compounds, producing numerous unmarketable polyesters and leaving other samples for future analysis. Over his nine years with DuPont, he published 62 technical papers, 31 of which collectively presented his theory of polymer generation and his vocabulary for this new form of chemical synthesis. He also filed for 69 patents, receiving more than 50.

In 1936 the National Academy of Sciences inducted Carothers into its fellowship, the first industrial chemist in its ranks. On February 21, 1936, Carothers married Helen Everett Sweetman, a fellow DuPont employee. She became pregnant the next year; however, Carothers did not live to see the birth of their child. The death of his favorite sister, Isobel, triggered a downward spiral in his manic-depressive tendencies. A colleague noticed him taking home a vial of poison cyanide but didn't realize that he intended to commit suicide. On April 29, 1937, Carothers ingested the poison and died in Philadelphia.

Two of Carothers's creations appeared posthumously: On November 27, 1937, his wife gave birth to their daughter, Jane; and in

September 1938, DuPont first marketed nylon, which had been called Tiber 66 as an untested sample in Carothers's lab. He had combined adipic acid with hexamethylene-diamine to create a dual-stranded polymer with six carbons on each side (hence the numeration). Nylon has since transformed the world with its versatility, and DuPont has continued to earn billions of dollars each year on the product of Carothers's genius.

⊠ Carr, Emma Perry
(1880–1972)
American
Chemist

Emma Perry Carr helped establish the study of spectroscopy in the United States, founding a program in spectroscopic studies at Mount Holyoke College, in South Hadley, Massachusetts, where she worked for her entire professional career. There she instituted innovative pedagogical techniques, including the collaborative approach to research. The American Chemical Society honored her as its first recipient of the Garvan Medal for women chemists.

Carr was born on July 23, 1880, in Holmesville, Ohio. Her mother, Anna Mary, was a housewife who raised five children (Emma was her third). Her father, Edmund Cone Carr, was a physician. Carr attended high school in Coshocton, Ohio, and after graduation, she matriculated at Ohio State University, where she was in a tiny minority as a female student. After her freshman year, she transferred to Mount Holyoke College, where she spent the next two academic years, from 1898 to 1899. She supported herself over the following three years working as an assistant in the college's department of chemistry and saved up enough money to complete her studies at the University of Chicago in 1905, graduating with a bachelor of science.

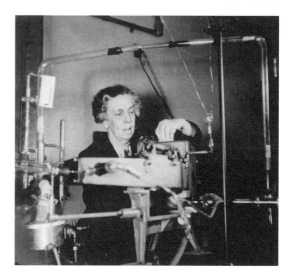

Emma Perry Carr, who helped establish the study of spectroscopy in the United States, when she founded a program in spectroscopic studies at Mount Holyoke College in Massachusetts *(The Mount Holyoke College Archives and Special Collections)*

Carr returned east to teach at Mount Holyoke from 1905 through 1907, and thereafter she pursued doctoral studies under Julius Stieglitz at her alma mater. Two scholarships—the Mary E. Woolley Fellowship and the Lowenthal Fellowship—supported her dissertation research in physical chemistry. In 1910 she became the seventh woman to earn a Ph.D. from the University of Chicago. She again returned to Mount Holyoke, this time as a full professor. In 1913 the college promoted her to head of the department of chemistry, a post she retained for the rest of her career. In this position she inspired generations of students with her meticulous experimental skills, as well as her inventive teaching and organized administrative skills.

Carr conducted spectroscopic studies on the configuration of electrons in organic molecules. She helped pioneer this line of investigation in the United States, as one of the country's first and foremost practitioners. In 1918 she published her first scholarly paper on the topic,

"The Absorption Spectra of Some Derivatives of Cyclopropane." The next year she traveled to Queens University in Belfast, Ireland, to learn techniques from leaders in the field of ultraviolet spectroscopy in order to establish a spectroscopy program at Mount Holyoke. She thus helped promote Holyoke, a women's college, as one of the preeminent institutions for the study of spectroscopy.

Carr moved on to focus her attention on the ultraviolet spectrum, specifically investigating the relationship between alkene's heat of combustion and that of the corresponding saturated hydrocarbon. She presented her findings at a September 1929 meeting in Minneapolis, Minnesota, and subsequently published her results in an important paper comparing the heats of combustion to ultraviolet spectra. She introduced an innovative model for research by encouraging collaboration between students and professors, thus allowing for the cross-fertilization of ideas between diverse levels of expertise.

Collaborative research under Carr's supervision involved first the preparation of highly purified hydrocarbons by means of vacuum spectrography, then a determination of their ultraviolet absorption spectra in both the liquid and the vapor phases. This research transformed the scientific understanding of carbon-carbon double bonds. It was also picked up by the petroleum industry, prompting Carr to deliver papers to the Petroleum Chemistry Section of the American Chemical Society in the 1940s.

After a long and illustrious career Carr retired from Mount Holyoke in 1946. She had distinguished herself throughout her career, winning the first Garvan Medal, in 1937, awarded by the American Chemical Society to the woman who most distinguished herself in the field of chemistry. She set a high benchmark for future recipients. She also received honorary degrees from several institutions: Allegheny College in 1939, Russell Sage College in 1941, and Hood College in 1957.

Carr had never married, and her failing health forced her family to have her admitted into a Presbyterian home for senior citizens in Evanston, Illinois. In an act of support and repayment for her years of devotion to the school, Mount Holyoke contributed to the maintenance of her care and medical expenses, while her nephew, James Carr, attended to her. On January 7, 1972, she succumbed to death after suffering heart failure.

Carver, George Washington
(1865?–1943)
American
Chemist

George Washington Carver's name is closely associated with his innovations involving the peanut: He developed more than 300 uses for the versatile legume. He also innovated applications for many other agricultural products, especially the sweet potato and the pecan. His life and success represented an ideal for African Americans in post-Reconstruction America, as he rose from the shackles of slavery to international prominence and appreciation that transcended the color of his skin.

Carver himself was uncertain of his date of birth, as were many African Americans born into slavery; he alternately dated it in 1864, "near the end" of the Civil War, or "just as freedom was declared." Linda McMurray, in her 1981 biography of Carver, surmises that he was probably born in the spring of 1865. His mother, Mary, had been purchased in 1855, at the age of 13, by Moses Carver, a prosperous slaveowner, although he sympathized with the Union cause. Mary bore an older son, Jim, and perhaps twin daughters who died in infancy.

Carver's paternity is uncertain, though the 1870 census lists him as "negro," as opposed to "mulatto" (the latter was usually used to indicate the progeny of white slave masters with their black female slaves). Carver believed his father to be a slave named Giles from a neighboring farm who died in a logging accident before he was born. While still an infant, Carver and his mother were kidnapped and removed to Arkansas; Moses Carver's neighbor, John Bentley, arranged for the return of Carver at the end of 1865, but his mother never returned, so Moses and his wife, Susan, raised Carver.

Carver was sent to Neosho, Missouri, to commence school, then moved to Ft. Scott, Kansas, in 1877 but moved away two years later after the lynching of a black man. In 1880 he ended up in Minneapolis, Kansas, where he graduated from high school. In 1882 he was swindled out of property in Minneapolis. In 1885 Highland College refused him admission on account of his race. In 1887 he homesteaded in Ness County, Kansas, where he planted corn on the advice of others, despite his intuition that the land would only support grasses such as wheat; his corn crops failed two years in a row. In 1890 he moved to Winterset, Iowa, where he worked washing and ironing clothing. He matriculated at Simpson College in Iowa later that year, and the next year he transferred to Iowa State College of Agriculture and Mechanic Arts.

Carver published his first scientific paper, "Grafting the Cacti," in 1893. The following year he earned his bachelor of science degree, and in 1896 he earned his master of science degree in botany. He joined the faculty of Iowa State, in charge of the systematic botany department's bacterial laboratory. On the strength of a recommendation by James Wilson, who chaired the school's agricultural department (he called Carver his ablest student in cross-fertilization and plant propagation), Carver landed an instructorship later in 1896 at the Tuskegee Normal and Industrial Institute in Alabama, where he remained for the rest of his career.

Over the next half century Carver generated a prodigious amount of research, develop-

ing 118 products synthesized from agricultural materials, including a synthetic rubber, instant coffee, fuel briquettes, meat tenderizer, and Worcestershire sauce. He derived 75 products from pecans, devised 108 applications for sweet potatoes, and produced more than 500 dyes and pigments from 28 different plants. He secured three separate patents between 1925 and 1927 for his process of producing paints and stains from soybeans.

More than anything else, though, Carver concentrated on the peanut. He commenced his peanut studies in 1903, and through the rest of his career, he developed more than 325 peanut-related products. In 1921 he made a presentation on peanuts to the United States Congress that impressed the legislature immensely and helped establish his reputation beyond the barriers of race. In fact, he broke down many of these barriers by lecturing throughout the South at colleges that barred blacks from their campuses.

Although the faculty at Tuskegee sometimes resented Carver's popular success, he gained wide appreciation for his innovations. Internationally, the Royal Society of Arts in London, England, inducted him into its membership in 1916. In 1923 the National Association for the Advancement of Colored People granted him its Springarn Medal for contributions to the advancement of his race. In 1928 Simpson College granted him an honorary doctorate.

In 1939 Carver opened a museum to display his work at Tuskegee. The year before, his health had begun to fail, and in 1942 his health spiraled downward. On January 5, 1943, Carver died of anemia at Tuskegee Institute and was buried there alongside his friend and former Tuskegee president Booker T. Washington. On July 14, 1943, President Franklin Delano Roosevelt apportioned $30,000 for a national monument to Carver, the first such recognition of an African American in the history of the United States.

⊠ Cavendish, Henry
(1731–1810)
English
Physicist, Chemist

Henry Cavendish was an eccentric scientist who published his findings only sporadically, thus leaving some of his significant discoveries unknown until long after his death. He did publish his discovery of hydrogen and carbon dioxide as substances distinct from air, as well as of combining hydrogen with air to create water with no other by-products. His name graces perhaps his most famous work, the Cavendish experiment, in which he calculated the approximate mean density of Earth. Most of Cavendish's experimentation on electricity and heat, however, remained unpublished until some 70 years after his death.

Cavendish was born on October 10, 1731, in Nice, France. Both of his parents hailed from noble lines: His mother, Lady Anne Grey, was the fourth daughter of the duke of Kent, and his father, Lord Charles Cavendish, was the fifth son of the second duke of Devonshire. He attended Hackney Seminary in London, starting in 1742, then matriculated to Peterhouse College of the University of Cambridge in 1749. He studied there until 1753, but his refusal to state adherence to the Church of England prevented him from receiving a degree.

With his brother, Frederick, Cavendish toured Europe and studied physics and mathematics in Paris. Upon his return to England, he took up residence in Clapham and Bloomsbury in London, where he lived frugally despite the fact that he had inherited a fortune from his uncle. He did spend his wealth, however, on scientific apparatuses and texts, amassing a vast library and laboratory that he made available to the public, despite his infamous aversion to social encounters with anyone other than scientific colleagues. This neurosis expressed itself most radically with women, so he avoided con-

Henry Cavendish, whose Cavendish experiment calculated the mean density of Earth *(E. F. Smith Collection, Rare Book & Manuscript Library, University of Pennsylvania)*

tact even with his female servants (he wrote notes containing his requests).

In 1766 he published a paper entitled "Three Papers containing Experiments on Factitious Airs," in which he identified what he called "inflammable air" (later termed *hydrogen* by Antoine Lavoisier) and "fixed air" (or carbon dioxide) as substance distinct from air. The next year he published an analysis of London city water combined with "fixed air," resulting in the suspension of calcareous matter; in other words, he discovered calcium bicarbonate. In 1783 he demonstrated that the composition of air was constant no matter the geographic region or altitude sampled. The next year he published "Experiments on Air," describing his ignition of hydrogen and air to produce water, with no other apparent by-products nor any loss in weight; he thus proposed that water is a compound.

The Cavendish experiment, which earned him a degree of renown, involved the measurement of the attraction between a large and a small lead ball using a Cavendish balance to read the relative torsion. By this means, he arrived at the gravitational constant (G), allowing him to calculate the mean density of Earth, which was about five-and-a-half times that of water. The methods of this experiment stood unimproved for more than a century.

Cavendish's investigations concerning heat and electricity lingered unpublished in his notebooks during his lifetime, though he discovered phenomena that anticipated the work of MICHAEL FARADAY and others. He discovered specific inductive capacity, compared the electrical conductivities of equivalent solutions of electrolytes, and expressed a version of Ohm's law. In keeping with his eccentricity, he measured the strength of electrical currents not with instrumentation but rather by evaluating relative degrees of pain upon shocking himself.

Recognition of his significance as a scientist came early—the Royal Society inducted him into its fellowship in 1760—and continued until late in his career—the Institute of France named him as one of only eight foreign associates in 1803. Cavendish died on February 24, 1810, in London. He was buried in Derby, at All Saints' Church, which maintains a monument to him.

Cavendish never married and hence had no direct heirs. However, he left more than 1 million pounds to diverse relations. One of these, the seventh duke of Devonshire, endowed a laboratory at the University of Cambridge in his benefactor's name, establishing the Cavendish Laboratory in 1871. In 1879, almost 70 years after Cavendish's death, James Clerk Maxwell examined his notebooks and published a collection of his previously unknown papers and findings.

⊠ Clark, Josiah Latimer
(1822–1898)
English
Chemist, Electrical Engineer, Civil Engineer

Josiah Latimer Clark contributed to the advancement of science in diverse ways: He invented the Clark cell, innovated the pneumatic transfer system, and founded a company that laid thousands of miles of telegraph cable. While his developments helped promote the dissemination of information, he also served as a repository of information, collecting texts important to the history of scientific thought as well as the ephemera that defined scientific thought in his day—the journals and reports issued by his contemporaries.

Clark was born on March 10, 1822, in Great Marlow, England. His brother, an engineer named Edwin, helped establish him in the scientific field by offering him a job on the Britannia Tubular Bridge in Wales in 1847. His work on this project prompted his promotion to assistant engineer on the Menai Straits Bridge project the next year. Clark went on to distinguish himself as an engineer, concentrating his efforts on developing applications for the emerging field of electrical engineering. In addition to his work, Clark, who never married, was a great lover of books and gardening.

As an electrical engineer, Clark invented the double-bell insulators that safeguarded telegraph wires from overload, an innovation that used to adorn all utility poles. He further advanced telegraphy by founding a company to manufacture telegraph cable. The company produced and laid more than 100,000 miles of submarine cable, creating a global telegraphic network.

Clark advanced the short-range transfer of physical objects by inventing the pneumatic transfer tube system. This system, which persists at bank drive-up windows, revolutionized the trading of stocks by allowing transactions to occur at some distance. In 1854 the city of London installed the first pneumatic mail system, a $1^1/_2$-inch-wide tube running the 220 yards between the London Stock Exchange and the Central Telegraph Office. A steam engine powered the pneumatic transfer of message cylinders at 20 feet per second, thereby relieving the traffic on the overloaded telegraph lines.

Within six years, the system expanded into the R. S. Culler/R. Sabine radial pneumatic telegraph and mail system throughout London. Similar systems sprouted around Europe: the Berlin stock exchange installed his system in 1865, and the city of Paris established a pneumatic mail system in 1868. Most major European cities followed suit: Hamburg, Vienna, Prague, Munich, Rome, Naples, Milan, Marseilles, and even Rio de Janeiro in Brazil all installed the Clark pneumatic system to transfer information faster than previously possible.

Clark applied his practical skills to the field of electrochemistry by inventing the Clark cell, familiar to all second-year chemistry students, who re-create his experiment to learn hands on its multifold thermodynamic implications. The Clark cell consists of a zinc amalgam anode and a mercury cathode both submerged in a solution saturated with zinc sulphate. Upon heating the cell generates a constant electromotive force (emf) of 1.4345 volts at 15° Celsius.

Clark effected his lasting impact on science not only through his own experimentation and innovations but also as a bibliophile ravenously collecting and meticulously chronicling the texts that defined the development of scientific thought. His collection of nearly 6,000 items included some 200 texts dating back to the advent of scientific thought and reporting on such topics as the seemingly magical powers of lodestar, the mechanics of the mariner's compass, and early theories attempting to explain electricity and magnetism. The early collection boasted works by seminal thinkers from the likes of Pliny to Descartes, but Clark also chronicled the writings of his near contemporaries: Franklin, Priestly,

Aepinus, and Coulomb, as well as the electrophysical treatises by Volta and Maxwell, for example.

In addition, Clark recorded the current development of his field, pulling together more than 100 sets of American and European electrotechnical periodicals dating from the 1840s to the turn of the century, as well as journals covering telephony, electroplating, and electromagnetic motors. Of course he kept abreast of his own field of specialization, telegraphy. He also collected such obscure documents as regulations, reports, trade catalogs, price lists, and annual reports of telegraphic companies.

Clark died on October 30, 1898. His library fell into the hands of the American electrical engineer, Schuyler Skaats Wheeler, who purchased the collection in its entirety when it went up on the auction block in 1901. Wheeler transported it across the Atlantic to New York City, where he donated it to the American Institute of Electrical Engineers. His deed of gift stipulated the cataloging of the collection, which was to be made available to the public in a reference library. Andrew Carnegie matched Wheeler's donations to fund the cataloging effort and added another $1 million on top of that to construct a library to house the collection.

In 1913 the Wheeler Collection joined other collections to form the core of the Engineering Societies Library. In early 1995 this library was split up, with the bulk of the collection transferring to the Linda Hall Library in Kansas City, while the Wheeler Collection remained (as stipulated by its donator) in New York City, housed in the Rare Book Collection of the New York Public Library.

⊠ **Cori, Gerty Theresa Radnitz**
(1896–1957)
Austro-Hungarian/American
Biochemist

Gerty Cori became the first American woman and third woman ever to win the Nobel Prize in

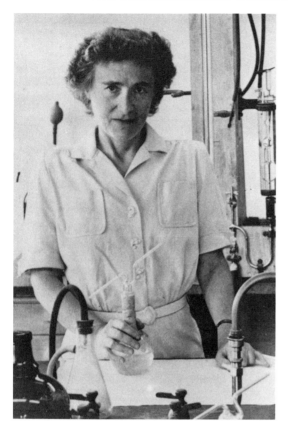

Gerty Theresa Radnitz Cori, the first American woman to win the Nobel Prize in science *(Bernard Becker Medical Library, Washington University School of Medicine)*

medicine or physiology, which she shared with her husband, Carl Ferdinand Cori, and the Argentine physiologist Bernardo A. Houssay for their work on carbohydrate metabolism. Investigating the mechanism whereby the carbohydrate glycogen breaks down into sugar glucose, the Coris discovered that the process involved more than simple hydrolysis (reaction with water) but rather relied on a series of reactions with enzymes. Together the Coris elucidated the entire process, now known as the Cori cycle in their honor.

Cori was born Gerty Theresa Radnitz on August 15, 1896, in Prague, then part of the

Austro-Hungarian Empire (now the Czech Republic). She was the eldest of three daughters born to Martha Neustadt and Otto Radnitz, a sugar refinery manager. She was tutored at home before entering private school. She attended secondary school at the Tetschen Realgymnasium, where she studied for the *matura* exam for entrance into university, which she passed.

In 1914 Radnitz matriculated at the University of Prague (also known as Ferdinand University) at the urging of her maternal uncle, a professor of pediatrics there. She studied medicine, collaborating with Carl Ferdinand Cori on studies of human complement, an antibody binder that controls immune responses. She earned her medical degree in 1920, and on August 5 of that year, she married Cori. The couple moved to the University of Vienna, where she worked as an assistant at the Karolinen Children's hospital conducting research on congenital myxedema, a malfunction of the thyroid gland.

Discouraged by the political and economic upheaval in postwar Europe (Gerty suffered from malnourishment for lack of food), the Coris immigrated to the United States in 1922. The New York State Institute for the Study of Malignant Diseases (later renamed the Roswell Park Memorial Institute) hired both Coris, Gerty as an assistant pathologist, and her husband as a biochemist. Her first publication in English tracked the growth of paramecia in reaction to a thyroid extract and, more important, marked the commencement of the carbohydrate metabolism studies that would define her career. Joined by her husband (against others' advice in an age when nepotism rules sought to partition marital bonds from the professional realm) when he could spare time from his official duties, they studied how insulin epinephrine affects the absorption of carbohydrates from the intestines, as well as studying glycerin formation and degradation.

In 1928 the Coris became naturalized citizens of the United States. Three years later, in

1931, the Washington University School of Medicine in St. Louis, Missouri, hired the Coris. There Carl assumed professorships in biochemistry and pharmacology and eventually chaired the latter department, while Gerty languished in a low-paying research position. Despite discrepancy in their official titles, however, they were equals in the laboratory, where they continued their collaborative research on sugar metabolism.

In their research the Coris sought to uncover the mechanism whereby the body converts glycogen, a carbohydrate polymer stored by the body, into glucose, another sugar polymer that feeds the body. The scientific community assumed that this process hinged on hydrolysis, or reactions with water molecules, but the Coris' research revealed that an enzyme (phosporylase) converted the glycogen into an intermediary compound (glucose-1-phosphate) made up of glucose and what is known as the Cori ester (a phosphate). They later discovered the next step was production of another intermediary compound (glucose-6-phosphate), formed by the action of the enzyme phosphoglucomutase, followed by the creation of glucose, which in turn transforms into lactic acid. This entire process was dubbed the Cori cycle in honor of its discoverers.

In the midst of this research, Cori gave birth to their only child, Carl Thomas, in 1936. In 1943 Washington University finally acknowledged her status as more than a mere assistant by promoting her to the rank of research associate professor of biochemistry. Four years later, in 1947, the university promoted her to a full professorship. Incidentally, that year she became the first American woman to win the Nobel Prize in medicine or physiology, an honor she shared with her husband and the Argentine physiologist Bernardo A. Houssay, who conducted carbohydrate metabolism studies on the function of the pituitary gland. Cori was only the third woman to win the Nobel Prize, after MARIE CURIE and IRÈNE JOLIOT-CURIE; interestingly, all three women shared the prize with their husbands.

After winning the Nobel, Cori turned her attention to the study of glycogen storage disorders, discovering that these inherited childhood diseases subdivided into two categories: those involving a surplus of glycogen and those involving abnormal glycogen, both of which depended on the action of enzymes. Throughout this post-Nobel period, Cori gathered more awards and honors, including the 1947 Squibb Award in endocrinology, which she shared with her husband; the 1948 Garvan Medal for women scientists from the American Chemical Society; the 1948 Women's National Press Award; and the 1950 Sugar Research Prize of the National Academy of Sciences, which also inducted her into its prestigious fellowship. In 1952 President Harry S Truman appointed her to the National Science Board of the National Science Foundation.

Unfortunately, in the summer of 1947, Cori, had discovered that she suffered from a rare bone-marrow disease, myelofibrosis; however, she persisted in her studies for another decade. She died of kidney failure on October 26, 1957, in St. Louis. In his eulogy famed newscaster Edward R. Murrow praised her dedication, intellectual integrity, courage, and professionalism in the service of science. Perhaps a more lasting testament to her significance as a scientist is the fact that six future Nobel laureates trained in her laboratory.

⊠ Cornforth, Sir John Warcup
(1917–)
Australian
Biochemist

John Cornforth shared the 1975 Nobel Prize in chemistry with VLADIMIR PRELOG for their independent research on the stereochemistry of enzymatic reactions. Cornforth made what the Royal Swedish Academy of Science called "an outstanding intellectual achievement" despite the fact that he was almost completely deaf since his 20s.

John Warcup Cornforth was born on September 7, 1917, in Sydney, Australia. He was the second of four children born to Hilda Eipper, the descendant of a German minister who had settled in New South Wales in 1832; Cornforth spent some of his childhood in rural New South Wales, at Armidale, and the other part in Sydney. His father was an English-born graduate of Oxford. When John was 10, his hearing began to deteriorate due to otosclerosis. At the age of 14, Cornforth set up a makeshift chemistry laboratory at home. He studied chemistry at the Sydney Boys' High School under Leonard Basser, who encouraged him to pursue the subject as a career in which he could communicate through his research.

At the age of 16, Cornforth matriculated at the University of Sydney, where he earned the University Medal in organic chemistry upon his graduation with first-class honors in 1937. He remained at the university to conduct graduate research under J. C. Earl on chemical compounds found in Australian plants, earning his master of science degree in 1938. After a year of postgraduate research, he was one of two Australian students to win the 1951 Exhibition Scholarship for study at Oxford; the other was Rita Harradence, an organic chemist from Sydney. After marrying in 1941, the couple collaborated throughout their careers (she aids in his communication), and together they have three children.

At Oxford Cornforth researched the laboratory synthesis of steroids under the organic chemist SIR ROBERT ROBINSON to earn his doctorate in 1941. During World War II, while working as a Medical Research Council (MRC) research scholar from 1942 through 1946, he identified the amino acid D-penicillamine as the major component of penicillin, and he was the first to synthesize it. He also contributed to the writing of the 1949 text *The Chemistry of Penicillin*. After the war he remained with the MRC at the National Institute for Medical Research, first in Hampstead, then in Mill Hill, London.

During those years Cornforth conducted research on the chemotherapy of tuberculosis and leprosy; on the process by which enzymes transform acetic acid into squalene, then cholesterol in the liver; and on the synthesis of cortisone sialic acid and squalene. However, he made his lasting impact on scientific history with his research into the stereochemistry of enzymatic reactions, prompted by Alexander Ogston in a brief note in 1948. In collaboration with his wife and George Popják, Cornforth conducted this research using isotopes of hydrogen to determine the possible mechanisms by which the reactions take place.

In 1962 Shell Research, Ltd., hired Cornforth and Popják as codirectors of its Milstead Laboratory of Chemical Enzymology at Sittingbourne in Kent, where he continued his stereochemical research. He focused on mevalonic acid, the parent of most steroids as well as of terpenes, which is the foundation of the flavor and odor components of plants. The departure of Popják from Milstead in 1968 left Cornforth as the sole director of the laboratory. Cornforth continued his line of stereochemical investigation, now collaborating with Herman Eggerer on the chiral methyl group.

Cornforth held concurrent visiting professorships at the University of Warwick from 1965 through 1971 and at the University of Sussex from 1971 through 1975. On October 17, 1975, the Royal Swedish Academy of Sciences announced its awarding of the Nobel Prize in chemistry to Cornforth and Prelog on December 12. That year, Cornforth departed his position at Milstein and accepted the University of Sussex's offer of the Royal Society research professorship, a post he retained until 1982, when he retired to emeritus status.

Throughout his career, several prestigious organizations inaugurated Cornforth into their membership: The Royal Society inducted him into its fellowship in 1953; in 1977 the Australian Academy of Science made him a coordinating member; and in 1978 both the American Academy of Sciences and the Royal Netherlands Academy of Science made him a foreign associate. The Royal Society of London granted him three of its most prestigious awards: the 1968 Davy Medal (with Popják), the 1976 Royal Medal, and the 1982 Copley Medal. The same year he received the Nobel, he was named Australian Man of the Year, and two years later, Queen Elizabeth II knighted him Sir John Cornforth. And in 1991 his homeland of Australia further named him Companion of the Order of Australia.

Francis Harry Compton Crick, whose role in determining the structure of deoxyribonucleic acid (DNA) won him, James Watson, and Maurice Wilkins the 1962 Nobel Prize in medicine or physiology *(Courtesy of The Salk Institute)*

⊠ Crick, Francis Harry Compton
(1916–)
English
Molecular Biologist

Francis Crick won the 1962 Nobel Prize in medicine or physiology with James Watson and Maurice Wilkins for their roles in determining the structure of deoxyribonucleic acid (DNA), the building blocks of life and the key to genetic transmission. He subsequently conducted exhaustive research on DNA before moving on to brain studies in an attempt to determine the mechanics of consciousness.

Francis Harry Compton Crick was born on June 8, 1916, in Northampton, England. His mother was Anne Elizabeth Wilkins, and his father, who managed a shoe and boot factory, Harry Crick. Crick attended Northampton Grammar School and then received a scholarship to Mill Hill School so his family moved to North London. He enrolled at University College at age 18 and completed the curriculum in three years, earning a second-class honors degree in physics, with a minor in mathematics, in 1937.

Crick remained at University College to study the viscosity of water at high temperatures under Edward Neville de Costa Andrade, but the advent of World War II interrupted his studies. He contributed to the war effort as a civilian worker at the British admiralty, designing mines to destroy ships beginning in 1940. That year he married Ruth Doreen Dodd, who gave birth to their son, Michael, during an air raid in London on November 25, 1940. (Michael went on to become a scientist.) Later in the war, Crick designed weapons for scientific intelligence at the admiralty headquarters in Whitehall. There, he met Odile Speed, an art student serving as a naval officer. Crick divorced his first wife in 1947 and married Speed in 1949. The couple eventually had two daughters together, Gabrielle and Jacqueline.

In 1947 Crick commenced doctoral studies on the physical properties of cytoplasm in cultured chick fibroplast cells under Arthur Hughes at the Strangeways Laboratory in Cambridge. In 1949 he transferred to the Cavendish Laboratory in Cambridge supported by a scholarship with the Medical Research Council Unit to study protein structure under the chemists (and future

Nobel laureates) MAX FERDINAND PERUTZ and John Kendrew.

In the fall of 1951, at Cavendish Laboratory, Crick shared an office with James Watson, a 23-year-old visiting fellow from the United States who was a member of the illustrious phage group (short for bacteriophages, or bacterial viruses). The two conspired to crack the code of DNA; however, DNA is too small for microscopic research, and academic policies limited X-ray crystallographic research to King's College Laboratories in London, where ROSALIND FRANKLIN and Maurice Wilkins were conducting just such research. Franklin and Wilkins stonewalled Crick's suggestions of pursuing the structure of DNA by building models (as American chemist LINUS CARL PAULING had done with protein molecules), preferring empirical research over theoretical research.

Crick and Watson built models of their own but could not confirm their results without access to X rays of DNA. Watson attended a November 1951 lecture by Franklin in which she proposed a "big helix" structure, but Watson neglected to take notes and misremembered vital information. After building a vastly flawed model, Crick and Watson were ordered to desist in their DNA research, which they did until January 1953, when Pauling proposed a model that later turned out to be flawed. Watson then visited Wilkins to request a look at the King's College DNA X rays, which Franklin had refused to share with Wilkins in protest of his sexist treatment of her. Wilkins, however, had secretly copied her photographs, and showed them to Watson, who sketched the image in the margin of a newspaper on the train ride back to Cambridge.

Crick and Watson set about building models, abiding by chemical rules as well as the X-ray photos, which suggested a helical structure, made up of two chains of DNA. On March 7, 1953, they succeeded in making a model that fit all the known constraints. They announced the Watson-Crick model (the former won a coin toss to determine primacy of name order) in a paper straightforwardly titled "A Structure of Deoxyribonucleic Acid" that appeared in the April edition of *Nature*. It humbly began, "We wish to suggest a structure for the salt of deoxyribose nucleic acid (DNA)."

Crick spent the 1953–1954 academic year on leave of absence at the Protein Structure Project at Brooklyn Polytechnic in New York. Upon his return, he submitted his dissertation, entitled "X-ray Diffraction: Polypeptides and Proteins," to earn his Ph.D. in 1954. Crick remained at the Cavendish Laboratories, conducting research with Vernon Ingram on the relationship between DNA and genetic coding. In 1957 he started collaboration with Sydney Brenner to extrapolate the amino acid sequence in proteins based on the sequence of DNA bases. In 1959 he served as a visiting lecturer at Rockefeller Institute for Medical Research (now the Rockefeller University), in New York City, as well as serving as a visiting professor at Harvard for that year and the next.

In 1962 Crick shared the Nobel Prize in medicine or physiology with Watson and Wilkins for their DNA breakthrough. Franklin had died of cancer in 1958, preventing her from being named with them, though her role in this discovery was duly acknowledged. Crick later contributed to the creation of a DNA/RNA dictionary before moving on to study the structure and function of histones, or proteins associated with chromosomes. In 1966 he published *Of Molecules and Men*.

In 1976 the Salk Institute for Biological Sciences in La Jolla, California, appointed Crick as its Kieckhefer Professor, an endowed chair. There Crick conducted research on the brain, specifically considering the mechanism of consciousness. He published three more books: *Life Itself* (with Leslie Orgel) in 1981; *What Mad Pursuit: A Personal View of Scientific Discovery* in 1988; and *The Astonishing Hypothesis: The Scien-*

tific Search for the Soul in 1994. Although he has not made any significant breakthroughs in his consciousness researches, he continued to pursue them in hopes of further elucidating the secrets of life.

⊗ **Crutzen, Paul J.**
(1933–)
Dutch
Atmospheric Chemist, Meteorologist

Paul Crutzen shared the 1995 Nobel Prize in chemistry for his identification of nitrogen oxides as catalytic converters of ozone in the stratosphere, a phenomenon that depletes the ozone layer, which is crucial to human survival on Earth. He shared the prize with MARIO MOLINA and FRANK SHERWOOD ROWLAND for their identification of chlorofluorocarbons (CFCs) as the main threat to the ozone layer.

Paul Josef Crutzen was born on December 3, 1933, in Amsterdam, the Netherlands. His mother was Anna Gurek; his father, Joseph C. Crutzen, was a waiter who was often unemployed during the Nazi occupation of Holland, which also interrupted Crutzen's elementary schooling. After World War II he qualified to attend high school at a time when not all children could. Too poor to afford a university education, he attended a two-year college in Amsterdam to earn his civil engineering degree in 1954.

Crutzen worked for the next four years at the Bridge Construction Bureau of the City of Amsterdam, and during that period also fulfilled his two-year military service obligation to his country. While on vacation in Switzerland in 1954, he met Tertu Soininen, a Finnish student, and the couple married in 1958. They moved to Gävle, Sweden, where he worked for a year in the House Construction Bureau and indulged his love of skating. The couple eventually had two daughters, Illona and Sylvia.

In 1959 the family moved to Stockholm, where Crutzen had landed a job as a computer programmer (with no experience in the field) in the Department of Meteorology at the Stockholm Högskola (renamed Stockholm University two years later). He conducted graduate studies at the university simultaneously, earning his *filosofie kandidat* (the equivalent of a master's of science) in 1963.

Crutzen continued with doctoral studies under Bert Bolin, focusing his research on the distribution of different forms of oxygen in the stratosphere, mesosphere, and lower thermosphere. He discovered that much of the chemistry of the stratosphere was based on guesswork, so he wrote his dissertation on the catalytic depletion of ozone with nitrogen oxide, among other things. He submitted his thesis, "Determination of Parameters Appearing in the 'Dry' and 'Wet' Photochemical Theories for Ozone in the Stratosphere," to earn his *filosofie licentiat* (the equivalent of a Ph.D.) with highest distinction in 1968, but he waited two years to publish his findings in a paper entitled "The Influence of Nitrogen Oxides on the Atmosphere Ozone Content."

Crutzen served a two-year postdoctoral fellowship from the European Space Research Organization at the Clarendon Laboratory in the Department of Atmospheric Physics at St. Cross College of the University of Oxford in England, where he continued to study the effects of nitrogen oxides (and hydrogen oxides) on atmospheric ozone under John Houghton and R. P. Wayne. He returned to the University of Stockholm in 1971 and composed his dissertation, entitled "On the Photochemistry of Ozone in the Stratosphere and Troposphere and Pollution of the Stratosphere by High-Flying Aircraft." He earned his *filosofie doctor* (the equivalent of a doctor of science degree) with highest distinction in 1973.

From 1974 through 1980 Crutzen worked at the National Center for Atmospheric Research

(NCAR) in Boulder, Colorado, first as a research scientist in the Upper Atmosphere Project until 1977 and then as a senior scientist and director of the Air Quality Division. During his early tenure he also consulted for the Aeronomy Laboratory of the Environmental Research Laboratories at the National Oceanic and Atmospheric Administration (NOAA), also in Boulder. Later in his tenure at NCAR he moonlighted as an adjunct professor in the Atmospheric Sciences Department at Colorado State University in Fort Collins. His research during this period focused on the atmospheric effects of the burning of savanna grasses and agricultural waste in the tropics, mainly in Brazil.

In 1980 the Max Planck Institute for Chemistry in Mainz, Germany, appointed Crutzen as the director of its Atmospheric Chemistry Division. At this time the journal *Ambio* commissioned Crutzen and John W. Birks, a chemistry professor from the University of Colorado at Boulder on sabbatical in Mainz, to consider the effects of nuclear war on Earth's atmosphere. By 1982 the pair had ascertained that the actual nuclear explosions would pale in their destructive capabilities compared to the havoc wreaked by carbon soot sent into the atmosphere by the ensuing fires, which would block 99 percent of the sunlight from reaching Earth's surface. They coined the term *nuclear winter* to describe this phenomenon, and *Discover* magazine named Crutzen its 1984 Scientist of the Year in recognition of the significance of this theory. At that time he was serving a two-year term as the executive director of the Max Plank Institute for Chemistry, from 1983 through 1985. The next year he published a book-length report of his and Birks's findings, *Environmental Consequences of Nuclear War 1985.*

When Susan Solomon, a senior scientist at NAOO and a former research student of Crutzen's, verified the existence of a hole in the ozone layer above Antarctica in 1986, Crutzen was one of the first scientists to identify aerosol

as one of the causes. He based this on his work from the 1970s, in which he identified that carbonyl sulfide produced in the soil or perhaps in the ocean rises to the stratosphere, where it is oxidized to sulfur dioxide and then to sulfuric acid, a phenomenon that leads to the depletion of ozone from the atmosphere.

The sum of Crutzen's work on the threats to the atmosphere earned him the 1995 Nobel Prize in chemistry. Crutzen divided the latter part of his career between multiple institutions: While maintaining his appointment at the Max Planck Institute for Chemistry, he also served as a part-time professor in the Department of Geophysical Sciences at the University of Chicago from 1987 through 1991, a part-time professor at the Scripps Institution of Oceanography of the University of California at La Jolla from 1992 on, and a part-time professor at the Institute for Marine and Atmospheric Sciences of Utrecht University in his homeland from 1997 through 2000. In November 2000 he retired to emeritus status.

⊠ **Curie, Marie**
(1867–1934)
Polish/French
Chemist, Physicist

Marie Curie was the first woman ever to win a Nobel Prize, which she received in physics in 1903 with her husband, Pierre Curie, and her doctoral adviser, ANTOINE-HENRI BECQUEREL, for their discovery of radioactivity and the element radium. She became the first scientist to win a second Nobel, when she received one in chemistry in 1911 for her isolation of metallic radium and complete elucidation of the element.

Curie was born as Marya Skłodowska on November 7, 1867, in Warsaw, Poland. She was the youngest of five children born to Bronsitwa Boguska, a pianist, singer, and teacher who died of tuberculosis when Skłodowska was 10 years

Marie Curie was the first woman ever to win a Nobel Prize and the first scientist to win the prestigious prize twice. She is shown here with her husband, Pierre Curie, with whom she shared the 1903 Nobel Prize in physics. *(AIP Emilio Segré Visual Archives)*

old. Thereafter, her father, Ladislas Skłodowski, a professor of physics and mathematics, raised the family. He educated his youngest daughter, introducing her to physics experimentation early on. She also attended high school graduating at the head of her class when she was merely 15.

For the next eight years Skłodowska worked as a tutor and governess to finance study in Paris.

During this period she studied in the "floating university," an underground movement of Polish professors who taught outside the Russian-controlled curriculum. In November 1891 she traveled to Paris, where she enrolled at the Sorbonne as "Marie," the French version of her name. In 1893 she graduated at the top of her class with a degree in physical sciences. The next year she graduated second in her class with master's degree in mathematical sciences and remained in Paris to work for a French industrial society.

A Polish friend introduced Skłodowska to Pierre Curie, a professor at the École de Physique et Chimie, and the couple, drawn together by a common humanitarian and scientific dream, married on July 26, 1895. Supported by an Alexandrovitch Scholarship (which she later repaid to allow others to benefit as she did), Marie Curie commenced doctoral studies on Becquerel's rays, discovered in 1896 by her thesis adviser, Henri Becquerel. First she renamed this phenomenon *radioactivity*, then she set to work on uranium in a makeshift laboratory housed in a shed behind the École de Physique et Chimie.

Within two months she observed two key aspects of radioactivity: first, the greater the amount of uranium in her sample, the more intense the rays, in exact proportion; second, the rays did not alter when she altered the uranium sample in any way (by temperature, exposure to light, combination with other elements, etc.). Tests of other minerals (chalcocite, uranite, and pitchblende) yielded higher-than-expected readings of radioactivity, suggesting the existence of a more powerful source of radioactivity than uranium. While conducting this research in 1897, Curie gave birth to Irène (a future Nobel laureate).

In the spring of 1898 Curie commenced her efforts to isolate the radioactive element in pitchblende; in the summer, her husband put his research on hold to join her in refining 8 tons of pitchblende to isolate a minuscule sample of radioactive material. In that sample the Curies identified two new radioactive elements: polonium, discovered in July 1898 and named after her homeland, and the even stronger radium, announced on December 26, 1898.

In 1900 Curie accepted a position at the École Normale Supérieure for girls in Sèvres, near Paris, as its first female teacher, while she continued her radium studies. By March 1902 the Curies had isolated enough radium (a decigram) to confirm its existence, determine its atomic weight, and describe some of its properties. In 1903 Curie became chief of the laboratory that she and her husband had established. In June of that same year she became the first woman to receive a doctorate in France, graduating summa cum laude. Also in 1903 the Curies received awards from around the world in recognition of their discovery of radium: the Davy Medal of the Royal Society; the Daniel Osiris Prize; and the Nobel Prize in physics, which they shared with Becquerel. Curie had become the first woman to receive the Nobel Prize. The next year she gave birth to a second daughter, Eve (who later wrote a biography of her mother). However, on April 19, 1906, Curie lost her husband when he absentmindedly stepped in front of a horse-drawn cart, which struck him dead.

In May 1906 the Sorbonne asked Curie to assume her husband's chair in physics, making her the first woman to hold a professorship in France. She continued her radium research, and in 1910 she and her assistant, André Debierne, succeeded in isolating metallic radium. In 1911

the Royal Swedish Academy of Sciences rewarded her work with an unprecedented second Nobel Prize, this time in chemistry. In 1914 the Sorbonne established a new laboratory for radioactivity studies, the Radium Institute, placing Curie at its head.

Curie spent World War I devising medical applications for radioactivity: She designed X-ray machines for the battlefield to help care for wounded soldiers and established 200 permanent X-ray radiological units in France and Belgium. In 1921, at the invitation of American journalist Marie Meloney, Curie toured the United States to raise funds for radioactivity research and returned with a gift from American women—a gram of radium. She toured the United States again in 1929 and returned this time with a gram of uranium (purchased with a $50,000 check given to her by President Herbert Hoover) for use in the newly established Curie Institute in her birthplace of Warsaw, Poland.

Upon her return Curie began to suffer from fatigue, dizziness, low-grade fevers, humming in the ears, and loss of eyesight; other scientists researching radioactivity suffered similar symptoms, suggesting toxicity in radioactive substances. Curie retired to a mountain sanitorium in Saint Cellemoz Haute Savoie, where she died of pernicious anemia on July 4, 1934. In 1995 the ashes of Pierre and Marie Curie were moved from a country cemetery to Paris, where they were interred in the Panthéon; she was the first woman so honored for her own accomplishments, not simply to rest aside her husband.

D

Dalton, John
(1766–1844)
English
Chemist, Meteorologist

At the beginning of the 19th century, John Dalton transformed the scientific understanding of molecules with his atomic theory, in which he proposed that atoms of different elements have different weights. He presented the first table of atomic weights as if an aside, appending it to the end of a paper in which he also proposed his law of partial pressures. His other scientific contributions—including his discovery of butylene, his determination of the composition of ether, and his conclusion of the magnetic nature of the aurora borealis—pale in comparison to the significance of his atomic theory.

Dalton was born on or around September 5, 1766, in Eaglesfield, near Cockermouth, in Cumberland, England. He was the third of six children born to Joseph Dalton, a handloom weaver who reared his family as Quakers. As a young student, Dalton studied under the Quaker meteorologist Elihu Robinson, who inspired the boy to record daily meteorological observations, a discipline he maintained through his entire life, resulting in the chronicling of more than 200,000 atmospheric observations in his notebooks, dating back to 1787.

Somewhat of a child prodigy, Dalton started teaching at the local school at the age of 12. Within three years he moved to the Quaker school in Kendal, where he and his brother served as assistant masters. The school appointed him its principal in 1785, a position he retained until 1793, during which time he collaborated with the blind mathematician and botanist John Gough. At the end of his tenure in Kendal, the Presbyterian college Manchester Academy appointed him as a tutor in natural philosophy and science on the strength of his reputation as a public lecturer. Also in 1793 Dalton published his *Meteorological Observations and Essays,* in which he explained the condensation of dew, tabulated water's vapor pressure at diverse temperatures, asserted that the aurora borealis was a magnetic phenomenon, and described how to construct and use basic meteorological apparatus. The next year he published the first detailed description of the condition that he and his brother suffered, color blindness (also known as Daltonism). Also in 1794 he joined the Manchester Literary and Philosophical Society, where he delivered more than 100 papers in his 50 years as a member.

In 1799 Dalton finally acknowledged that his teaching duties eclipsed his own researches, prompting him to resign in favor of freelancing as a private tutor. That year the Manchester Literary and Philosophical Society moved into a

The atomic theory of matter, developed by John Dalton, states that all elements are comprised of minute, indestructible particles called atoms. *(E. F. Smith Collection, Rare Book & Manuscript Library, University of Pennsylvania)*

new house and offered Dalton room for living, teaching, and conducting research. The next year, in 1800, the society appointed him its secretary. In October 1803 he read a paper on "The Absorption of Gases by Water and Other Liquids," which included his law of partial pressures (also known as Dalton's law), proposing that the pressure of mixed gases was equivalent to the sum of the pressures each gas would exert if it were alone in the same space. He later published this paper in 1805.

At the end of this paper Dalton humbly and almost incidentally added, "I shall subjoin my results, as far as they appear to be ascertained by my experiments." His addendum amounted to the first tabulation of atomic weights, an assertion of such significance that the Royal Institu-

tion of Great Britain invited him to expound his atomic theory further in lectures from 1803 through 1804. Based on Dalton's lost notebooks, Henry Roscoe surmised that Dalton arrived at his atomic theory in answer to the question why gases mix instead of separating into layers based on their densities. He resolved this quandary by proposing that the rate of proportion by which gases combine represents the relative atomic masses of those gases.

Dalton expounded his atomic theory more completely in his 1808 text *New System of Chemical Philosophy*. That year the society named him its vice president, and more than a decade later, in 1819, he ascended to the presidency of the organization. At about this time recognition of the significance of his scientific contributions came to roost: In 1816 the French Academy of Sciences named him a corresponding member (later honoring him as one of only eight foreign associates in 1830); in 1822 the Royal Society elected him into its fellowship and in 1826 awarded him its first Royal Medal.

In 1833 the British government granted him a pension, liberating him from the shackles of tutoring for a living, and in 1836 this pension doubled. Dalton died on July 27, 1844, in Manchester. Forty thousand mourners viewed his coffin, and 100 carriages accompanied his casket on the route to its burial. A German air raid in 1940 destroyed the Manchester Literary and Philosophical Society's house, where Dalton had boarded and stored his papers; in 1993 some of these papers were discovered and may be restorable of the fire damage they sustained.

⊠ Daniell, John Frederic
(1790–1845)
English
Chemist, Meteorologist

John Frederic Daniell invented a voltaic cell, now named the Daniell cell, that maintained its charge much longer than the existing electric cells at the time. This invention bolstered the telegraph industry through its infancy, allowing for sustained transmissions with its constant current. Earlier he had invented the dew-point hygrometer to measure atmospheric humidity. In testament to his unprecedented intelligence, the Royal Society inducted him into its ranks at a very early age, and King's College created a professorship in chemistry for Daniell, its first such chair, despite the fact that he had never attained a postsecondary education.

Daniell was born on March 12, 1790, in London, England. His father was a lawyer. Daniell's relatives gave him his first job at their sugar refinery and resin factory, where he was introduced to the chemical process. Chemistry lectures by William T. Brande inspired him to pursue his own chemical investigations. The excellence of this independent research brought him to the notice of the prestigious Royal Society, which inducted him into its fellowship in 1814, when he was a mere 23 years old.

Daniell conducted meteorological research, in addition to his chemical studies, and in 1820 he invented a device called a dew-point hygrometer to measure the humidity of the atmosphere. Three years later he published *Meteorological Essays,* a collection of his papers on Earth's atmosphere, the trade winds, and instructions for constructing meteorological instrumentation. What distinguished this text was Daniell's use of physical laws to explain atmospheric phenomena, as well as his meticulous exactitude in meteorological observations and measurements. On a practical note, his suggestion that the moisture of hothouses required monitoring led to a transformation in the management of hothouses. For its second edition, which came out in 1827, he revised this text to include a discussion on radiation.

In 1831 King's College in London appointed Daniell as its first professor of chemistry on the strength of his research and writings and despite the fact that he lacked academic credentials. In the mid-1830s, he turned his attention to electric cells in response to the demand for more consistent and longer-lasting power sources for the burgeoning telegraph industry. At the time telegraphy depended on the voltaic cell, invented by Alessandro Volta in 1797, which lost its potential once the energy had been drawn due to hydrogen bubbles gathering on the copper plate and creating resistance to the free flow of the circuit. Voltaic cells thus had an extremely brief shelf life, forcing telegram messages to remain exceedingly brief lest the energy supply fail mid-message.

In 1836 Daniell devised a new type of cell consisting of a negative zinc amalgam electrode immersed in a dilute solution of sulfuric acid contained in a porous pot, surrounded by a solution of copper sulfate contained in copper with a positive copper electrode immersed in it. The porous pot allowed hydrogen ions to pass through to the copper sulfate but prevented the mixing of the two electrolytes. This cell, now known as the Daniell cell, sustained a constant current over long periods of time and thus served as a perfect energy source for telegraphy. British and American telegraph companies employed the Daniell cell exclusively, though other constant current cells were developed thereafter (namely, Grove's nitric acid depolarized cell and Sand batteries). Interestingly, telegraph operators measured the cell's power by the degree of pain it induced upon contact with their nerves.

In 1839 Daniell attempted to fuse metals by means of a 70-cell battery; this arrangement, however, generated such a powerful electric arc that the ultraviolet rays damaged Daniell's vision, as well as harming the eyes of other observers, who walked away from the experiment with an artificial sunburn. He followed up on these investigations more carefully to demonstrate that a metal's ion, not its oxide, carries the electric charge in electrolysis of metal-salt solutions.

Daniell dedicated his 1839 book *Introduction to the Study of Chemical Philosophy* to the eminent chemist, MICHAEL FARADAY, who was a close friend. The Royal Society granted Daniell its Rumford Medal in 1832, and then after he had invented his eponymous cell, the society presented him its Copley Medal in 1837. Daniell, who never married, died on March 13, 1845, in London.

⊠ Davy, Sir Humphry
(1778–1829)
English
Chemist

Sir Humphry Davy is best remembered for his discovery of the psychoactive effects of what came to be known as laughing gas. However, his significance as an experimental chemist reached much beyond this one contribution. He discovered potassium, and he was the first to isolate sodium, magnesium, calcium, strontium, and barium. On the more practical side, he invented the miner's safety lamp.

Davy was born on December 17, 1778, in Penzance, in the Cornwall region of England. He was the eldest of five children born to a wood carver who died in 1794. He attended primary school at the Truro Grammar School, then embarked on a surgical/apothecarian apprenticeship under J. Bingham Borlase in order to support his family. His interest turned to chemistry, however, after reading *Traité Elémentaire*, which prompted him to test Antoine Lavoisier's experimental conclusions. Davy demonstrated a contradiction to the caloric theory by rubbing two ice cubes together; the fact that they melted disproved the theory of heat as a substance, suggesting instead that it was a product of motion. He also demonstrated that light was a constituent of oxygen, not a transformation of heat. He later contradicted Lavoisier by showing that hydrochloric acid contains no oxygen.

In 1798 Thomas Beddoes hired Davy as superintendent at his Medical Pneumatic Insti-

tution in Bristol, where Davy acted as a guinea pig by inhaling various gases to determine their potential therapeutic effects (and incidentally internalizing some of their detrimental effects, whether he realized it or not). His most infamous research involved nitrous oxide, inhaling 16 quarts in seven minutes, rendering him "completely intoxicated." In 1800 he published his findings in *Researches … concerning nitrous oxide.* That year he also commenced electrolytic research on voltaic action (recently discovered by Alessandro Volta in 1797), which Davy determined to be the result of chemical reaction. He took up this line of investigation later in his career.

Davy's nitrous oxide investigations impressed Count Rumford (Benjamin Thompson) so much that the count invited Davy to become an assistant lecturer at the Royal Institution in February 1801. Davy quickly rose through the academic ranks, as the institute promoted him to lecturer by June and professor of chemistry by May 1802. He established himself as a preeminent lecturer, casting a favorable light on the institute.

During a five-week period in 1806, when the Royal Institution relieved Davy of his academic and administrative responsibilities, he performed 108 experiments, predominantly returning to the electrolysis studies he had commenced at the turn of the century. He reported his results in the Bakerian Lectures to the Royal Society in 1806. The next year, he delivered another round of Bakerian Lectures in November 1807, this time reporting his discovery of potassium by electrolyzing molten potash and his isolation of sodium. Napoléon awarded him a 3,000-franc prize for the top galvanic electricity research of the year. The next year, 1808, he isolated magnesium, calcium, strontium, and barium, and in 1810 his demonstration of oxygen's absence from what was known as "oxymuriatic acid" led him to conclude that it was an element, which he named *chlorine* (after the Greek term *chlōros*, meaning "yellow-green").

On April 8, 1812, Davy was knighted, and three days later he married Jane Apreece, a Scottish widow who had inherited wealth. The couple spent the next several years touring Europe, accompanied by MICHAEL FARADAY, Davy's assistant ever since the untrained scientist had submitted detailed notes on Davy's lectures. Also in 1812 he published his *Elements of Chemical Philosophy*, followed the next year by his *Elements of Agricultural Chemistry*. Upon return from his continental tour, he invented the miner's safety lamp, which burned without exploding even in the presence of the volatile mixture of oxygen and methane, though he refused to patent the invention, leading to false counterclaims by locomotive engineer George Stephenson as its inventor.

When Davy returned from a second European tour from 1818 to 1820, the Royal Society elected him its president. The society had awarded him its 1805 Copley Medal and its 1818 Rumford Medal. Interestingly, Davy twice opposed Faraday's induction into the Royal Society (perhaps masking his jealousy of his former assistant's success), though Faraday eventually gained entrance into the fellowship in 1824.

In 1827 Davy fell ill, presumably in response to the damage he'd done to his body by inhaling gases at the Medical Pneumatic Institution. In 1829 he moved to Rome, where he suffered a heart attack. He subsequently died on May 29, 1829, in Geneva, Switzerland.

⊠ Debye, Peter
(1884–1966)
Dutch/American
Chemist, Physicist

Peter Debye distinguished himself first as a physicist by utilizing complex mathematical applications to solve physics problems. He determined the dipole moment of molecules, which is now measured in debyes in his honor. Later he

Peter Debye, for whom the base unit of the electric dipole moment is named *(E. F. Smith Collection, Rare Book & Manuscript Library, University of Pennsylvania)*

fused his physics research with chemistry to propose the Debye-Hückel theory of electrolytic dissociation. In 1936 he won the Nobel Prize in chemistry in recognition of these contributions.

Petrus (Pie) Josephus Wilhelmus Debije was born on March 28, 1884, in Maastricht, a city in the Limburg region of the Netherlands. His mother, Maria Anna Barbara Ruemkens, was a theater cashier, and his father, Johannes Wilhelmus Debije, was a foreman in a metalworks factory. His sister and only sibling was born in 1888. Debije had no intention of continuing his education after attending the Hoogere Burger School from 1896 through 1901, so he neglected the study of Greek and Latin necessary for entrance into Dutch university. After topping the graduation examinations for the region, though, he decided to continue, so he crossed the border to Aachen, Germany, where the Technische Hochschule had no classics requirement. There he anglicized his name to Peter Debye.

Debye served as an assistant to the theoretical physicist Arnold Sommerfeld from 1904 until his graduation in 1905 with a degree in electrical engineering. Thereafter he followed his mentor to the University of Munich, where Debye pursued doctoral studies while continuing to assist in Sommerfeld's lab. He presented his dissertation, a mathematical theory of rainbows with discussion of the effects of radiation on spherical particles, on July 1, 1908, and remained at the university as a *privatdozent*, or lecturer, after receiving his doctorate.

Debye made a name for himself by providing solutions to vexing scientific questions of his day. He deduced Max Planck's empirical radiation law (proposed a mere two months before) by experimenting with modes of vibration in a radiation cube. Next, Debye helped Albert Einstein resolve an inadequacy in his 1907 quantum theory of solids by mathematically calculating the vibration of waves in an elastic sphere. The University of Zurich in Switzerland recognized this feat by hiring Debye in 1911 to fill a one-year appointment in the prestigious chair of theoretical physics recently vacated by Einstein. While in Zurich, Debye conducted his famous work on the dipole moments of molecules, or their tendency to rotate in an external magnetic field due to their electrical polarity.

In 1912 Debye moved to the University of Utrecht in the Netherlands. While there, he married Mathilde Alberer, the daughter of his landlord back in Munich, on April 10, 1913. The couple subsequently had two children: Peter Paul Ruprecht, who was born in 1916, became a physicist and collaborated with his father, and Mathilde Maria Gabiele was born in 1921.

At the urging of German physicist David Hilbert, the University of Göttingen in Germany hired Debye as a professor of theoretical and experimental physics in 1914. Yet again Debye focused his attention on solving a scientific quandary: how to conduct X-ray diffraction studies by something other than the Laue-Bragg method, which required the use of large crystals. With his assistant, Paul Scherrer, he discovered that powdered lithium fluoride diffracts X rays; the discovery was subsequently named the Debye-Scherrer method.

The University of Zurich rehired Debye in 1920 as a professor of experimental physics and director of the Federal Institute of Technology's physics laboratory. There he extended his experimentation to include chemical applications. Attempting to explain electrolytic dissociation, he collaborated with Erich Hückel to produce a mathematical model that exactly describes the attraction and repulsion of ions in solution. The Debye-Hückel theory superseded the theory first propounded by SVANTE AUGUST ARRHENIUS. Debye also provided a mathematical solution to the Compton effect, or the diversion of X-ray wavelengths by electrons, in 1923.

In 1927 the University of Leipzig in Germany appointed Debye as a professor of experimental physics and director of its Institute of Physics. He remained there for seven years, until 1934, when he moved to a similar position as a professor of theoretical physics at the University of Berlin and the director of the newly established physics institute at the Kaiser Wilhelm Gesellschaft in Dahlen.

Debye was nominated for the Nobel Prize in physics 15 times since David Hilbert first threw his name in the hat in 1916, and he was nominated for the Nobel Prize in chemistry every year from 1927. Finally, in 1936 he won the Nobel Prize in chemistry for his work on dipole moments and on X-ray diffraction.

In 1939 Cornell University in Ithaca, New York, invited Debye to deliver the George Fisher

Baker chemistry lectures, and he stayed on there as a way of escaping Nazi Germany with his family. Cornell appointed him a professor of chemistry and head of the chemistry department in 1940. He became a United States citizen in 1946, and in 1948 Cornell named him its Todd Professor. He retired to emeritus status in 1952 but continued his scientific studies, producing 64 more publications thereafter.

Besides the Nobel Prize, Debye also won a plethora of other awards: the 1930 Rumford Medal of the Royal Society, the 1935 Lorentz Medal of the Royal Netherlands Academy of Sciences, the 1937 Franklin Medal of the Franklin Institute, the 1949 Faraday Medal and Gibbs Medal, the 1957 Kendall Award, the 1963 Nichols Medal, and the 1963 Priestley Medal of the American Chemical Society. He suffered a heart attack at Kennedy international airport in New York in April 1966; on November 2, 1966, another heart attack took his life in his home in Ithaca.

⊠ Diels, Otto
(1876–1954)
German
Chemist

Otto Diels collaborated with his former student KURT ALDER to identify one of the most ubiquitous reactions in nature—what became known as the Diels-Alder reaction. The pair received the 1950 Nobel Prize in chemistry jointly in recognition of this work, but Diels had already distinguished himself as an important chemist before his work with Alder: He identified carbon suboxide, a previously unknown compound; he oxidized cholesterol to create Diels acid; and he hydrogenated cholesterol to generate the Diels hydrocarbon.

Otto Paul Hermann Diels was born on January 23, 1876, in Hamburg, Germany. His mother, Bertha Dübell, was the daughter of a dis-

trict judge, and his father, Hermann Diels, was a professor of classical philology at the University of Berlin. All three of their sons followed their father's footsteps into academia: Diels's older brother, Ludwig, became a professor of botany at the University of Berlin; his younger brother, Paul, became a professor of Slavic philology at the University of Breslau.

Diels attended secondary school at the Joachimsthalsches Gymnasium in Berlin from 1882 to 1895, when he graduated to the University of Berlin. He studied chemistry under EMIL FISCHER, who won the Nobel Prize in chemistry in 1902. After serving his compulsory year of military service in 1896, he conducted his dissertation research on cyanuric compounds to earn his doctorate in 1899. He remained at the university into the next century, first as an assistant in Fischer's lab at the Institute of Chemistry and then as a lecturer as of 1904. That year he rose to international prominence by designing the gold medal–winning chemical apparatus display in Germany's chemistry exhibit in the Louisiana Purchase Exposition in St. Louis, Missouri.

In 1906 Fischer effected Diels's promotion to assistant professor. That year and the next Diels performed groundbreaking work by dehydrating diethylmalonic acid with phosphorous pentoxide to produce a malodorous gas that he subsequently realized to be an unidentified compound; he dubbed it carbon suboxide. Also in 1907 Diels published his classic textbook, *Einführung in die Organische Chemie*, which went through 22 editions and remained in print continuously through 1966. He later published an elementary inorganic chemistry lab manual in 1922.

In 1909 Diels married Paula Geyer, the daughter of a government official, and together the couple had three sons and two daughters. In 1913 the University of Berlin appointed Diels as the head of its chemistry division, and then in 1916 he moved to a professorship at Christian Albrecht University in Kiel, where he remained for the rest of his career as the director of its Institute of Chemistry.

Diels focused his research on the oxidation of cholesterol, a method by which he discovered what is called Diels acid, a dicarboxylic acid that has a high melting point. He subsequently shifted methods from oxidation to dehydrogenation of cholesterol, using selenium to remove the hydrogen by forming hydrogen selenide gas. By this process he identified a hydrocarbon that became known as Diels hydrocarbon, a basic component of cholesterol and many other natural compounds, such as steroids. He reported this work in 1927.

The next year he and Alder published their work on the reaction between acrolein and butadiene, a specific version of what became known as the Diels-Alder reaction. This process conjoins a dienophile (acrolein, in this case), or a double-bonded molecule, with a conjugated diene (butadiene, in this case), or a molecule containing two adjacent double bonds. Over the next 16 years, Diels published 33 more papers on this type of reaction.

World War II devastated Diels: In the winter of 1943–44, two of his sons (Hans Otto and Klaus) were killed within three months of each other on the Russian front; in 1944 Allied bomb raids destroyed both his laboratories and his home; and in 1945 his wife died. In the wake of these tragedies, Diels requested an early retirement, which the university granted in 1945. However, Diels remained on staff until his successor was named in 1948, and he continued to lecture until 1950. Thereafter, his health deteriorated due to arthritis, and on March 7, 1954, he died of a heart attack at the age of 78. Two years earlier he had received the Grosskreuz des Verdienstordens der Bundesrepublik Deutschland, and in 1931 he had received the Adolf von Baeyer Memorial Medal from the Society of German Chemists.

E

⊠ Eigen, Manfred
(1927–   )
German
Physical Chemist

Manfred Eigen opened up new worlds of chemical analysis by studying reactions that transpire in mere nanoseconds, a realm of inquiry inaccessible before Eigen innovated his method of disrupting compounds to observe their return to states of equilibrium. This process, which he called the "relaxation" technique, allowed Eigen to time the rates of chemical reactions, revealing much about the reactants in the process. For this work Eigen received the 1967 Nobel Prize in chemistry with SIR GEORGE PORTER and RONALD G. W. NORRISH for their similar work on fast chemical reactions.

Eigen was born on May 9, 1927, in Bochum, Germany. His mother was Hedwig Feld; his father, Ernst, was a chamber musician. Eigen attended secondary school at the Bochum Gymnasium for humanities, then he fought in an army anti-aircraft artillery unit at the end of World War II. He enrolled at the Georg-August University in Göttingen in 1945 to study chemistry and physics under Arnold Eucken, for whom he wrote his dissertation on the specific heat of heavy water and aqueous electrolyte solutions to earn his doctorate in natural sciences in 1951. The next year he married Elfriede Müller, and together the couple had two children: Gerald, born in 1952, and Angela, born in 1960.

Eigen remained at the university for two years as an assistant lecturer in physical chemistry under Ewald Wicke, after which the Max Planck Institute for Physical Chemistry hired him. There he commenced his studies using his "relaxation" technique of chemical analysis. Previously chemists had measured chemical attributes by observing chemical reactions; Eigen revolutionized this line of study by instead disrupting a chemical compound and then observing its return to equilibrium as a means of adducing its attributes.

Eigen's colleagues, Konrad Tamm and Walter Kurtze, conducted ultrasonic studies measuring distance through seawater that yielded different results from the same studies with distilled water; Eigen correctly theorized that magnesium sulfate in seawater absorbs some ultrasonic waves, hatching the notion that he could study the composition of chemical compounds by disrupting them with events such as ultrasonic bombardments, eliciting a reaction and then a quantifiable return to their normal state, or "relaxation." With Leo de Maeyer, he measured relaxation times down to the nanosecond, thus uncovering chemical processes that had previously been "invisible" to the scientific eye. He first reported on his technique in his 1954 paper published in the *Discussions of the Faraday Society*, "Methods for Investigation of Ionic Reactions in Aqueous Solutions with Half Times as Short as 10-9 Sec.: Application to Neutralization and Hydrolysis Reactions."

The institute promoted Eigen to the rank of research fellow in 1958 and four years later named him head of its Department of Biochemical Kinetics. In 1964 he headed the directorship of the Max Planck Institute, and the next year Cornell University named him Andrew D. White Professor at Large. He served a three-year term as managing director of the institute from 1967 through 1970, and at that same time the German Federal Republic appointed him member of its Scientific Council.

Recognition of his work's significance came relatively early to Eigen: The German Physical Society granted him its 1962 Otto Hahn prize, and the American Chemical Society awarded him its 1963 Kirkwood Medal and its 1967 Linus Pauling Medal. Also in 1967 he received the Nobel Prize in chemistry. The year before, three universities (Harvard University, Washington University, and the University of Chicago) bestowed honorary doctorates on him; also in 1966 the National Academy of Sciences in the United States appointed him a foreign associate. He went on to receive the Faraday Medal from the British Chemical Society in 1977.

Eigen applied his relaxation spectrometry on biochemical reactions in an effort to trace his way back to the most basic chemical reaction that created organic life, which he viewed as a chance occurrence. He discussed this theory in his 1981 book, *Laws of the Game: How the Principles of Nature Govern Chance*, coauthored by Ruthild Winkler. Two years earlier he discussed his theory of hypercycles—how nucleic acids organize themselves into proteins—in his text entitled *Hypercycles: Principles of the Self-Organization of Macromolecules* (coauthored by Peter

Schuster). Eigen's work has also proven instrumental in other scientific endeavors, such as radiation chemistry and enzyme kinetics.

⊠   Elion, Gertrude Belle
(1918–1999)
American
Biochemist, Pharmacologist

Gertrude Belle Elion developed a panoply of drugs to cure diseases and alleviate ailments as diverse as leukemia, gout, herpes, malaria, arthritis, and the rejection of organ transplants. Soon after she retired, her laboratory completed her work on azidothymidine (AZT), the first drug to counter the effects of autoimmunodeficiency syndrome (AIDS). In all, she held 45 patents for her innovations. For these innovations she shared the 1988 Nobel Prize in medicine and physiology with George Herbert Hitchings, her career-long collaborator, and the British biochemist Sir James Black.

Elion was born on January 23, 1918, in New York City, the first of two children (she had a brother). Her mother, Bertha Cohen, was a Russian immigrant, and her father, Robert Elion, was a Lithuanian immigrant who became a dentist. Elion excelled in school, skipping two grades. Her father's losses in the 1929 stock market crash prevented him from financing his daughter's college education, but her high marks in high school earned her a full scholarship to Hunter College, the women's arm of the College of the City of New York. In 1937, at the age of 19, she graduated summa cum laude from Hunter.

Despite this superlative distinction, graduate programs and industry jobs alike rejected Elion's applications, so she spent the next several years picking up jobs where she could: She taught biochemistry to nurses for one semester, worked in a friend's laboratory, and acted as a substitute science teacher in public schools. She

Recipient of the 1988 Nobel Prize in medicine, Gertrude Belle Elion knew from the age of 15 that she wanted to do cancer research. *(Courtesy Burroughs Wellcome Company)*

financed her own graduate study at New York University, where she was the only woman student in the chemistry department. She earned her master of science summa cum laude in 1941. The sexism of academia and industry forced her to work below her abilities as a quality-control chemist at the Quaker Maid Company, checking mayonnaise color and pickle acidity. A short stint as a research chemist at Johnson & Johnson ended when her division collapsed after she had been there only six months.

The departure of men to serve in World War II, however, opened up many of the positions appropriate to Elion's qualifications. On June 14, 1944, the Wellcome Research Laboratories of the Burroughs Wellcome Company (now Glaxo Wellcome) hired her to work in the lab of biochemist George Hitchings, commencing a collaborative relationship with him that lasted the rest of their careers. Her attempt to earn a doctorate was thwarted when the Brooklyn Polytechnic Institute (now Polytechnic University), where Elion took night classes for two years, required her to choose between her job and full-time study. As a single woman (she never married) supporting herself, she chose work over education. She worked her way up the ranks by filling the positions that Hitchings vacated on his rise up the corporate ladder.

In the early 1950s Hitchings and Elion, by then a senior research chemist, conducted research on purine biochemistry, focusing their investigations on the metabolic difference between normal cells and malignant cells. These studies resulted in the 1951 discovery of the compound 6-mercaptopurine (6MP, marketed as Purinethol), the first drug to counteract the effects of childhood leukemia. The Sloan-Kettering Institute (now the Memorial Sloan-Kettering Cancer Center) confirmed its efficacy, and the U.S. Food and Drug Administration (FDA) approved it for commercial distribution in 1953.

Continued pharmokinetic research yielded even more uses for 6MP and its derivatives. Elion derived azathioprine (marketed as Imuran), which suppressed the rejection of kidney transplants and treated rheumatoid arthritis. Elion and Hitchings's lab discovered a veritable factory line of new drugs: pyrimethamine (marketed as Daraprim and Fansidar) to treat malaria; trimethoprim (marketed as Bactrim and Septra) to fight urinary and respiratory tract infections, as well as Pneumocystis carinii pneumonia (which now plagues AIDS victims); allopurinol (approved by the FDA as Zyloprim in 1966) to address gout; and acyclovir (discovered in 1969 and approved by the FDA as Zorivax in 1982) to remedy the herpes simplex virus, as well as shingles and chicken pox.

Elion retired from her position as head of the Department of Experimental Therapy to emerita scientist status at Wellcome Research Laboratories in 1983. The next year her laboratory completed work that she had commenced in developing azidothymidine (AZT), the first drug to effectively medicate the swelling ranks of those suffering from the newly identified disease AIDS. Far from resting in retirement, Elion accepted an appointment as a research professor of medicine and pharmacology at Duke University, in Durham, North Carolina, where she advised third-year medical school students. The American Association for Cancer Research elected her as its president, and she also worked with the National Cancer Institute, the Leukemia Society of America, and the World Health Organization.

The Royal Swedish Academy of Sciences granted Elion its 1988 Nobel Prize in chemistry jointly with Hitchings and Black. Two decades earlier the American Chemical Society had recognized the significance of her work with its 1968 Garvan Medal, and in the wake of her receipt of the Nobel Prize, she was inducted into the National Academy of Sciences as well as the National Inventors' Hall of Fame and the National Women's Hall of Fame. In 1991 she won the National Medal of Science, and in 1997 she received the Lemelson-MIT Lifetime Achievement Award. Elion died on February 21, 1999.

F

Faraday, Michael
(1791–1867)
English
Chemist, Physicist

Michael Faraday is best remembered for his work with electricity, as he discovered electromagnetic induction and helped found field theory. He also helped popularize the understanding of science as a clear public lecturer and contributed directly to the field of chemistry by discovering benzene and liquefying chlorine.

Faraday was born on September 22, 1791, in Newington, Surrey, England, the third of four children. His mother was Margaret Hastwell; his father, James Faraday, was a journeyman blacksmith who worked sporadically due to ill health and died in 1809. Faraday received his early education through the Sandemanian sect, a dissenting Christian group, and he retained a healthy skepticism throughout his scientific career that was balanced by a faith in God and in his own ability to test his conclusions experimentally.

Faraday quit school at the age of 14 to apprentice as a bookbinder to a London bookseller named George Riebau; he continued his education as an autodidact, reading the books he bound. While rebinding a copy of the *Encyclopaedia Britannica*, he became entranced by the entry on electricity and thereafter submerged himself in the study of science. Toward this end he joined the City Philosophical Society in 1810 and attended public lectures of SIR HUMPHRY DAVY at the Royal Institution in 1812, meticulously taking notes that he subsequently bound and presented to the lecturer. Fortuitously for Faraday, the famous scientist was then suffering a temporary blind spell as the result of a laboratory explosion, so he required an assistant. He chose Faraday, hiring him on March 1, 1813.

From 1813 to 1815, Faraday accompanied Davy on a European tour that amounted to his scientific education at the hands of the scientific luminaries of the day. After his return to England, the Royal Institution appointed him as superintendent of the laboratory apparatuses and the meteorological collection on May 21, 1821. On June 2 of that year he married Sarah Barnard, who hailed from a prominent Sandemanian family. The couple had no children. Following up on the implications of Hans Christian Oersted's discovery of electromagnetism, Faraday discovered the principle of the electric motor on September 3 and 4, 1821.

Faraday was the first to liquefy chlorine, in 1823. Two years later he discovered and isolated benzene. Also in 1825 the Royal Institution appointed him director of laboratories. Around this same time period he discovered the first compounds of chlorine and carbon to be identi-

Best known for his work with electricity, Michael Faraday also contributed to the field of chemistry by discovering benzene and liquefying chlorine. *(Photo by John Watkins, courtesy AIP Emilio Segré Visual Archives)*

fied. In 1826 he founded the Royal Institution's Friday Evening Discourses and the Christmas Lectures for juveniles (both of which continue today); he delivered 123 of the former and 19 of the latter in his lifetime, thereby establishing his reputation as a clear lecturer and disseminator of scientific knowledge. In 1827 he published his first book, *Chemical Manipulation*, and in 1830 the Royal Military Academy in Woolwich appointed him a professor of chemistry.

Almost exactly a decade after his discovery of the electric motor principle, on August 29, 1831, he discovered electromagnetic induction. Over the next three years he developed his the-

ory of electrochemical action, introducing (along with William Whewell) such now-familiar electrical terms as *electrode*, *electrolyte*, *anode*, and *cathode*. He published his theories in the book *Experimental Researches in Electricity* in 1839, with subsequent editions appearing until 1855.

At the 1845 meeting of the British Association in Cambridge, a 21-year-old Scot named William Thomson (later known as Lord Kelvin) inquired if Faraday had considered the effect on light passing through an electrolyte. Faraday had, with disappointing results, which he confirmed upon subsequent experimentation. However, this line of investigation inspired him to try passing light through glass in an electromagnetic field, which rotated the light's plane of polarization. This magneto-optical effect is now known as the Faraday effect.

On November 4, 1845, Faraday studied the effect of electromagnetism on the glass itself and found that it, too, rotated its alignment. He repeated this experiment on other substances, finding that they also reacted to electromagnetism, a phenomenon he called "diamagnetism." Faraday presented these findings in an April 1846 lecture entitled "Thoughts on Ray-vibrations," which he subsequently generalized into what is now known as field theory. Although most scientists rejected this theory at the time, Thomson and James Clerk Maxwell confirmed it mathematically after Faraday's death, and it became a cornerstone of modern physics.

Faraday continued to write, publishing *Experimental Researches in Chemistry and Physics* in 1859 and *A Course of Six Lectures on the Chemical History of a Candle* in 1861. He refused the offer of knighthood, standing firm on his Sandemanian principles of humility. In the 1860s he retired from public life to the "grace and favour" house for special residents at Hampton Court, where he sank into senility. He died on August 25, 1867, at Hampton Court, in Surrey. Five days later, he was buried in the Sandemanian plot in Highgate Cemetery.

⊠   **Fischer, Emil**
(1852–1919)
German
Chemist

Emil Fischer received the 1902 Nobel Prize in chemistry for his disciplined studies of proteins and sugars. More an experimentalist than a theorist, he distinguished himself by making discovery after discovery through the systematic investigation of the composition of various chemicals. In addition to his research on amino acids and glucose, he synthesized more than 100 derivatives of purine, which he had discovered to be the base of uric acid.

Considered by many to be the father of biochemistry, Emil Fischer received the 1902 Nobel Prize in chemistry for his studies of proteins and sugars. *(E. F. Smith Collection, Rare Book & Manuscript Library, University of Pennsylvania)*

Emil Hermann Fischer was born on October 9, 1852, in the small town of Euskirchen, near Bonn, Germany. He was the youngest of six children born to Julie Poensgen and Lorenz Fischer. His father, an astute businessman who owned a grocery store, then a wool-spinning mill, and later a brewery, rejected his family's traditional Lutheranism by pronouncing himself an atheist. Ever a pragmatist, he steered Fischer clear of mathematics and physics in favor of chemistry, which stood to earn his boy a better living. His father had already vainly attempted to lure Fischer into the family business, until he realized that his only son had failed to inherit his business acumen: "The kid is too dumb to be a businessman; he should go to school."

After graduating at the top of his class from gymnasium in 1869, Fischer matriculated in the spring of 1871, at the University of Bonn, where he attended the lectures of August Kekulé and Rudolf Clausius. He transferred the next year to the University of Strasbourg to study under ADOLF VON BAEYER. He conducted his doctoral research on fluorescein, a dye derived from coal tar, which glows fluorescent yellow green in solution and hence could be used to trace water through a system. He earned his doctorate in 1874.

In 1875 Fischer followed his mentor, Baeyer, to the University of Munich, and within three years he became a *privatdozent* (lecturer). In 1879 the university named him an associate professor of analytical chemistry. The year before he had collaborated with his cousin Otto Phillip Fischer to demonstrate that rosaniline, an important dye first synthesized by August von Hofmann in 1862, is a derivative of triphenylmethane. This discovery fueled the German dyestuff industry's takeover of the world market and also led to the synthesis of novocaine.

Fischer also helped launch the German drug industry with his uric acid investigations, which he commenced in 1881. The following year he posited the chemical formulas for uric acid, caffeine, theobromine, xanthine, and guanine, and over the next two decades he succeeded in synthesizing caffeine and theophylline (later used in the motion-sickness drug Dramamine) in 1895 and uric acid in 1897. He ultimately realized that they all had a common oxygenated parent, which he synthesized in 1898 and dubbed *purine*. By the turn of the century, he had synthesized 130 derivatives (including hypoxanthine, xanthine, theobromine, adenine, and guanine) from purine. After the turn of the century he discovered the hypnotic and sedative 5,5-diethylbarbituric acid, the base compound from which he later synthesized barbiturates. (Fischer discovered phenobarbital in 1912.)

Back in 1882 the University of Erlangen named Fischer a full professor of chemistry and director of its Chemistry Institute. In a stuffy laboratory there he inhaled phenylhydrazine, not realizing that its toxicity would eat at his insides—his kidneys, liver, and lungs. When he recovered in 1885 from the first waves of ailments that would last the rest of his life, he moved to a professorship at the University of Würzburg. In 1888 Fischer married Agnes Gerlach, who gave birth to three sons in seven years before she died, in 1895.

Three years earlier, in 1892, the University of Berlin had appointed Fischer to the chair recently vacated by rosaniline discoverer von Hofmann and as director of its Chemistry Institute. The university granted him a free hand over the construction of new laboratory buildings; he rose to the occasion by designing ventilated labs that became the standard throughout the industry and the academy worldwide. He also instituted a new methodology for pursuing chemical investigations by delegating small teams of graduate students to perform experiments until they encountered deviation from expected results, at which point Fischer would collaborate with the students to discover the chemical explanation for the variation. In this way theory preceded practice by only one step.

In 1894 Fischer commenced research on sugars and enzymes that elicited his metaphorical explanation for the interaction between the two: An enzyme was a key that fitted only one lock, in the form of a specific sugar. Then in 1899 he started a methodological study (as he organized *all* his research) of proteins, wherein he synthesized many of the amino acids and after the turn of the century identified valine, proline, and hydroproline. His sugar studies combined with his protein investigations earned him the 1902 Nobel Prize in chemistry, as well as the 1909 Helmholtz Medal, after his continuing protein research resulted in the discovery of what he termed *peptides*, or strings of amino acid segments.

When the British blockaded Germany early in World War I, Fischer's country impressed him into service to synthesize the vital resources that it could no longer access on the open market. Fischer discovered how to synthesize saltpeter (potassium nitrate) and nitric acid, both essential ingredients in explosives; he also synthesized camphor to stabilize gunpowder and pyrites, a base of sulfur for explosives.

On the personal front Fischer lost two sons serving as medical officers to the war, as well as his own fortune in the inflationary aftermath of the Versailles Treaty. Some attribute his July 15, 1919, death to the cancer he had developed in his bowels, while others suspect suicide. His surviving son, who followed in his footsteps to become a distinguished biochemist, honored his father's lifework by arranging for the deposit of all his papers in the Emil Fischer Library, which the University of California at Berkeley proudly constructed for its October 9, 1952, opening.

⊠ Fischer, Ernst Otto
(1918–)
German
Inorganic Chemist

Ernst Otto Fischer established a new field of inorganic chemistry upon his discovery of sandwich compounds, so called because of the layers of carbon rings that surrounded metal atoms. Fischer explained the unpredictable stability of these compounds by experimentally confirming bonds between each carbon atom and the core metal atom. SIR GEOFFREY WILKINSON arrived at almost identical results from independent research, and some two decades later the pair shared the 1973 Nobel Prize in chemistry.

Fischer was born on November 10, 1918, in Solln, a suburb of Munich, Germany. His mother, Valentine Danzer, had two other children with Fischer's father, Karl Tobias, a professor of physics at the Munich Technische Hochschule. Fischer attended secondary school at the Theresien Gymnasium, graduating in 1937. His mandatory two years of service in the German military turned into full combat on the Polish, French, and Russian fronts during World War II. On leave from active duty during the winter of 1941–42, he initiated chemical studies at his father's institution, the Munich Technische Hochschule.

Upon his return to action Fischer was captured by Allied forces and held as a prisoner of war through the remainder of the conflict. After his release in the fall of 1945, Fischer returned to Munich for the reopening of the Technische Hochschule, from which he graduated in 1949. He continued as a research assistant under Walter Heiber, a founding father of metal-carbonyl chemistry (the fusion of metals with carbon-oxygen compounds). Fischer focused his doctoral research on carbon-nickel bonds and submitted his dissertation, "The Mechanisms of Carbon Monoxide Reactions of Nickel II Salts in the Presence of Dithionites and Sulfoxylates," to earn his Ph.D. in 1952.

Fischer remained at the Technische Hochschule, turning his attention to another metal-carbonyl compound, a combination of iron with two five-carbon rings that was first identified as dicyclopentadienyl iron in early 1952. Fischer and his colleagues examined the compound using X-ray crystallography, while the British chemist Geoffrey Wilkinson and his lab pursued the same line of study spectroscopically in the United States. Both teams ascertained independently that the carbon rings "sandwiched" the iron, which combined with each carbon atom in both rings to create an unexpectedly stable structure. Although Wilkinson published his results first, in the spring of 1952, Fischer's publication (with coauthor W. Pfab) of his results experimentally confirmed the pentagonal antiprism structure whereby the carbon atoms staggered themselves between the rings, explaining the compound's stability, whereas Wilkinson had only hypothesized. The two teams thus complemented each other synchronously.

After establishing the existence of sandwich compounds such as ferrocene (the name given to the carbonyl-iron compound), Fischer and his colleagues investigated other metal-carbonyl compounds, establishing the existence of such compounds as cobaltocene and nickelocene. They next turned their attention to dibenzenechromium, which they demonstrated to consist of a neutral chromium atom sandwiched between two neutral benzene molecules. Fischer collaborated with Walter Hafner to develop the Fischer-Hafner method of synthesizing this compound by means of a reductive ligation process activated by aluminum powder and halides.

The Technische Hochschule promoted Fischer to the rank of assistant professor in 1954. In 1957 the University of Munich offered him a full professorship at its Institute for Inorganic Chemistry, and two years later it elevated him to the rank of senior professor. In 1964 he returned to the renamed Technische Universität to succeed his mentor, Hieber, as director of the Inorganic Chemistry Laboratory. That same year Fischer coauthored a short article announcing the preparation and isolation of stable carbene complexes, later named Fischer carbene complexes. In 1966 he coauthored with H. Werner the text *Metal (pi)-Complexes*.

In 1973 Fischer and Wilkinson shared the Nobel Prize in chemistry for their simultaneous discovery and verification of so-called sandwich compounds and for their subsequent work establishing this field of science. Not one to rest on his laurels, Fischer followed up on his receipt of the Nobel by announcing his success at synthesizing the first carbene complexes and the first carbyne complexes (metal atoms connected to yet another layer of carbons). As with the discovery of double-layered metal-carbonyl compounds, the discovery of triple-layered compounds opened up a whole new subfield of research. He continued to contribute to his field by publishing his accumulated wisdom in the 1983 text *Transition Metal Carbene Complexes* (cowritten by Karl Heinz Dotz et al.).

Besides the Nobel Prize, Fischer received numerous other honors, including the 1957 Göttingen Academy Prize and the 1959 Alfred Stock Memorial Prize of the Society of German Chemists. He capped off his distinguished career by disseminating his influence and knowledge through a series of visiting positions: as the 1969 Firestone Lecturer at the University of Wisconsin at Madison, a 1971 visiting professorship at the University of Florida at Gainesville, the 1973 Arthur D. Little Visiting Professorship at the Massachusetts Institute of Technology, and as the Visiting Distinguished Lecturer at the University of Rochester in New York. A lifelong bachelor, Fischer continued to explore organometallic compounds in the late 1990s.

⊠ Fischer, Hans
(1881–1945)
German
Chemist

Hans Fischer established the study of pyrrole chemistry, or the investigation of pigmentation in natural substances such as blood, bile, and plant leaves. He incorporated the factory-line concept to his laboratory, generating more than 60,000 microanalyses necessary for the synthesis of 130 porphyrins. In 1929 he succeeded in synthesizing hemin, the pigment in the hemoglobin of human blood. In recognition of this accomplishment the Royal Swedish Academy of Sciences conferred on him its Nobel Prize in chemistry in 1930.

Fischer was born on July 27, 1881, in Höchst-am-Main, Germany. His mother was Anna Herdegen; his father, Eugen Fischer, was a dye chemist at Meister, Lucius, and Brüning and became the laboratory director at the Kalle Dye Works in Biebrich soon after Fischer's birth. Fischer attended primary school in Stuttgart and secondary school in Wiesbaden, graduating in 1899. He then studied chemistry under Theodore Zincke at the University of Marburg, where he earned his doctorate in 1904. He then studied medicine at the University of Munich, earning his license to practice in 1906 and his degree in 1908.

Fischer commenced his research career studying peptides and sugars at the First Berlin Chemical Institute under EMIL FISCHER (no relation), who had won the 1902 Nobel Prize in chemistry. In 1910 Hans Fischer returned to Munich to study the bile pigment bilirubin (commencing a career-long pursuit of pigment investigations) under Friedrich von Müller at the Second Medical Clinic. In 1912 he became a lecturer in internal medicine, and the following year he became a lecturer in physiology at the Physiological Institute, where he worked under Otto Frank.

In 1916 Fischer replaced ADOLF OTTO REINHOLD WINDAUS as professor of medical chemistry at the University of Innsbruck, in Austria, and two years later he moved to a similar position at the University of Vienna. In the intervening year he had a kidney removed in response to a case of tuberculosis he contracted when he was 20. During this time he also witnessed his father's falling irretrievably into a crevice while mountain climbing in the Austrian Alps.

After World War I ended, Fischer returned again to Munich in 1921 to direct the Institute of Organic Chemistry at the Technical University, where he established an extensive and efficient laboratory. There he recommenced his pigment studies, focusing on the blood hemoglobin pigment hemin, which contains over 70 atoms. In the course of this study Fischer removed iron from the compound, forming porphyrins, which he further reduced into their constituent pyrroles (five-atom closed-ring compounds consisting of four carbon atoms and one nitrogen atom).

Fischer streamlined the experimental process for pyrrole chemistry investigations by coordinating simultaneous experiments on different phases of the research. He increased efficiency by relegating the labor- and time-intensive Gatterman aldehyde synthesis process to "cooks," thereby freeing the time of his graduate students for more specialized research. By applying this division of labor to the experimental process, Fischer was able to produce more than 60,000 microanalyses in his laboratories over the years. This work yielded the first synthesis of porphyrin in 1926 and the synthesis of hemin in 1929. Fischer won the 1930 Nobel Prize in chemistry for his synthesis of hemin.

In 1935 Fischer married Wiltrud Haufe, who was 30 years his junior. Between 1934 and 1940 he published (with coauthor Hans Orth) three volumes of *Die Chemie des Pyrrols*, the definitive text on pyrrole chemistry. During this time period he also pursued research on chlorophyll, the plant pigment related to hemin, and once he had commenced his team on this line of investigation, he returned to his research on bilirubin, the bile pigment. Fischer's laboratory eventually synthesized some 130 porphyrins, including the elusive bilirubin in 1944.

World War II devastated Fischer, as he witnessed the bombing of his laboratories. Suffering from depression, he committed suicide on Easter Sunday, March 31, 1945. He left behind a legacy of accomplishment, though. He had received the Liebig Memorial Medal in 1929 and the Davy Medal in 1937, in addition to an honorary doctorate from Harvard University in 1936. Furthermore the lines of investigation that he set in motion continued after his death, and in 1960 the synthesis of the pigment chlorophyll was finally accomplished.

⊠ Flory, Paul
(1910–1985)
American
Physical Chemist

Paul Flory helped establish the field of polymer chemistry, though his influence reaches far beyond the bounds of one subfield of chemistry. His approach to research, a combination of hypersimplification and commonsense application of the most logical solution to problem solving, have exerted even more influence on the world of science than any one of his numerous discoveries. He won the 1974 Nobel Prize in chemistry in recognition of his far-reaching contributions to science generally and to polymer chemistry specifically.

Paul John Flory was born on June 19, 1910, in Sterling, Illinois. Both his parents were first-generation college graduates from German families who had immigrated to the United States six generations earlier. His mother, Martha Brumbaugh, was a schoolteacher who gained her training at Manchester College; his father, Ezra Flory, worked in the clergy and in education. Flory graduated from Elgin High School to his mother's alma mater in North Manchester, Indiana, where he studied under the chemistry professor Carl W. Holl. The school's limited curriculum took him a mere three years to complete in 1931; similarly, he completed Ohio State University's requirements for a master's degree in a mere three months. After switching to physical chemistry for his doctoral studies, he produced a dissertation under the guidance of Herrick L. Johnston on the photochemical dissociation of nitric oxide, earning his Ph.D. from OSU in 1934.

In the depths of the Great Depression, the central research department of the DuPont Company offered Flory a secure position as a research chemist in the laboratory of WALLACE HUME CAROTHERS studying macromolecules and polymers. Flory made the first of his many intuitive leaps of faith in asserting that the size of a polymer does not in the slightest affect its rate of chemical reactivity, which he confirmed experimentally and theoretically with his "most probable distribution" approach to statistical mechanics. Also during this period Flory married fellow DuPont employee Emily Catherine Tabor, and together the couple had two daughters, Susan and Melinda, and one son, Paul John Jr. In the wake of Carother's suicide, Flory left DuPont to accept a position as a research associate at the University of Cincinnati.

Flory spent World War II conducting research on synthetic replacements for rubber at the Esso (now Exxon) Laboratories of the Standard Oil Development Company and at the Goodyear Tire and Rubber Company. During this time he developed a hypothesis independently from but simultaneously with M. L. Huggins—the renowned Flory-Huggins theory—concerning polymer-solution thermodynamics.

After the war the head of Cornell University's chemistry department, PETER DEBYE, invited Flory to deliver the George Fisher Baker Lectures in chemistry, which resulted not only in his joining the Cornell faculty but also in the publication in 1953 of the compiled transcripts as *Principles of Polymer Chemistry*, still considered the bible of the field. Just as he had earlier pointed out that immersing polymers in solvent caused them to expand for easier observation, he now asserted that specific temperatures facilitated the study of polymers, which he called the "theta point" but is universally referred to as the Flory temperature.

Also while at Cornell, he published one of the first papers ever on liquid crystals.

The Mellon Institute of Industrial Research enticed Flory to Pittsburgh in 1957 to serve as its executive director of research. There he attempted to apply the same organizational scheme to administration as he had applied to his research, but the position ultimately proved frustrating given that chemicals behave more predictably, or perhaps more logically, than humans. When Stanford University offered him a chemistry professorship in 1961, he jumped at the chance to return to teaching and conducting research. Within half a decade, the university established an endowed chair for him as the first J. G. Jackson–C. J. Wood Professor of Chemistry.

In 1969 Flory published his second influential text, *Statistical Mechanics of Chain Molecules*, which was translated into Russian as well as Japanese. A critical mass of recognition showered on him in 1974, as he received the National Medal of Science, the Priestley Medal of the American Chemical Society, and the Nobel Prize in chemistry. The next year he retired to emeritus status, but he remained active by exerting his influence as a Nobel laureate to petition for human rights. He served on the Committee on Human Rights of the National Academy of Sciences from 1979 through 1984. Flory died on September 8, 1985, at his weekend home in Big Sur, California, where he was writing a paper to present at the national meeting of the American Chemical Society.

⊠ **Franklin, Rosalind**
(1920–1958)
English
Chemist, X-ray Crystallographer

Rosalind Franklin played a crucial but underappreciated role in the discovery of the structure of deoxyribonucleic acid (DNA). She provided the key that unlocked the secret: an X-ray crystallographic photograph of DNA that her supervisor J. D. Bernal called one of "the most beautiful X-ray photographs of any substance ever taken." Maurice Wilkins, Franklin's adversarial colleague at King's College in London, showed her photograph without her permission or knowledge to Cambridge researchers FRANCIS HARRY COMPTON CRICK and James Watson. The photo served as the missing link in their pursuit, and the pair published an elucidation of DNA's structure soon thereafter. Franklin did not share the 1962 Nobel Prize in medicine and physiology with Watson, Crick, and Wilkins, because she had died four years earlier, although many speculate that had she lived, the entrenched sexism of the time might have deprived her of Nobel laurels anyway.

Franklin was born on July 25, 1920, in London, England. Her mother, Muriel, was a volunteer social worker who raised five children, and her father, Ellis, was a successful banker who believed that women's pursuit of professional careers destined them to unhappiness. Franklin, who attended one of the few London girls' schools that taught physics and chemistry, decided at age 15 that she would become a scientist. Despite paternal objections, she matriculated in 1938 at Newnham College, Cambridge's school for women, and she graduated in 1941 with a degree in chemistry.

After holding a graduate fellowship for the next year, Franklin contributed to the cause in World War II as an assistant research officer for the British Coal Utilization Research Association (CURA), where she elucidated the structures of carbon and graphite. She utilized some of this research in her doctoral dissertation to earn a Ph.D. in physical chemistry from Cambridge University in 1945. She subsequently spent three years in Paris, studying X-ray diffraction at the French government's Laboratoire Central des Services Chimiques de L'État, where she conducted X-ray crystallographic studies of colloidal carbon.

In 1950 she returned to England, and the following year King's College of the University of London hired Franklin as a research associate under John Randall, who assigned her to study DNA; Randall had also assigned Maurice Wilkins, who was away at the time, to study DNA. Upon his return Wilkins assumed Franklin to be his subordinate instead of his equal, but, in fact, it was Franklin whom Randall charged with the task of elucidating DNA's structure due to her superior X-ray crystallographic skills.

Franklin commenced her investigations by taking X-ray photographs of DNA crystals, the traditional "dry" method. In May 1952 she experimented with an untested method by photographing "wet" samples of DNA, or those prepared with water molecules instead of in their crystalline form. However, she put aside these photos, not yet far enough into her study to recognize their significance. In January 1953 Wilkins showed Watson his appropriated copy of Franklin's "wet" photo, which clearly showed DNA's double-helical structure with an external "backbone." Watson and Crick used this knowledge as the foundation for their construction of a DNA model that abided by all known chemical and physical rules. The April 25, 1953, edition of *Nature* published the Watson-Crick model of DNA's structure, along with a supporting article by Franklin and Wilkins. The discovery thrilled Franklin, ignorant of her own essential role in its development.

Franklin moved from Randall's lab to J. D. Bernal's at the University of London's Birkbeck College, where she conducted X-ray crystallographic studies of the tobacco mosaic virus (TMV) that revealed its structure to be not solid, as was assumed, but rather a hollow tube. She also identified ribonucleic acid (RNA), a cousin of DNA as the foundation of TMV. As she was conducting X-ray crystallographic studies on the poliovirus in 1956, she learned of her ovarian cancer, which progressed over the next two years. On April 16, 1958, she died in London.

Watson's published account of the events leading up to the discovery of DNA's structure, *The Double Helix* (1968), presented Franklin in an unfavorable light. Her friend, Anne Sayre, sought to right this wrong by searching exhaustively through the existing records (which she made available to the public) to paint a more accurate portrait of Franklin in her 1975 book, *Rosalind Franklin and DNA*. This account made it clear that Franklin played a key role in the discovery of the structure of DNA and also revealed the brutal politics that suppressed recognition of her contribution in the first place, thereby redressing this wrong for the record of scientific history.

⊠ Friedel, Charles
(1832–1899)
French
Chemist, Metallurgist

Frenchman Charles Friedel's name is inextricably linked to that of the American James Mason Crafts for their joint discovery in 1877 of a set of hydrocarbon reactions. The Friedel-Crafts alkylation, as it was dubbed, proved vital to petrochemical advances that launched an entire industry, while also providing the key to advances in academic research. Friedel also distinguished himself as a mineralogist, serving as the curator of the mineral collections of the Écoles des Mines for two decades before taking on a professorship as well. An affable man professionally, in his personal life he held strong Protestant beliefs, pitting him against the tide of religious and scientific thought in France at the time.

Friedel was born on March 12, 1832, in Strasbourg, France. He received his education there as well, studying under LOUIS PASTEUR, an experience that inspired him to pursue further studies and eventually a profession in chemistry. He then moved to Paris, where he studied at the Sorbonne under C. A. Wurtz, whose chair he later filled at what is also known as the University of Paris. Friedel married twice and had six children.

In 1856 the Écoles des Mines hired Friedel as its curator for its mineral collections, testament to not only his expertise in his field of mineralogy (one of his two specialties) but also his administrative skills. Two decades later the school added teaching to his responsibilities when it appointed him to a professorship in the mineralogy department. In his mineralogy studies, Friedel focused on the synthesis of minerals by heating mineral-water solutions and placing them under pressure in order to elicit reactions that result in new combinations. In 1862 he discovered secondary propyl alcohol, thus verifying Hermann Kolbe's prediction of its existence.

In 1884 Friedel's mentor, C. A. Wurtz, passed away, leaving open his chair in the organic chemistry department at the Sorbonne. Friedel accepted the position and retained this professorship until his death, 15 years later. In this position he transitioned from his first field of scientific specialization—mineralogy—into his other field—organic chemistry—and overlapped them in his investigation of silicon compounds. Following in the footsteps of Wurtz, Friedel studied the synthesis of ketones, as well as the synthesis of secondary propanol, and extended on Wurtz's studies with his own research.

Friedel's organic chemistry research had commenced, however, before he held a position in that field. In the 1870s he commenced his trans-Atlantic collaboration with the American chemist James Mason Crafts conducting organosilicon chemistry studies. They immortalized themselves with their 1877 discovery of alkylation, an electrophilic process whereby alkyl halide reacts with aromatic hydrocarbons—catalyzed by the presence of a Lewis acid (often anhydrous aluminum chloride)—and thereby creates alkyl benzenes. This process became known as Friedel-Crafts alkylation. At the same time the pair discovered the acylation of hydrocarbons, a similar process.

Their discovery found immediate academic applications and has become an integral component of the chemistry curriculum. However, it was the application of this reaction in the petrochemical industry that cemented the significance of the Friedel-Crafts process, which has generated numerous advances in the field. Friedel died on April 20, 1899, in Montauban, France.

⊠ **Fukui, Kenichi**
(1918–1998)
Japanese
Theoretical Chemist, Engineer

Kenichi Fukui shared the 1981 Nobel Prize in chemistry with Roald Hoffman for their independent but complementary research on chemical reactivity. Fukui supplied in-depth mathematical and experimental confirmation of the frontier orbital theory of reactivity, and Hoffman provided an elegant explanation for visualizing the process.

Fukui was born on October 4, 1918, in Nara on the island of Honshu, Japan, the eldest of three sons. His mother was Chie Fukui; his father, Ryokichi Fukui, was a foreign trade merchant and factory manager. On the advice of Professor Gen-itsu Kita, he attended Kyoto Imperial University in the Department of Industrial Chemistry. He earned his bachelor of arts degree in engineering in 1941 and proceeded to work on the synthesis of fuels at the Japanese Army Fuel Laboratory.

In 1943 Fukui returned to Kyoto Imperial University (where he spent the rest of his career) as a lecturer in the fuel chemistry department of the Engineering School. Within two years the university had promoted him to an assistant professorship while he continued to conduct doctoral research. In 1947 he married Tomoe Horie, and together the couple had two children: a son named Tetsuya and a daughter named Miyako.

He earned his doctorate in chemical engineering in 1948, and three years later he became a full professor in physical chemistry.

Fukui focused his research on aromatic hydrocarbons, asking why molecular reagents attacked certain substrates more than others. He noticed that the density of orbitals, or electrons orbiting atomic nuclei, helped determine the rate that atoms reacted with other chemicals: The densest concentration of orbitals stood ready to offer electrons for reactions, while the lowest density of orbitals stood ready to accept electrons to initiate a reaction. Fukui called these two types of orbital HOMOs (highest-occupied molecular orbitals) and LUMOs (lowest-unoccupied molecular orbitals).

In 1952 Fukui published a paper simultaneously with R. S. Mulliken, one of the founders of molecular orbital theory, whose piece noted that bases with high concentrations of occupied molecular orbitals tended to overlap with acids with concentrations of unoccupied orbitals. He called the former "donors" and the latter "acceptors" and named his observation the overlap-orientation principle. Fukui applied this same logic to his theory on frontier orbitals (or orbitals that are unpaired and thus can accommodate reactions), hypothesizing that HOMOs react most readily with LUMOs, then providing mathematical and experimental substantiation of this notion of electron swapping, or delocalization.

Fukui's theory did not receive the attention it deserved, partly due to the arcane mathematics involved and partly due to the proprietary status of Western opinion in the scientific community. In 1965, though, the American scientists R. B. Woodward and Roald Hoffmann published several brief communications that illustrated the frontier orbital theory of chemical reactivity gracefully, approaching the idea from a graphical perspective unexplored by Fukui. The Japanese approach complemented the American approach synergistically, offering future researchers a complete map to the processes of molecular reactions—what became known as the principle of the conservation of orbital symmetry.

In the 1970s Fukui expanded his examination of chemical reactions to consider the paths of interaction. He demonstrated how the geometry of molecules determined their reactivity. Also in the 1970s Fukui assumed more responsibility within his institution and his profession: He became the dean of the engineering faculty at Kyoto from 1971 through 1973 and served as the vice president of the Chemical Society of Japan from 1978 to 1979 and as its president from 1983 to 1984.

In 1981 Fukui received the Nobel Prize in chemistry in conjunction with Hoffman; Wood-ward most likely would have received a part in the prize if he had not died two years earlier. Besides the Nobel Prize, Fukui received many other honors, including the 1962 Japan Medal of Honor and the 1981 Order of Culture, as well as foreign membership in the Royal Academy and the National Academy of Sciences.

In 1982 Fukui retired to emeritus status at Kyoto University in order to assume the presidency of the Kyoto Institute of Technology. He retired to emeritus status there in 1988 in order to take on the directorship of the Institute for Fundamental Chemistry, a position he held for the next decade, until his death on January 9, 1998.

G

⊠ **Gay-Lussac, Joseph-Louis**
(1778–1850)
French
Chemist

Joseph-Louis Gay-Lussac made his name as an extremely precise experimentalist who avoided theorizing, preferring instead to extrapolate his actual results into generalized conclusions. Such was the case with his most famous proposition, the law of combining volumes, which held that gases mixed at standard ratios. He conducted other important work as well: He was the first to isolate boron, and he discovered the properties of iodine and cyanogen.

Gay-Lussac was born on December 6, 1778, in Saint-Léonard, in the Haute-Vienne region of France. His grandfather was a physician, and his father, Antoine Gay, was a lawyer and judge who added the name of family property to his surname to distinguish himself from others in the region with the same last name. During the French Revolution, Antoine Gay was arrested, so he sent his son to Paris at the age of 14 to continue his education at a boarding school.

After Gay-Lussac finished his studies at the École Polytechnique and the École des Ponts et Chaussées, he served as an assistant to C. L. Berthollet, who hosted the Arcueil Society (which included such luminary chemists as Antoine Lavoisier) at his country house in the Arcueil region of the same name. In the winter of 1801–02, Gay-Lussac conducted research on the expansion properties of various gases. Independent of JOHN DALTON and Jacques Charles, he devised a law that determined equal volumes of gas expanded at equal rate, which he published in 1802. This law is now called Charles's law after chemist Jacques Charles, who had discovered it 15 years earlier than Gay-Lussac did, though he did not publish his results, which were inferior to Gay-Lussac's precise measurements due to the latter's purging of water from his apparatus.

In 1804 Gay-Lussac ascended 23,000 feet in a balloon to gather air samples to ascertain whether its composition differed at various altitudes; it did not, he discovered. His results contradicted those of Alexander von Humboldt, who did not respond with venom but rather rewarded his junior's precision by inviting Gay-Lussac to accompany him on a European tour in 1805, taking terrestrial magnetism measurements on their way from Rome, Italy, through Switzerland to Berlin, Germany. Upon his return to France, Gay-Lussac held several academic positions from 1806 until 1808, when he became a professor of physics at the Sorbonne, a position he retained for the next two and a half decades. A year after this appointment, in 1809, he became a professor of chemistry at the École Polytechnique.

Joseph-Louis Gay-Lussac, known for his precise experiments, proposed the law of combining volumes, which holds that gases mix at standard ratios. *(E. F. Smith Collection, Rare Book & Manuscript Library, University of Pennsylvania)*

On December 31, 1808, Gay-Lussac announced his law of combining volumes at the meeting of the Société Philomatique in Paris: "The compounds of gaseous substances with each other are always formed in very simple ratios of volume," he stated. Based on his atmospheric air research, he compared the weight proportions of elements in three different oxides of nitrogen, noticing that these corresponded approximately to $1:\frac{1}{2}$, 1:1, and 1:2. John Dalton vehemently refused to accept this law, in part because it refuted his own theories; however, his objections proved unfounded, and the law stands, though it has undergone some corrections.

Between 1813 and 1814, Gay-Lussac conducted a detailed study of iodine, which had been discovered in 1811 by B. Courtois. SIR HUMPHRY DAVY contested Gay-Lussac's priority in his iodine studies; in the two prominent scientists' boron studies, Gay-Lussac (and his collaborator L. J. Thenard) isolated the element nine days prior to Davy's team, but Davy beat Gay-Lussac to the punch in publishing the results. In other chemical studies Gay-Lussac investigated the substance Prussian blue to gain a comprehension of cyanogen and cyanides. In 1818 he became superintendent of the French government's gunpowder factory.

In 1827 Gay-Lussac contributed to the chemical industry by innovating improvements to the sulfuric acid production process: He figured out a way to recycle nitrogen monoxide after oxidizing it from sulfuric trioxide. The absorption towers used in the process are now known as Gay-Lussac towers, which were used well into the 20th century for sulfuric acid production. His other innovations include a portable barometer, an improved pipette and burette (names he introduced), and a process for titration of silver with chloride.

In 1829 Gay-Lussac became the chief assayer of the French mint, and after resigning his post at the Sorbonne in 1832, he took up a chemistry professorship at the Jardin des Plantes. During this late period of his career he entered the political arena, gaining election to the Chamber of Deputies in 1831, and to the Senate in 1839. Gay-Lussac, who never married, died on May 9, 1850, in Paris.

⊠   Gilbert, Walter
(1932–   )
American
Molecular Biologist, Chemist

Walter Gilbert was corecipient with FREDERICK SANGER of the 1980 Nobel Prize in chemistry for their independent but simultaneous discovery of DNA (deoxyribonucleic acid) sequencing. PAUL BERG, a biochemist who pioneered gene splicing, also received the 1980 prize. Leading up to his DNA sequencing, Gilbert had verified the existence of lac repressors, a trigger mechanism for DNA.

Gilbert was born on March 21, 1932, in Cambridge, Massachusetts. His mother, Emma Cohen, was a child psychologist, and his father, Richard V. Gilbert, was a professor of economics at Harvard University. The family moved to Washington, D.C., when Gilbert was seven, and there he became interested in science. He ground his own telescopic glass at the age of 12 and studied nuclear physics independently at the Library of Congress while a senior at Sidwell Friends High School. After graduating he matriculated at Harvard in 1949, majoring in chemistry and physics. In 1953 he graduated summa cum laude and married Celia Stone, with whom he later had two children, a son and a daughter. He remained at Harvard to earn his master's degree in physics in 1954.

Gilbert traveled to England to pursue doctoral studies in theoretical physics under Abdus Salam at Cambridge University, where he also worked with the famous geneticists FRANCIS HARRY COMPTON CRICK and James Watson. For his dissertation he mathematically predicted the actions of elementary particles in "scattering" experiments. He earned his Ph.D. in mathematical physics in 1957 and returned to Harvard to work under Julian Schwinger as a National Science Foundation postdoctoral fellow. Within two years Harvard had appointed him to an assistant professorship.

In 1960 Watson, who had moved from Cambridge to Harvard, included Gilbert in his team of investigators using pulse labeling to locate messenger ribonucleic acid (mRNA), which transfers information from DNA to ribosomes. Proof of the existence of messenger RNA came from another laboratory, however, at the California Technological Institute. Gilbert then collaborated with the German postdoctoral fellow Benno Muller-Hill in attempting to isolate the lac repressor, which hypothetically acted as a kind of "on/off" switch for genetic transferal. Gilbert's team tracked the process in *Escherichia coli* bacteria by introducing "tagged" (radioactive) lactose, which elicited the anticipated result: The lac repressor bonded to the lactose, confirming the existence of the former.

In 1964 Harvard tenured Gilbert as an associate professor of biophysics, and four years later it promoted him to a full professorship. In 1972 Harvard named him to an endowed chair as the American Cancer Society Professor of Molecular Biology. By then he had located the site of lac

Walter Gilbert shared the 1980 Nobel Prize in chemistry with Frederick Sanger for their independent but simultaneous discovery of deoxyribonucleic (DNA) sequencing and with Paul Berg for his discovery of recombinant DNA. *(Courtesy The Biological Laboratories, Harvard University)*

repression on the lac operon, and he used this gene in his next line of investigation—the sequencing of DNA. Gilbert, in collaboration with his graduate student Allan Maxam, tagged the DNA radioactively (a similar tactic as in his lac repressor research), then passed broken fragments of the DNA strands through gel by zapping it with an electrical current, a process called gel electrophoresis. X-ray photographs of the segments revealed their chemical composition, allowing Gilbert and Maxam to identify the chemical sequence of DNA. Biochemist Frederick Sanger innovated a similar procedure at about the same time, hence the awarding of the 1980 Nobel Prize in chemistry to Gilbert and Sanger, as well as to Berg.

Starting in 1978 Gilbert began to investigate the possibility of approaching the new field of biotechnology from a corporate perspective. By 1981 he had resigned from his Harvard professorship to help found the biotech firm Biogen N.V., for which he served as chair of the scientific directors and then as chief executive officer. Within three years the inherent conflict of interest between pure science and capitalistic science frustrated him, and in 1985 he returned to Harvard, which had named him the H. H. Timken Professor of Science in the cellular and developmental biology department. Two years later he assumed the chair of the department as the Carl M. Loeb University Professor.

Gilbert has spent the later part of his career working on the human genome project, an attempt to map the entire DNA scheme of humans. Although the project received criticism for treading into God's territory, Gilbert defended its importance for the potential to discover new cures for disease. He convinced Congress to budget $2 million annually to support the project in his laboratories at Harvard through a National Institutes of Health grant. He even started a company to copyright the genetic code.

Besides the Nobel Prize, Gilbert has received ample recognition for his career. He shared the 1979 Albert Lasker Basic Medical Research Award with Sanger and won Columbia University's 1979 Louisa Horwitz Gross Prize, as well as the 1980 Herbert A. Sober Memorial Award of the American Society of Biological Chemists.

⊠ Good, Mary Lowe
(1931–)
American
Chemist

Mary Lowe Good's career has encompassed the heights of the field of chemistry, as well as its breadth: She has held an endowed chair in academia, served as a senior vice president in industry, headed a department of the federal government, and presided over the American Chemical Society. In 1997 she won the highest honor bestowed by that society, its Priestley Medal, adding it to a cornucopia of other honors she has received throughout her illustrious career.

Good was born Mary Lowe on June 20, 1931, in Grapevine, Texas. Her mother, Winnie Mercer, was a schoolteacher, and her father, John W. Lowe, was a superintendent of schools. When she was still young, the Lowe family moved to Arkansas, where she had her first exposure to chemistry as a teenager in their basement darkroom. She did not study chemistry in high school, however, due to curricular limitations of the school district. When Lowe entered the University of Central Arkansas, she intended to study home economics, but her freshman chemistry course, taught by an enthusiastic older professor, changed her mind. She attended classes year-round to finish her undergraduate course of study in three years (she had to make way for her three younger siblings to attend college), graduating in 1950 at the age of 19 with a bachelor of science in chemistry, with minors in physics and mathematics.

The University of Arkansas at Fayetteville offered Lowe a fellowship for graduate study in inorganic chemistry and radiochemistry, so she earned her master's degree in 1953 and her doctorate in 1955. While at the university, she met Billy Jewel Good, a physics graduate student, and the couple married on May 17, 1952. Together they had two sons, Billy John and James Patrick.

In 1954 Good accepted a position as instructor of chemistry at Louisiana State University (LSU) at Baton Rouge, where her husband completed his doctorate in physics. By 1956 the university had promoted her to assistant professor, and in 1958 it transferred her to its New Orleans campus as an associate professor. There she became a full professor in 1961. In 1970 she published her book *Integrated Laboratory Sequence: Volume III—Separations and Analysis*. LSU named her the Boyd Professor of Chemistry in 1974, and in 1978 she returned to LSU's Baton Rouge campus as the Boyd Professor of Material Science in the division of engineering research.

During the 1970s Good worked in executive positions at the American Chemical Society (ACS), serving on the board of trustees from 1972 to 1980 and chairing the board for the last three of these years. This position charged her with responsibility for a $150-million budget, preparing her perfectly for her next professional step from academia into industry. In 1980 Universal Optical Products (UOP), a division of the Allied Signal Corporation, hired her as vice president and director of research. In 1985 she became president and director of research for Signal Research Center, and the following year she was named president of engineered materials research for Allied Signal. In 1988 she ascended to the rank of senior vice president for technology at Allied Signal. Meanwhile, she had maintained her affiliation with the ACS and in 1987 became the second woman to serve as the society's president.

Good's extensive service to both academia and industry qualified and prepared her perfectly for her next role: In 1993 President Bill Clinton appointed her as undersecretary of technology in the Department of Commerce. In this position she held ultimate responsibility over the Technology Administration, the National Institute of Standards and Technology, and the National Technology Information Service.

In 1997 Good became the first woman to win the Priestley Medal, the ACS's oldest and most prestigious award. This added to the plethora of other distinctions she had won over the years: the 1969 Agnes Faye Morgan Research Award, the 1973 Garvan Medal, the 1975 Herty Medal, the 1983 American Institute of Chemistry Gold Medal, and the 1992 National Science Foundation Distinguished Public Service Award, as well as the American Association for the Advancement of Science Award. In addition she was named the 1982 Scientist of the Year by *Industrial Research & Development* magazine.

However, Good is perhaps more proud of the awards she has given, not received. For a decade, from 1978 to 1988, she served as chairman of the Amelia Earhart Awards Committee of Zonta International, a women's service organization, and oversaw the awarding of scholarships to women pursuing graduate study in aerospace science. At the beginning of her tenure, she had difficulty locating women qualified for the prize, but by the end of her term, her committee had its choice between a host of superb candidates. Not only did Good help choose these candidates, but also the success of her career served as a catalyst for the opening up of the sciences to women.

⊠ **Graham, Thomas**
(1805–1869)
Scottish
Physical Chemist

Thomas Graham was a brilliant experimentalist who extrapolated the results of his investigations into generalized laws. His most famous proposition, which quantified the rule governing the intermixture of gases, was named after him—Graham's law. He left his homeland early in his career to install himself in London, where he actively involved himself in the affairs of the Royal Society of London (which awarded him amply for his accomplishments) and inhabited a chair at University College. Shy and unassuming, he did not distinguish himself before the lectern, lacking the charisma to communicate his ideas socially; however, he excelled at testing and manipulating his ideas in the laboratory and communicated well enough in writing to disseminate the fruits of his labor.

Graham was born on December 21, 1805, in Glasgow, Scotland. His father was a successful textile manufacturer who planned a ministerial life for his son. Graham matriculated at the University of Glasgow in 1819, at the age of 14. Thomas Thomson's lectures impressed a fascination in chemistry upon the youth, who reportedly in turn impressed his senior with the question that he would later set out to prove: "Don't you think, Doctor, that when liquids absorb gases, the gases themselves become liquids?" After earning his master's degree in 1826, Graham moved to the University of Edinburgh for two years of study under Thomas Charles Hope.

In 1830, after tutoring mathematics and chemistry and teaching science to laborers upon his return to Glasgow, Graham became one of the first professors appointed to a chair in chemistry at the Andersonian College (later known as the Royal College of Science and Technology). While there, in 1833, he published his first statement on what is now known as Graham's law in his paper entitled "On the Law of the Diffusion of Gases." In it he held that gases intermix spontaneously in volumes governed by the square root of their respective densities. The Royal Society of Edinburgh awarded him its Keith medal later that year. Also in 1833 Graham pub-

lished the results of his phosphate compound research, determining through the hydration of phosphoric acid that three, not two, salts of phosphate existed.

In 1834 the Royal Society of London inducted him into its fellowship, and three years later Graham went down to London to accept the society's 1837 Royal Medal. Instead of returning back home, he stayed in London, where he landed a professorship at the newly founded University College (now the University of London) and remained there the rest of his career. In 1841 he became president of the Chemical Society of London, which was the first national chemical society and which he helped found. (France followed suit by founding a similar society in 1857, Germany in 1867, and the United States in 1876.)

Besides phosphoric compounds and the diffusion of gases, Graham concerned himself with colloids, inventing an entirely new, logical vocabulary to describe and discern the substances. He coined the terms *gel*, *sol*, and *colloids*, among others, clarifying the field for future scientists but confusing his contemporaries. He also invented a "dialyzer" to separate colloids from crystalloids. This technique of dialysis remains in use at hospitals, most important to purify the blood of patients suffering from kidney failure. Graham is considered by many to be the father of colloidal chemistry.

In 1854, after a two-decade tenure at University College, Graham took up Isaac Newton's old post as master of the mint. Instead of using the position as an excuse to rest on his laurels, he set about to reorganize the institution. At the same time he continued to conduct original research. In 1857 he published his account of coffee adulteration in a paper entitled "Report on the Mode of Detecting Vegetable Substances mixed with coffee for the Purpose of Adulteration," which appeared in the *Journal of the Chemical Society of London*. He also investigated the formation of alcohol in the bread-making process, generating enough of the by-product to ignite gunpowder.

The Royal Society recognized the significance of Graham's scientific contributions with numerous honors. It awarded him a second Royal Medal in 1863, having given him the Copley Medal the prior year. Other societies honored him as well. The year before, in 1862, the Paris Academy of Sciences (whose formation he had instigated) awarded him its Prix Jecker. He continued to work at the mint until his death on September 16, 1869. The master of the mint position was eradicated after his tenure, as nobody could possibly fill his shoes.

Honors continued to grace Graham posthumously. In June 1872 James "Paraffin" Young presented the city of Glasgow a statue of his mentor, Graham, sculpted by William Brodie. The University of Strathclyde in Glasgow further acknowledged Graham's significance by naming the Thomas Graham Building after him, allowing his legacy to continue visibly to this day.

⊠ **Grignard, François-Auguste-Victor**
(1871–1935)
French
Organic Chemist

Grignard reagents, named after their discoverer, Victor Grignard, proved instrumental in conjoining different chemicals into compounds. His process proved inexpensive, simple, and extremely flexible in generating a wide range of chemical compounds. This discovery earned him not only a post as Commander in the Legion of Honor but also the Nobel Prize in chemistry in 1912.

François-Auguste-Victor Grignard was born on May 6, 1871, in Cherbourg, France. His mother was Marie Hébert. His father, Théophile Henri, was a sail maker and foreman for the military arsenal. Grignard attended the Collège de Cherbourg from 1883 through 1887, when he

passed the baccalaureate. Financial constraints prevented him from attending the École Polytechnique or the École Normale Supérieure, the preeminent French scientific institutions in Paris, so he matriculated at the École Normale Spéciale at Cluny, a teacher training school, in 1889. However, the school closed its doors within two years, transferring its students on scholarship to the Faculty of Sciences at nearby Lyons. Unfortunately Grignard failed his mathematics examination for his *licence*.

Grignard spent a year in compulsory military service from 1892 to 1893 and upon his discharge finally earned his *licence* in mathematics. He then rejoined the science faculty in Lyons as an assistant in the general chemistry laboratories of Philippe A. Barbier, who also mentored his doctoral research. Barbier had been experimenting with organometallic compounds, or compounds of organic radicals attached by a metal, and he prompted his student to pursue similar studies for his dissertation. Grignard's challenge consisted in performing the combination while avoiding spontaneous combustion at room temperature; he achieved this objective by using anhydrous ether as a stabilizing agent. In a famous experiment he added magnesium shavings in ether to methyl iodide, which became known as Grignard reagent.

In the summer of 1901 Grignard presented his dissertation, in which he listed 29 compounds (including phenylisobutyl carbinol, dimethylphenyl carbinol, and dimethylbenzyl carbinol) he had prepared with his reagents, to the science department at Lyons. The key to the importance of his method was its universality. Although scientists already knew how to synthesize organometallic compounds, the processes were extremely complicated and volatile. Grignard's reagents, on the other hand, generated stable compounds that did not spontaneously combust. Other scientists followed up on his research, and by 1905, 200 papers discussed organomagnesium synthesis. Almost a half century after his dissertation, some 4,000 scientific papers concerned themselves with his results.

Grignard stayed on at the science faculty in Lyons, where he served as a laboratory instructor from 1898 until he became a lecturer in 1906 and an associate professor in 1908. In 1909 he joined the faculty of the Nancy Chemical Institute and rose to the rank of professor of organic chemistry soon thereafter. In 1910 he married the widow Augustine Marie Boulant, a childhood friend, and together the couple had two sons, Robert Paindestre and Roger Grignard.

In the first decade of the new century Grignard received numerous awards from the Académie des Sciences: the Prix Cahours in 1901 and 1902, the Berthelot Medal in 1902, and the Prix Jecker in 1906. In 1912 the Royal Swedish Academy of Sciences awarded him its Nobel Prize in chemistry, which he shared with Paul Sabatier. The academy recognized Grignard's significant contribution to science with his discovery of Grignard reagents. Two years

later he served in the French military as a corporal with sentry duty guarding a railroad bridge in Normandy. He spent the remainder of World War I conducting research in the chemical divisions division, including the preparation of dichloroethyl sulfide, or mustard gas.

In 1919 Grignard returned to Lyons and filled the chair of general chemistry vacated by Barbier upon his retirement. In 1921 he took over the directorship of the School of Industrial Chemistry. In 1929 the Faculty of Sciences appointed him as its dean, a position he retained until his death on December 13, 1935, after a six-week illness.

Before his death Grignard had managed to complete two volumes of a projected 15-volume series on organic chemistry. His son Roger, who followed in his footsteps to become a chemist, in collaboration with Grignard's former student Jean Colgone, published two more volumes of this work posthumously by gathering together his lectures on organic chemistry.

H

Haber, Fritz
(1868–1934)
German
Chemist

Fritz Haber led a controversial and tragic life. He devised the most practical method for synthesizing ammonia from its constituent elements, a technique that proved instrumental not only in the manufacture of agricultural fertilizers but also in the production of explosives. Haber also pioneered the first use of poisonous gas on the battlefield, a development he believed would hasten the end of World War I, but which only proved to entrench the deadlock, as the French retaliated with their own use of chemical warfare. Some scientists and humanists protested his receipt of the 1918 Nobel Prize in chemistry, arguing that his development of ammonia synthesis and chemical warfare amounted to crimes against humanity. He later proved his humanism by resigning his cherished post at the helm of the Kaiser Wilhelm Institute in protest over the Nazi discrimination against Jews.

Haber was born on December 9, 1868, in Breslau, in Prussia, Germany (now Wrocław, Poland). His mother, Paula Haber, died in childbirth, the first of many tragedies to visit Haber's life. Haber's father, dye and pigment importer Siegfried Haber, remarried in 1877. His new wife, Hedwig Hamburger, gave birth to three daughters and also raised Haber with more compassion than his father could muster. Haber sought refuge in the literature of Goethe, introduced to him at the St. Elizabeth Gymnasium.

In 1886 Haber commenced his university studies, which he conducted according to the custom of the time by attending several institutions: the University of Berlin for one semester; the University of Heidelberg, where he studied for a year and a half under ROBERT WILHELM BUNSEN; and the Charlottenberg Technical College in Berlin, where he spent a year and a half after fulfilling his obligatory military service. He earned his Ph.D. in 1891.

This nomadism continued in Haber's early professional career, as he cycled through three industrial jobs before his final failure in his father's business. Haber imported his knowledge of industry, however, back to academia, and later in his career he built bridges between these two disparate but interdependent sectors of society. First he served as junior assistant to Ludwig Knorr at the University of Jena for a year and a half, then moved in 1894 to the Fredericiana Technical College in Karlsruhe, where he served as an assistant in the Department of Chemical and Fuel Technology.

During his tenure there Haber published the first of his influential books, *Experimental Studies of*

Fritz Haber and his team of scientists invented chemical weapons and poison gas to aid the German war effort during World War I. *(E. F. Smith Collection, Rare Book & Manuscript Library, University of Pennsylvania)*

the Decomposition and Combustion of Hydrocarbons, an 1896 text on petroleum refining that earned him a promotion to the rank of *privatdozent* (lecturer). Two years later he published *Outline of Technical Chemistry on a Theoretical Basis*, which elicited a further promotion to associate professor. He received international recognition for the publication of his third book, *The Thermodynamics of Technical Gas Reactions*, in 1905.

That same year Haber commenced his ammonia research, and the next year the new Institute of Physical Chemistry and Electrochemistry in Karlsruhe appointed him its director and a professor. In 1907 his nemesis, WALTHER HERMANN NERNST, publicly challenged Haber's findings on ammonia, prompting Haber to adopt Nernst's method of experimentation under increased atmospheric pressure in order to prove his own calculations correct. In the process Haber discovered the most efficient means of synthesizing ammonia on an industrial scale, while also discovering a replacement for iron as a superior catalyst for the reaction. This discovery immediately aided in the production of fertilizers beginning in 1910 but soon proved more significant in the manufacture of nitric acid for use in explosives.

In 1912 the newly established Kaiser Wilhelm Institute of Physical Chemistry and Electrochemistry installed Haber as its director, a position created expressly for him. However, World War I erupted, drawing him into the service of his country to oversee the production of ammonia for fertilizer and explosives. He also developed chlorine gas, which was first employed on April 11, 1915, when the German military dropped 5,000 cylinders of the poisonous chemical near Ypres, Belgium, killing some 150,000.

Haber's wife, the chemist Clara Immerwahr (who married him in 1901 and bore a son, Hermann, in 1902), condemned what she considered to be a perversion of science toward destructive ends. Despite the fact that she committed suicide in 1915 in protest of his actions, Haber headed the Chemical Warfare Service in 1916. Haber remarried in 1917. His second wife, Charlotte Nathan, bore him a son and a daughter, but his devotion to science eclipsed his domestic affections, resulting in the dissolution of their marriage a decade later.

The Royal Swedish Academy of Sciences postponed its awarding of the 1918 Nobel Prize in chemistry to Haber until November 1919. Clearly, his innovation of ammonia synthesis techniques significantly transformed the world; however, opponents of his receipt of the prize claimed that the negative consequences of his discovery (essentially, the elongation of the war) outweighed any positive contributions he made through it. He received the Nobel despite the protests.

Haber tried to salvage his country from the devastating effects of inflation in the wake of onerous reparations demanded in the Treaty of Versailles by sending expeditions to scour the ocean for gold, but his calculations proved extremely optimistic and the harvest was scant. He remained at the head of the Kaiser Wilhelm Institute until 1933, when he resigned in protest of the National Socialist Party's freeze on hiring Jews. Haber himself was a lapsed Jew and found it unconscionable to hire anyone other than the best candidate for a position, regardless of creed, so he quit and immigrated to Britain.

There chemist and future president of the Israeli state Chaim Weizmann invited Haber to become the director of the physical chemistry department at the Daniel Sieff Research Center in Rehovot in what is now Israel. Haber suffered a heart attack before his departure, and in Basel, Switzerland, on his way to fill the position, another heart attack killed him on January 29, 1934.

⊠ **Hahn, Otto**
(1879–1968)
German
Radiochemist

Otto Hahn inadvertently discovered nuclear fission while duplicating IRÈNE JOLIOT-CURIE's experiment bombarding uranium with neutrons. The reaction generated not only an element that Hahn identified as barium, which splits uranium's atomic weight in half, but also unthinkable amounts of energy, as demonstrated by the detonation of atomic bombs through fission over Hiroshima and Nagasaki in World War II. Hahn also discovered numerous radioactive elements, such as radiothorium, radioactinium, and mesothorium, as well as the first nuclear isomer, uranium Z.

Hahn was born on March 8, 1890, in Frankfurt am Main, Germany. In 1897 he graduated from secondary school in Frankfurt and began university studies in Munich and Marburg. He conducted his doctoral research on organic chemistry under Theodor Zincke, submitting his dissertation in 1901 to earn his Ph.D.

Hahn remained at his alma mater, the Chemical Institute at Marburg, as an assistant for two postdoctoral years. In the fall of 1904 he traveled to London to work at University College under SIR WILLIAM RAMSAY. While preparing pure radium salts that year, Hahn discovered radiothorium, a new radioactive substance. The next year he crossed the Atlantic and went to Montreal, Canada, to work under LORD ERNEST RUTHERFORD at the Physical Institute at McGill

Otto Hahn, who inadvertently discovered nuclear fission *(Max-Planck-Institut für Physik, courtesy AIP Emilio Segrè Visual Archives)*

University. He made yet another discovery, of radioactinium, and he proceeded to study the alpha rays of his two discoveries with Rutherford.

In 1906 Hahn returned to Germany, installing himself at the Chemical Institute of the University of Berlin to work under EMIL FIS-CHER. Within a year he was appointed as a university lecturer. Also in 1907 he discovered yet another radioactive by-product of thorium, mesothorium. Hahn then commenced his 30-year collaboration with Lise Meitner, when she moved from Vienna to Berlin in 1907. In 1912 he became a member of the Kaiser Wilhelm Institute for Chemistry in Berlin-Dahlem. The following year he married Edith Junghans, with whom he had one son, Hanno, who was born in 1922 and died in 1960 in an accident.

It wasn't until after Hahn returned from his service in World War I, in 1918, that he made his next major discovery: With Meitner he discovered the longest-lived isotope of protactinium. They followed this achievement with the discovery in 1921 of uranium Z, the first nuclear isomer (or radioisotope that differs in energy content and half-life from other radioisotopes whose nuclei contain the same subatomic particles) of a radioactive atom. In 1928 the Kaiser Wilhelm Institute named Hahn its director; in 1933 he took a leave to serve as a visiting professor at Cornell University in Ithaca, New York.

In 1938 Meitner, a Jew, fled the anti-Semitism of Nazi Germany, so Hahn carried on his radiochemical research with Fritz Strassman. That year Irène Joliot-Curie discovered artificial radioactivity by bombarding uranium with neutrons to create a radionuclide of thorium. When her husband, Pierre, told Hahn his hypothesis—that the radionuclide they had created was really a transuranium element, such as lanthanum—the incredulous Hahn enlisted Strassman to re-create Joliot-Curie's experiment to disprove this hypothesis.

What Hahn and Strassman discovered astounded them: The reaction split the element in half to create barium atoms, which weigh just half as much as uranium. Furthermore, calculations suggested that a nuclear reaction such as this, which was unheard of, should generate as much as a hundredfold more energy than any known nuclear reactions. Experiments confirmed this prediction, much to their amazement. In a letter to Copenhagen, Denmark, dated December 19, 1938, Hahn pleaded with Meitner, "Perhaps you can suggest some fantastic explanation." His longtime partner confirmed what Hahn suspected—that they had opened up the Pandora's box of a reaction so strong, it defied comprehension—and she suggested it be called *nuclear fission*.

Hahn and Strassman published their results in the January 6, 1939, issue of the journal *Naturwissenschaften*, announcing that their findings were "in opposition to all the phenomena observed up to the present in nuclear physics." They didn't hazard an explanation, but let the results speak for themselves. The Danish physicist Niels Bohr immediately grasped the implication that a fission chain reaction could be sustained, which he communicated to Albert Einstein. Einstein wrote a letter to President Franklin D. Roosevelt, advising him to conscript the foremost scientists in a coordinated effort to harness the potential energy of fission to defend against the spread of fascism.

Before the demonstration of fission's destructive capabilities at Hiroshima and Nagasaki in Japan, Hahn received the 1944 Nobel Prize in chemistry for his discovery of fission. Hahn, who was being held by the British in hopes of extracting from him the status of the German bomb-building efforts, responded to the Royal Swedish Academy of Science that he was "regrettably unable to attend this ceremony." Also in 1944 Hahn ascended to the presidency of the Max Planck Institute in Göttingen when it subsumed the Kaiser-Wilhelm Institute.

Although Strassman and Meitner did not share the Nobel Prize with Hahn, they did share the 1966 Enrico Fermi Award with him Hahn devoted the latter part of his career to trying to temper the buildup of nuclear weaponry. He died on July 28, 1968, in Göttingen.

⊠ **Harden, Arthur**
(1865–1940)
English
Biochemist

Arthur Harden helped establish the modern field of biochemistry with his research on fermentation, which held profound implications for the understanding of the processes of metabolism. Harden relied more on perspiration than inspiration, allowing his meticulous experimentation to guide him to a conclusion instead of hypothesizing his intended outcome and then trying to achieve that result. In accepting the Noble Prize in chemistry in 1929, he acknowledged that science owed less recognition to him personally than to the field of biochemistry for rearing its head through his experiments.

Harden was born on October 12, 1865, in Manchester, England. He was the third of nine children born to Eliza MacAlister of Paisley, Scotland, and Albert Tyas Harden, a Manchester merchant. He grew up in an austere household that frowned upon the theater, though Harden found refuge in the literature of Charles Dickens. He attended Victoria Park School as a youngster and Tettenhall College in Staffordshire as a teenager, from 1876 to 1881.

At the age of 16, Harden matriculated at Owens College (later known as the University of Manchester), where he studied chemistry under Henry Roscoe. In 1885 he earned a degree in chemistry with first-class honors. The next year he received a Dalton scholarship that financed his graduate study at the University of Erlangen in Germany under ERNST OTTO FISCHER. He wrote his dissertation on synthetic organic chemistry, earning his doctorate in

1888. Thereafter he returned to his alma mater in Manchester as a junior lecturer under Roscoe's successor, H. B. Dixon, who later promoted Harden to senior lecturer.

Harden distinguished himself first as a teacher and, to a greater extent, as a writer. His love of the history of chemistry prompted him to excavate JOHN DALTON's notebooks chronicling the development of his atomic theory. Harden's studies resulted in the 1896 revisionist text *A New View of the Origin of Dalton's Atomic Theory*, which he coauthored with his former mentor, Roscoe. The following year he collaborated with F. C. Garett on the textbook *Practical Organic Chemistry*. That same year he revised and edited Roscoe's *Treatise on Inorganic Chemistry*.

Also in 1897 Harden accepted a professorship of chemistry at the British Institute of Preventative Medicine (renamed the Jenner Institute in 1898 and the Lister Institute in 1903). Almost immediately, however, the institute ceded responsibility for educating medical students in chemistry to the medical schools of the University of London, thereby freeing Harden to devote himself to research. He responded by focusing on fermentation studies, following up on EDUARD BUCHNER's discovery of extracellular yeast fermentation that year.

Harden elucidated the entire process of fermentation, thereby helping establish the field of biochemistry by opening the door to future scientists to further his discoveries and establish the mechanics of metabolism. Among his achievements, Harden first identified the sugar glycogen as the agent that survived the extraction from yeast cells and activated fermentation with Buchner's zymase (enzymes). Even more significant was the discovery by him and his assistant William Young of a cofactor, which Harden called "coferment," that was necessary for fermentation. Furthermore, Harden and Young discovered the existence of phosphate in the ash of heated coferment, thereby identifying phosphate as an important active agent in the fermentation process.

In 1900 Harden married Georgina Sydney Bridge, the eldest daughter of Cyprian Wynard Bridge of Christchurch, New Zealand. The couple had no children, and she died in 1928.

In 1905 the head of the biochemistry department departed from the Lister Institute, prompting it to combine the chemistry department that Harden then headed with the biochemistry department and placing Harden in the chairman's seat for the new department. In 1911 Harden helped found the Biochemical Society, and the following year Harden became the coeditor of the *Biochemical Journal*. Over the next quarter century, it is estimated that he read about 18 million words in editing the journal.

Also in 1912 Harden accepted a professorship in biochemistry at the University of London, though he did not abandon the Lister Institute: He served as the university department's acting director during World War I, when the head departed for the front. The war interrupted his fermentation studies, as he focused his attention on the war effort by studying the effects of nutrition on the development of the common trench diseases of beriberi and scurvy.

Harden received the Nobel Prize in chemistry in 1929 for his fermentation research performed more than two decades earlier. In 1935 the Royal Society, which had inducted him as a member in 1909, awarded him its Davy Medal. The next year Harden was knighted. He died at his home in Bourne End on June 17, 1940. The Biochemical Society, which he helped found, established an annual conference named after him with the money inherited from Harden's estate.

⊠ **Hassel, Odd**
(1897–1981)
Norwegian
Physical Chemist

Odd Hassel established the three-dimensionality of molecular geometry, a discovery that elicited

DEREK H. R. BARTON's realization of the correspondence between molecular function and structure. Barton published his theory of conformational analysis, which relied on Hassel's experimental confirmation of his hypotheses, in 1950, but it was not until 1969 that the Royal Swedish Academy of Sciences acknowledged the significance of their work with the Nobel Prize in chemistry.

Hassel and his twin brother, Leif, were born on May 17, 1897, in Christiania (now Oslo), Norway. His father, the gynecologist Ernst Hassel, died when Hassel was only eight years old, leaving his mother, Mathilde Klaveness, to raise her four sons and one daughter. Hassel entered the University of Oslo in 1915 to study chemistry, writing his thesis on the kinetics of nitro-compound reduction to earn his *candidatus realium* degree in 1920.

Upon his graduation at age 23, Hassel embarked on a tour of the continent, conducting research in France, Italy, and Germany. While working in Kasimir Fajan's laboratory in Munich, Germany, Hassel discovered absorption indicators, dyes used to measure silver and halide ions with greater accuracy than previously possible. In Berlin he helped determine the crystal structures of such inorganic substances as bismuth and graphite. After receiving his doctorate in 1924 from the University of Berlin, FRITZ HABER nominated Hassel for the Rockefeller Fellowship, which supported the young scientist's continued stay in Berlin while conducting research at the elder scientist's institution, the Kaiser Wilhelm Institute. There Hassel learned the new technique of X-ray crystallography, a method of determining atomic structure by focusing radiation on pure crystals of a substance.

Hassel returned to the University of Oslo in 1925 as an instructor, and the next year he rose to the rank of associate professor of physical chemistry. He focused his research on the question of ring-shaped carbon molecules, specifically cyclohexane, which he suspected filled three dimensions instead of the commonly held belief of only two. He identified the geometrical impossibility of these molecules existing on one plane, as the determined number of bonds between the carbon and hydrogen atoms prevents it from maintaining flatness.

If not flat, two configurations could account for their depth: Either atoms are raised off the plane on one side only (called the "boat" configuration, after its graphical resemblance to the bow of a canoe), or they depart from the plane on both sides (the "chair" configuration, after its resemblance to a reclining easy chair). In 1930 Hassel confirmed by means of X-ray crystallography that the cyclohexane molecule sits in the chair position, thus proving its three-dimensionality. After this discovery he switched from X-ray crystallography to electron diffraction, a method that identified the dipole moment and worked with liquids and gases, which X-ray crystallography could not do. It was not until 1942 that he demonstrated that the molecule oscillates back and forth between the boat and chair configurations, preferring the latter; the following year he published his elucidation of the theory of three-dimensionality, complete with experimental verification.

In the beginning of World War II, in 1940, Germany invaded Norway, and after three years of occupation, the Nazis closed the University of Oslo, sending its faculty to a concentration camp in Grini, near Oslo. Hassel continued his research during his interment and forged lasting relationships with fellow imprisoned scientists Per Andersen and Ragnar Frisch. They gained their freedom in November 1944.

News of Hassel's experimental confirmation of molecular three-dimensionality was slow in disseminating, as he refused to publish his results in German-language journals. An English translation of his 1943 paper summarizing his three-dimensional theory did not come until 1945, and it did not reach Derek Barton until the next year. In 1949 Barton combined the implications of his research on molecular characteristics with those of Hassel's three-dimensionality to demonstrate that certain molecular functions coincided with certain geometric configurations. This line of study became known as *conformational analysis* after Barton published his four-page paper introducing and confirming his theory, which leaned heavily on Hassel's experimentation.

Hassel acted as the Norwegian editor of the *Acta Chemica Scandinavica* from its founding in 1947 until 1956, publishing more than 65 of his own English-language papers in its pages. In the 1950s Hassel changed the course of his research from carbon rings to the crystal structures of charge-transfer complexes, a process by which one part "donates" an electron to the other part, which "accepts" it.

Hassel retired from the University of Oslo in 1964, though he remained active in his field. Not until 1969 was his groundbreaking work of the 1940s recognized with the awarding of the Nobel Prize in chemistry to Hassel and Barton for their joint establishment of the new field of conformational analysis. Hassel had already received the Fridtjof Nansen Award in 1946 and the Gunnerus Medal from the Royal Norwegian Academy of Sciences, as well as the Guldberg-Waage Medal from the Norwegian Chemical Society in 1964. His twin brother died in 1980, taking part of Hassel's spirit to the grave. Hassel, who never married, survived only another year, dying in Oslo on May 15, 1981, just two days before their 84th birthday.

⊠ **Haworth, Sir Walter**
(1883–1950)
English
Chemist

Walter Haworth was the first chemist ever to synthesize vitamin when he produced vitamin C in vitro in 1933. This accomplishment freed humans from dependence on fresh produce to avoid scurvy, as ascorbic acid could now be generated in the lab. He spent his earlier career investigating carbohydrate chemistry, making important discoveries regarding monosaccharides, disaccharides, and polysaccharides. He developed the formula for determining the molecular structure of the monosaccharide glucose, later named the Haworth formula. For these achievements he shared the 1937 Nobel Prize in chemistry with PAUL KARRER, who had done similar work on vitamins A and B.

Walter Norman Haworth was born on March 19, 1883, in the town of Chorley in Lancashire, England. He was the fourth child born to Hannah and Thomas Haworth. While working in the linoleum factory his father managed, Haworth realized that the chemistry of dyes interested him more than the production process, so he abandoned the industry for the academy. His father reacted to this slight by withholding financial support for his schooling, but Haworth managed to pass the Manchester University entrance examination in 1903 after private tutoring. He studied terpenes under the chemistry department chairman, William Perkin Jr. In 1906 he graduated with first-class honors in chemistry.

Haworth spent the next three years as an assistant to Perkin while conducting doctoral research. In 1909 he received a scholarship that allowed him to travel to the University of Göttingen in Germany to study under OTTO WALLACH, who won the Nobel Prize in chemistry in 1910 for his work on terpenes. Haworth finished his Ph.D. in Göttingen and used the second year of his scholarship to complete another doctorate, a Doctor of Science, from the University of Manchester in organic chemistry in 1911.

Haworth commenced his professional career as a senior demonstrator at Imperial College of Science and Technology in the South Kensington region of London, a post he retained only one year. In 1912 he accepted a lectureship at United College of the University of St.

Andrews. There Haworth dropped his terpene research in favor of carbohydrate chemistry research with the two top experts in the new field, Thomas Purdie and James Colquhoun Irvine. World War I interrupted their investigations for several years, and soon after it ended, Haworth resumed his carbohydrate studies independently when he moved to the University of Durham as a professor of chemistry in 1920. Two years later he married Violet Chilton, daughter of Sir James Dobbie, and together the couple had two sons.

In 1925 Haworth identified mistakes in the existing formula for the monosaccharide glucose and corrected it to generate a new model for the structure of glucose dubbed the Haworth formula: a ring of five carbon atoms and one oxygen atom, with an extra carbon atom attached. He subsequently identified the structures of other monosaccharides, including mannose, galactose, and fructose. He then moved to investigating disaccharides (two simple sugar units conjoined), identifying the structures of maltose, cellobiose, and lactose, among others. In 1926 the University of Birmingham named him its Mason Professor, and in 1929 he published *The Constitution of Sugars*, which quickly became considered a classic text defining its field.

In 1932 Albert Szent-Györgyi discovered hexuronic acid in oranges and cabbages and in the adrenal gland. After consulting with Haworth, the pair agreed that the agent was structured identically to vitamin C, so they named the discovery *ascorbic acid* (for its anti-scurvy properties). By the next year Haworth had succeeded in identifying ascorbic acid's structure (which was flat) and producing it synthetically. The international scientific community hailed the significance of Haworth's work by granting him the 1937 Nobel Prize in chemistry, making him the first British organic chemist to receive the Nobel in chemistry.

Haworth suffered a downturn in his health over the next three years, but he recovered enough to serve as chairman of the British Chemical Panel for Atomic Energy during World War II. His return to health allowed him to serve in other significant capacities in his field: From 1943 to 1946 he served as the dean of the University of Birmingham faculty, and from 1944 through 1946 he filled the presidency of the British Chemical Society.

Haworth received ample recognition of the significance of his career, winning a host of prestigious honors: the 1933 Longstaff Medal of the British Chemical Society and the 1934 Davy Medal and the 1942 Royal Medal of the Royal Society, which had inducted him as a member in 1928. He was knighted in 1948. Two years later, on his 67th birthday, he suffered a fatal heart attack in his Birmingham home.

⊠   **Herzberg, Gerhard**
(1904–1999)
German/Canadian
Physical Chemist

Gerhard Herzberg helped found the study of molecular spectroscopy, a process that separated molecular spectra into distinct lines, as a prism separates light into the spectrum of colors. By analyzing these lines, Herzberg drew conclusions about the structure of molecules, such as nitrogen, oxygen, and hydrogen. Later in his career Herzberg developed spectroscopic techniques for studying free radicals, or short-lived molecules generated in certain chemical reactions. For this pioneering work in molecular spectroscopy, Herzberg received the 1971 Nobel Prize in chemistry.

Herzberg was born on December 25, 1904, in Hamburg, Germany. His mother was Ella Biber, and his father, the comanager of a small shipping company, was Albin Herzberg. In 1924 Herzberg enrolled at the Technische Universität at Darmstadt, where he studied physics under Hans Rau to earn his bachelor of science in engi-

Gerhard Herzberg, who received the 1971 Nobel Prize in chemistry for his pioneering work in molecular spectroscopy *(AIP Emilio Segrè Visual Archives, Segrè Collection)*

neering in 1927. The next year he submitted a dissertation on the interaction of electromagnetic radiation with matter for his doctorate in engineering physics.

In 1928 Herzberg conducted postdoctoral research with James Franck, Max Born, and Walter Heitler at the University of Göttingen. In collaboration with the latter he demonstrated that the nuclei of nitrogen follow not Fermi statistics, as was commonly believed, but Bose statistics—an early indication that atomic nuclei are devoid of electrons. He spent another postdoctoral year at the University of Bristol in England, before submitting the fruit of his labors—theoretical work on molecular orbitals—for conferment of *habilitation* status, qualifying him for an unsalaried lectureship back at Darmstadt in 1930.

The year before Herzberg had married Luise H. Oettinger. The couple eventually had two children together, Paul and Agnes. When the National Socialist Party (Nazis) rose to power in 1933, one of the laws enacted forbade faculty members with Jewish wives from teaching at universities; this directly affected Herzberg, whose wife was Jewish. That same year W. T. Spinks, a physical chemist from the University of Saskatchewan, had studied with Herzberg; upon his return to Canada, Spinks secured a visiting professorship for the German refugee in 1935 through Carnegie Foundation funding that had been established specifically to aid scholars in exile. When a position in the physics department opened up, Herzberg became a permanent member of the faculty.

In 1936 Herzberg published a series of his Göttingen lectures on spectroscopy in German, and the next year *Atomic Spectra and Atomic Structure* came out in an English translation. In 1939 he published the first tome of a four-volume series (published over four decades), entitled *Molecular Spectra and Molecular Structure*. Simultaneous to his writing, Herzberg worked to improve the equipping of the university's laboratories. In 1939 he succeeded in purchasing a high-speed spectrograph through an American Philosophical Society grant.

In 1945 Herzberg became a naturalized citizen of Canada, after which he served three years as a professor at the University of Chicago, conducting spectroscopic research at its Yerkes Observatory. While there, he discovered new bands in the oxygen spectrum, which were called Herzberg bands in his honor. In 1947 the National Research Council (NRC) of Canada appointed him as principal research officer. In 1949 the NRC promoted him to the position of director of its physics division. Over the next two decades he established the NRC as the "foremost center for molecular spectroscopy in the world," according to the Royal Swedish Academy of Sciences.

In 1969 the NRC appointed Herzberg its first Distinguished Research Scientist, relieving him of all administrative responsibilities so that he could focus his attention on research. In 1971 he published his tract on free radicals, or molecules that appear during certain chemical reactions and then quickly disappear. Also in 1971 the Royal Swedish Academy of Sciences awarded him the Nobel Prize in chemistry. However, the death of his wife that year counterbalanced his elation with sorrow. A year later he married Monika Tenthoff.

In 1974 the NRC honored the scientist by founding the Herzberg Institute of Astrophysics, where he continued to conduct astronomic spectroscopic research. Many other organizations honored Herzberg as well: The American Chemical Society granted him its Willard Gibbs Medal and its Linus Pauling Medal, the Canadian Association of Physicists awarded him its Gold Medal, and the Royal Society of London conferred its Royal Medal upon him. Herzberg died on March 3, 1999.

⊠ Heyrovský, Jaroslav
(1890–1967)
Czechoslovakian
Physical Chemist

Jaroslav Heyrovský invented polarography, a process for measuring the chemical composition of solutions. This automated procedure decreased the analysis time from hours to milliseconds and made it possible to analyze several compounds in a solution simultaneously, as each element generates its own specific polarographic wave, captured on photographic film for further examination. For his research, and findings, Heyrovský was awarded the 1959 Nobel Prize in chemistry.

For his invention of polarography, Jaroslav Heyrovský won the 1959 Nobel Prize in chemistry. *(E. F. Smith Collection, Rare Book & Manuscript Library, University of Pennsylvania)*

Heyrovský was born on December 20, 1890, in Prague, Austria-Hungary (now the Czech Republic). His mother was Klára Hanl; his father, Leopold Heyrovský, was a professor of Roman law at what was then called the Charles-Ferdinand University in Prague. The name was later changed to Charles University. Heyrovský attended secondary school at the Akademicke Gymnasium and matriculated in 1909 to Charles University, where his father was rector. He studied physics under František Záviška and Bohumil Kučera. At the suggestion of chemist Bohumil Brauner, Heyrovský transferred to University College in London to study under SIR

WILLIAM RAMSAY. He received his bachelor of science there in 1913.

Heyrovský commenced doctoral studies on the electrode potential of aluminum under Ramsay's successor at University College, Frederick G. Donnan, who suggested that he employ a dropping mercury electrode when the passivation of the aluminum electrode posed problems. The eruption of World War I while he was visiting home detained him in Prague, though he managed to continue his doctoral research at Charles University in the laboratory of J. S. Štěrba-Böhm. He also served as a dispensing chemist and a roentgenologist in a military hospital. In 1918 he submitted his dissertation, entitled "The Electro-Affinity of Aluminum," to earn his Ph.D. from Charles University.

Serving as an assistant to Brauner, Heyrovský continued his aluminate studies toward a habilitation thesis that earned him a docent position in physical chemistry in 1920. The following year the University of London awarded him a Doctor of Science. Then in 1922 Charles University appointed him to an associate professorship and the head of the chemistry department. That same year he published a paper in the Czech journal *Chemicke Listy* describing his discovery of polarography.

Heyrovský had invented the process of polarography during his dissertation research in response to a problem with his experiments with aluminum, as an oxide coating built up on the electrodes. His solution involved measuring the applied voltage on the surface tension of mercury droplets falling from a glass capillary, which he called its electrocapillary curve. By chance, one of the members of his doctoral examination committee, Kučera, was conducting mercury surface tension research, so he used the oral examination as an opportunity to direct Heyrovský to work with Dr. Šimunek, who was doing even more specialized mercury investigations. This work resulted in Heyrovský's discovery that the comparison between stepped current and voltage curves in a dropping mercury electrode yielded results that very accurately identified the chemical compounds being studied. In collaboration with his Japanese assistant, Masuzo Shikata, he developed a photographic process that automatically recorded the current-voltage curves. In 1925 he published three more papers on the technique, instrumentation, and theory of conducting polarography with a dropping mercury electrode.

The year before, in 1924, Charles University had promoted Heyrovský to the title of extraordinary professor and appointed him director of its newly established Institute of Physical Chemistry. In 1926 he married Marie Kořánova, who assisted him in his scientific writings. The couple had two children—daughter named Jitka, who became a biochemist, and a son named Michael, who entered his father's field of polarography. Also in 1926 the university promoted him to a full professorship, freeing him to lecture at the University of Paris under a Rockefeller fellowship that year. In 1933 he served a six-month stint as the Carnegie Visiting Professor at three California institutions (the University of California at Berkeley, Stanford University, and the California Institute of Technology). The following year he traveled to St. Petersburg, USSR, to deliver the Mendeleev Centenary lecture.

Heyrovský raised the prominence of Czech scientific research by founding a journal, *Collection of Czechoslovak Chemical Communications*, under the auspices of the Royal Bohemian Society of Science. In order to disseminate information to the international scientific community, he translated papers into English while his associate, Emil Votoček, translated them into French (German and Russian translations were later added). Similarly, he promoted the use of polarography by assembling (with the assistance of his wife) exhaustive bibliographies of works discussing the process. Polarography became an established method of analysis: It is applied in cancer diagnoses, in measuring the blood lead

content, and even in determining the vitamin C content in fruits and vegetables. The field of polarography also proliferated, spawning such offspring as oscillographic polarography (which Heyrovský himself established), pulse polarography, chronopotentiometry, alternating current (AC) polarography, cyclic voltametry, and amperometric titrations.

In 1950 Charles University founded the Polarographic Institute, naming Heyrovský its director. The next year he won the Czech State Prize, and in 1955 he received the Order of the Czechoslovak Republic. Four years later he won the 1959 Nobel Prize in chemistry. In 1964 the Czechoslovak Academy of Sciences honored him by renaming the Polarographic Institute the J. Heyrovský Institute of Polarography. Three years later, on March 27, 1967, Heyrovský died in Prague at the age of 76.

⊠ **Hill, Henry Aaron**
(1915–1979)
American
Organic Chemist

Henry Aaron Hill specialized in researching polymers, large molecules made up of smaller molecules. He focused his attention on both natural polymers, such as rubber and resins, and synthetic polymers, such as plastic. He rose through the ranks in industry, eventually establishing a company in collaboration with others before founding his own research firm. He also served his field on the National Commission on Product Safety under President Lyndon B. Johnson and on the American Chemical Society's board of directors as its first African-American president.

Hill was born on May 30, 1915, in St. Joseph, Missouri. He attended Johnson C. Smith University in Charlotte, North Carolina, where he studied chemistry. He earned his bachelor of science in 1936, then spent a year pursuing graduate studies at the University of Chicago before transferring to the Massachusetts Institute of Technology (MIT). There he conducted some of his doctoral research under James Flack Norris. He submitted his dissertation, "Test of the Van't Hoff's Principle of Optical Superposition," earning his Ph.D. from MIT in 1942. He married the next year, and he and his wife had one child.

After sending out 54 letters of application, most of which met with rejections based more on his race than his ability, Hill finally received an offer for a position as head of chemistry research at Atlantic Research Associates in Newtonville, Massachusetts, in 1942. The next year the company promoted him to research director, and two years after that, in 1945, the North Atlantic Research Corporation (as it had been renamed) promoted him to a vice presidency in charge of research. In addition to his administrative duties, Hill contributed research to innovating water-based paints, rubber adhesives, and synthetic rubbers.

In 1946 the Dewey & Almy Chemical Company hired Hill as a group leader to manage team research on polymers. During his six years there he formed relationships with colleagues who later joined with him as partners in founding a new research company, National Polychemicals, that prepared chemical intermediaries for polymer production. For four years from 1952, Hill served as an assistant manager for National Polychemicals, based in Wilmington, Massachusetts. The company promoted him to a vice presidency in 1956, and he retained the position for five years.

Finally, in 1961 Hill fulfilled his goal of independence by founding his own company, Riverside Research Laboratory, and served as its president. Hill successfully carved a niche by offering not chemical products but chemical expertise on fabric flammability, as well as consulting on resins, plastics, and rubber. As head of his own company, Hill garnered respect instead of suffering the prejudice he had endured lower

on the corporate ladder. Other chemists and companies sought his advice, and his opinion carried weight, backed as it was with years of success in the industry (National Polychemicals, which he had cofounded prior to Riverside Research Laboratory, grossed annual sales of more than $10 million in 1971). In fact, in 1976 the Rohm & Hass Company requested he direct its operations for one year.

For 38 years of his career, Hill worked actively as a member of the American Chemical Society (ACS). In 1963 he served as chairman of the Northeast section, and in 1968 he chaired the society's Committee on Professional Relations, which developed comprehensive standards and guidelines for hiring and employment in the chemical field. That same year President Lyndon B. Johnson appointed Hill to the National Commission on Product Safety, charged with improving the safety of products while holding companies liable for safety lapses. Hill also chaired the compliance committee of the National Motor Vehicle Safety Advisory Council and worked on the Information Council on Fabric Flammability.

Hill served on ACS's board of directors for seven years, starting in 1971. In 1977 he also served as the society's first African-American president. In this position of influence he helped establish the society's Norris Award in honor of his MIT mentor, James Flack Norris, "the first big man ... who was more interested in my ability to learn chemistry than in the identity of my grandparents." Hill remained active in his profession up until he suffered a heart attack and died on March 17, 1979.

Hinshelwood, Sir Cyril
(1897–1967)
English
Chemist

Cyril Hinshelwood focused his research on the kinetics of chemical reactions, shedding light on the molecular interaction between the two most basic natural elements—hydrogen and oxygen, which combine to make water. He later focused on the kinetics of bacterial growth. Hinshelwood shared the 1956 Nobel Prize in chemistry with NIKOLAI SEMENOV.

Cyril Norman Hinshelwood was born on June 19, 1897, in London, England. He was the only child of Ethel Smith and Norman MacMillan Hinshelwood, a chartered accountant who shortly thereafter moved his family to Canada. Just before the death of Hinshelwood's father, in 1904, his mother moved with her son back to England. Hinshelwood attended the Westminster City School before receiving the Brackenbury Scholarship to Balliol College of Oxford University, which he had to defer to serve as an explosives chemist at the Queensferry Royal Ordnance Factory in Scotland during World War I. He was promoted to assistant chief chemist in 1918; shortly thereafter he returned to complete his scholarship at Balliol, earning both his master of arts and doctor of science in a mere one year.

Balliol appointed Hinshelwood as a fellow upon graduation in 1920, and the next year he also entered the fellowship of Trinity College as a tutor. He commenced his research career measuring the rate of decomposition under diverse conditions (differing temperatures, pressures, and catalysts) for solid mixtures in the presence of oxidants such as potassium permanganate and ammonium dichromate. He then turned his interest from interactions between solids to gaseous reactions.

From 1923 on through the rest of his research career, he investigated kinetics, or molecular motion, in chemical reactions. He collaborated with W. H. Thompson on the study of propionaldehyde, which decomposed more quickly at high pressures than predicted by the Lindemann theory. Hinshelwood accordingly supplemented the theory by taking into account the internal energy released by polyatomic molecules upon activation, or what he called "quasi-unimolecular" reactions.

In 1926 Hinshelwood published the classic text *Kinetics of Chemical Change* (which went through three subsequent printings by 1940). The next year, in 1927, he commenced investigations on one of the most basic of chemical interactions, that of hydrogen and oxygen, which becomes explosive under low pressure due to a branching chain reaction akin to those observed by Nikolai Semenov between phosphorus and oxygen. The Royal Society recognized the significance of Hinshelwood's findings by inducting him into its fellowship in 1929.

Hinshelwood turned his attention from gaseous reactions to liquid reactions in the 1930s. He became increasingly interested in the chemical reactivity of bacterial growth. He summarized these findings in his 1946 text *The Chemical Kinetics of the Bacterial Cell*, which he followed up with *Growth, Function, and Regulation in Bacterial Cells*, coauthored by A. C. R. Dean in 1966. The king of England, George VI, further recognized the importance of Hinshelwood's discoveries by knighting him in 1948. This recognition freed him to wax philosophical about his field in the 1951 text *Structure of Physical Chemistry*.

Having made his name through experimentation and practical applications, Hinshelwood spent the latter part of his career as an ambassador of his field. He served in prominent positions at several organizations: president of the Chemical Society from 1946 to 1948; foreign secretary of the Royal Society from 1950 to 1955 and its president from 1955 to 1960; and president of the Faraday Society from 1961 to 1962. He retired from his chair at Oxford University in 1964 but continued to serve as a senior research fellow at Imperial College while also holding a chair at Queen Elizabeth College in his hometown of London.

Accolades showered Hinshelwood's career. Aside from the 1956 Nobel Prize in chemistry, the Société Chimique de France awarded him its Lavoisier Medal in 1935, and Oslo University bestowed on him its Guldberg Medal in 1952. The Royal Society granted him its 1942 Davy Medal, its 1947 Royal Medal, and its 1962 Copley Medal. He was also elected into the Order of Merit in 1960. The year before, in 1959, he grieved the death of his mother, with whom he had lived throughout his life as an inveterate bachelor. He held diverse interests, including Chinese pottery and languages. Hinshelwood died on October 9, 1967, in London.

⊠ Hodgkin, Dorothy Crowfoot
(1910–1994)
English
X-ray Crystallographer

Dorothy Hodgkin pioneered the field of X-ray crystallography, using this method to ascertain the structures of several important substances, including penicillin, vitamin B_{12}, and insulin. In the process, she overcame both social and technological restrictions, such as the limited laboratory space and equipment provided to women scientists at the time and the limitations of analog computers until the advent of electronic computers. In honor of the significant advances in the understanding of chemical structure she had elicited, Hodgkin received the 1964 Nobel Prize in chemistry.

Hodgkin was born Dorothy Crowfoot on May 12, 1910, in Cairo, Egypt. Her mother, Grace Mary Hood, was an expert in Coptic textiles, an amateur botanist, and a nature artist who illustrated *Flora for the Sudan*. Her father, John Winter Crowfoot, was an archaeologist, classical scholar, and an inspector for the British Ministry of Education, a post that required extended stays worldwide. Crowfoot lived in England with her paternal grandmother in Beccles, where she attended private school. She began growing crystals at this time, and her

Dorothy Crowfoot Hodgkin pioneered the use of X-ray crystallography, which she used to understand the structures of penicillin, vitamin B_{12}, and insulin. *(AIP Emilio Segrè Visual Archives, Physics Today Collection)*

mother gave her Sir William Henry Bragg's *Concerning the Nature of Things* for her 16th birthday, sparking her lifelong interest in crystallography. She enrolled at the Sir John Leman School in 1921, graduated in 1928, and spent the following summer in Palestine on one of her father's archaeological digs, an experience that almost swayed her interest away from crystals.

In October 1928 Crowfoot entered Somerville College for women at Oxford University, where she read chemistry and physics while serving as a research assistant under H. M. Powell. She earned her bachelor's degree in 1932. Her aunt, Dorothy Hood, who financed her undergraduate education, continued to provide for Crowfoot during her postgraduate study at Cambridge University until Somerville College offered her a two-year research fellowship. The first year of her fellowship, she remained at Cambridge, conducting X-ray crystallographic studies on the structure of vitamin D and sex hormones under John D. Bernal and Isidor Fankucken; the second year her fellowship required her to return to Somerville, which was sorely underequipped for crystallographic research. Crowfoot conducted her experimentation on makeshift equipment in makeshift spaces in the basement of the university museum, until she received a Rockefeller Foundation grant (thanks to SIR ROBERT ROBINSON) to purchase two X-ray tubes, two oscillation cameras, and a Weisenberg camera, the minimal equipment requirements for her research.

Crowfoot submitted a dissertation on the chemistry and crystallography of sterols to earn her doctorate from Cambridge in 1937. That spring she met the historian and African studies specialist Thomas L. Hodgkin (whose great uncle had identified what became known as Hodgkin's disease), and the couple married on December 16, 1937. Their separate professional commitments forced them to live apart until 1945, when he finally secured a position at Oxford. Despite this separation, Dorothy and Thomas Hodgkin enjoyed a happy marriage, and together they had three children—Luke (born in 1938), Elizabeth (born in 1941), and Toby (born in 1946).

Oxford appointed Hodgkin as a university lecturer and demonstrator while she continued her crystallographic research. She analyzed the crystal structures of the halogen derivatives of cholesterol and subsequently studied zinc insulin (prepared by Robinson) and then penicillin (from a sample sent by Ernst Chain). Throughout World War II she continued her studies on penicillin hydrochloride in conjunction with her graduate student Barbara-Low Rogers. They utilized an abandoned IBM punch-card machine to determine the chemical structure of penicillin hydrochloride in 1945. Hodgkin also identified a penicillin analog, cephalosporin C, which proved much easier to synthesize in large-scale production and which subdued penicillin-resistant infections.

In 1948 Hodgkin commenced analysis of the vitamin B_{12} molecule, which appeared to be an insurmountable task due to its immense molecular weight (1,355) and structural complexity. Hodgkin harnessed the same punch-card machine to begin B_{12} calculations, but over the six years of research, electronic computers became available, and with them she completed the determination of the chemical structure of vitamin B_{12}, which she published in 1957.

That same year she finally gained financial security when Oxford appointed her to a readership (equivalent to a full professorship in the United States), and the next year the university finally outfitted her in a proper laboratory. In 1960 the Royal Society named Hodgkin its Wolfson Research Professor. Four years later the Royal Swedish Academy of Sciences awarded Hodgkin its 1964 Nobel Prize in chemistry, the first such awarded to a British woman, for her elucidations by X-ray crystallography of penicillin vitamin B_{12}, and insulin. The following year she became the second woman (after Florence Nightingale) to receive the British Order of Merit.

Hodgkin had set out to uncover the structure of insulin directly after she deduced the structure of vitamin B_{12}. By 1969 she had analyzed 70,000 crystallographic X rays to decipher the 777 atoms of insulin, demonstrating their three-dimensional configuration as 51 amino acids. In 1977 she retired from her professorship with the Royal Society but remained active in her field by serving as the chancellor of Bristol University, a position she retained until 1988, the same year she updated her insulin analysis thanks to advanced computer applications. She died at her home in Shipston-on-Stour, Warwickshire, England, on July 30, 1994.

⊠ Hoobler, Icie Gertrude Macy
(1892–1984)
American
Biochemist

Icie Gertrude Macy Hoobler pioneered the study of nutrition from a scientific and chemical perspective, focusing specifically on the effects of nutrition on women, mothers, infants, and children. Although she experienced sexual discrimination throughout her career, she persisted in paving the way for countless women scientists who followed in her footsteps.

Hoobler was born Icie Gertrude Macy on July 23, 1892, in Gallatin, Missouri, where her parents were farmers. Macy's interest in the sciences was inspired by her observations of the food chain in nature and how those at the top of this chain, humans, sometimes suffered needlessly from malnutrition due to poverty. She attended the Central College for Women in Lexington, Missouri, where she earned her associate's degree in English in 1914 (one source reports that she received her A.B. degree from Randolph-Macon Woman's College). She next earned a Bachelor of Science in chemistry from the University of Chicago in 1916. Her adviser, Jules Steiglitz, recommended her for a chemistry assistantship at the University of Colorado at Boulder, where she earned a master's degree two years later.

In 1918 Macy matriculated at Yale University to pursue doctoral studies in physiological chemistry. Early on she attended a lecture on the dairy industry in which the American biochemist Lafayette B. Mendel exhorted women scientists especially to conduct research on human nutrition in general and focus specifically on pregnant women and malnourished infants. (Mendel became famous for training 45 women Ph.D.s out of 124 of his doctoral students altogether.) Inspired, Macy devoted herself to the study of the health of mothers, infants, and children. She focused her doctoral research on the nutritional value of cottonseeds, discovering that they contained a poison, gossypol, that made laboratory animals sick. Based on this work, she earned her doctorate in 1920.

Macy took up a chemistry assistantship at the Western Pennsylvania Hospital in Pittsburgh. Within a few months, however, she developed acute nephritis (inflammation of the kidney) as the nearest women's restroom was in another building a half-block away from where she worked. She almost resigned due to this and other instances of sexual discrimination, but the director of the hospital's board of trustees intervened and, upon hearing of her mistreatment, chided her supervisors and ensured her that she would receive the respect and privileges accorded to all staff members. Her research there focused on fetal analysis for calcium and magnesium, and she compared the composition of the bloods of mothers and their newborn babies.

Macy spent a year recuperating from her nephritis at the University of California at Berkeley, where she not only taught her appointed course on food chemistry but also picked up several other teaching assignments. The president of the university also asked her to serve as an inspector of institutions for the state of California educational system. While at Berkeley, she collaborated with Agnes Fay Morgan. Then, in 1923 the Merrill-Palmer School for Motherhood and Child Development appointed her director of the Nutrition Research Project, which was run in conjunction with the Children's Hospital of Michigan in Detroit.

In 1931 Macy's laboratory moved to be under the auspices of the Children's Fund of Michigan, with Macy serving as director of its research division. However, her research remained unchanged, as she continued to focus her studies on women's metabolism during reproductive cycles, the makeup of mother's milk, chemical changes throughout childhood growth, childhood nutrition, and the chemistry of red blood cells. In 1938 Macy married Dr. B. Raymond Hoobler, who supported her continued professional activities wholeheartedly (she chose not to heed his urging to keep her own name). Sadly, he died five years later (he was 20 years her senior), in 1943.

In 1945 Hoobler published the book that established her lasting reputation, *Hidden Hunger*, coauthored by H. H. Williams. This period represented the peak of her career, as she won the Garvan Award, the highest honor accorded women by the American Chemical Society, the very next year. Far from resting on her laurels, however, she continued to contribute to the knowledge of her field: Also in 1946 she published the second volume of *Nutrition and Chemical Growth in Childhood, Original Data;* she had published the first volume, *Evaluation,* in 1942. She published the third volume, *Calculated Data,* in 1951. In 1952 she received the Osborne and Mendel Award from the American Institute of Nutrition

In 1954 Hoobler retired to a research consultantship with the Merrill-Palmer Institute, a position she retained for the next two decades. In 1957 she and H. J. Kelly published *Chemical Anthropology.* Throughout her career she oversaw her laboratory's publication of more than 300 journal articles. In 1982 she collaborated with H. H. Williams and A. G. Williams to write her autobiography, *Boundless Horizons: Portrait of a Pioneer Woman Scientist.* Two years later, on January 6, 1984, Hoobler died in her birthplace of Gallatin, Missouri.

⊗   **Hückel, Erich**
(1896–1980)
German
Physical Chemist

Erich Hückel collaborated with his mentor, PETER DEBYE, to develop the Debye-Hückel theory of strong electrolytic dissociation, mathematically accounting for the instances not covered by SVANTE AUGUST ARRHENIUS's theory of electrolytic dissociation. Hückel also performed independent research on aromaticity, identifying the planar structure of benzene and its related compounds. He devised Hückel's rule, which posited a mathematical formula for calculating the numbers of electrons in aromatic compounds.

Erich Armand Arthur Joseph Hückel was born on August 9, 1896, in Berlin-Charlotten-

berg, Germany. His father was a medical doctor, and his brother, Walter Hückel, became a famous organic chemist. When Hückel was three years old, his family moved to Göttingen. In 1914 he entered the University of Göttingen to study physics. World War I interrupted his studies for two years, as he contributed to the war cause by working on aerodynamics under L. Prandtl in the university's Applied Mechanics Institute. In 1918 he returned to his university studies in mathematics and physics. He wrote his dissertation on the diffraction of X rays by liquids under Debye to earn his doctorate in experimental physics in 1921.

Hückel remained at the University of Göttingen for two years of postdoctoral research, working first under the mathematician David Hilbert then under the physicist and Nobel laureate Max Born, as a physics assistant in both instances. In 1922 he moved to the Zurich Technische Hochschule in Switzerland to rejoin his mentor, Debye, under whom he worked as an unpaid lecturer. In 1923 Hückel and Debye revised Svante Arrhenius's theory of electrolytic dissociation in solutions, which posited the equilibrium between undissociated solute molecules and the ions from dissociated molecules, whose positive and negative charges counterbalanced each other; however, this theory failed to account for certain strong electrolytic solutions.

Debye and Hückel solved this dilemma by suggesting that these strong electrolytes dissociate completely into ions, thus disrupting the equilibrium of the negative and positive charges described by Arrhenius that applies to most situations. The pair devised mathematical formulas of the electrical and thermodynamic properties of these strong electrolytes to prove their theory. In 1925 Hückel married Annemarie Zsigmondy, daughter of RICHARD ADOLF ZSIGMONDY, who won the Nobel Prize in chemistry that same year.

In 1928 Hückel won a Rockefeller Foundation Fellowship, which he used to study under Frederick Donnan for a brief stint and then under

Niels Bohr in Copenhagen, Denmark. Afterward he served as a fellow of the Notgemeinschraft der Deutschen Wissenschaft at the University of Leipzig. In 1930 the Technische Hochschule in Stuttgart appointed him to a position in chemical physics. That year he commenced his studies of benzene and its related aromatic compounds (such as pyridine), which are generally five- and six-membered rings. He collaborated with his brother, Walter, on these studies.

Hückel established that aromatic molecules must have one 2π orbital on each member of the ring, and six π electrons (or an aromatic sextet) in the cyclic arrangement of 2π orbitals. This configuration accounts for the planar structure of aromatic compounds. He subsequently devised Hückel's rule, a mathematical formula for calculating the number of π electrons in aromatic compounds: $(4n + 2)$, where n is a whole number. Nonaromatic compounds can be calculated by the formula $4n$ ($n = 1$); for example, benzene itself has six π electrons, whereas cyclooctatetrene has eight π electrons and thus is not an aromatic compound. The American chemist William van Doering later formulated a version of this rule as well.

In 1937 Hückel moved to the University of Marburg, which had appointed him a professor of theoretical physics. He stayed there through the remainder of his career, retiring in 1962. His other researches involved double- and triple-bond compounds, or unsaturated compounds; he also investigated free radical compounds, which contain a nonbinding, or "free," radical. Hückel died in 1980.

As a creative aside, though the lines are sometimes attributed to his brother, Hückel composed the following verse (translated from the German by Felix Bloch) in honor of Austrian physicist Erwin Schroedinger's research wave function.

> Erwin with his psi can do
> Calculations quite a few.
> But one thing has not been seen:
> Just what does psi really mean?

I

Ingold, Sir Christopher
(1893–1970)
English
Organic Chemist

Sir Christopher Ingold, who was known as "the prof" among his peers in the field, is considered one of the fathers of modern organic chemistry. His theories, considered "heretical" when he first proposed them in the early 1930s, became the cornerstones of organic chemistry, and the terminology he introduced is now standard. Ingold advanced the field with his use of reaction kinetics to study the mechanisms of nitration, displacement, and elimination reactions. He also conducted exacting spectroscopic studies on the first excited states of benzene. These studies pioneered organophysical chemistry, or the application of physical methods to organic chemistry.

Christopher Kelk Ingold was born on October 28, 1893, in London, England. Within a few years, however, his family moved to Shanklin on the Isle of Wight due to his father's failing health. Ingold attended Sandown Grammar School, before matriculating at Hartley University College (later renamed the University of Southampton), where he earned his bachelor of science in 1913.

Ingold spent the next five years at Imperial College of the University of London, investigat-

ing spiro-compounds of cyclohexane under J. F. Thorpe. In 1918 he left to spend two years as a research chemist at the Cassel Cyanide Company in Glasgow, Scotland. He returned to Imperial College in 1920 as a lecturer in organic chemistry. There he met Edith Usherwood, a fellow chemist, and the couple married soon thereafter.

In 1924 Leeds University hired Ingold as a professor of organic chemistry. In 1926 he advanced the notion of mesomerism, a concept allowing for the coexistence of a molecule as a hybrid of two equally possible structures. After six years at Leeds he joined University College London to fill the chair vacated by SIR ROBERT ROBINSON, a position Ingold retained for the remainder of his career. In 1932 he started publishing some of his ideas about reaction mechanics, which are now fundamental to his field.

In 1934 Ingold wrote his famous review on the electronic theory of organic reactions, a tract that became known as the "English heresy." Within five years, however, not only this reaction mechanism theory but also his accompanying terminology (such as *electrophilic, nucleophilic,* and *inductive*) became standard in the field. At University College, Ingold collaborated extensively with E. D. Hughes, conducting the above studies, as well as investigations of elimination reactions and the mechanisms of prototropic and anionotropic systems. Ingold is also known for many other contributions to the field of organic chemistry: He solved the mystery of the Walden inversion; studied the reversibility of the Michael reaction, relating this to prototropy; investigated the mutarotation of sugars; and in collaboration with VLADIMIR PRELOG, specified the sequence rule.

Ingold gathered together a career's worth of research in his field to publish his comprehensive reference book, *Structure and Mechanism in Organic Chemistry*, in 1953 (the second edition, which appeared in 1969, ran more than 1,200 pages). From 1952 through 1954, he served as the president of the Royal Society of Chemistry (RSC), an organization that honored him extensively throughout his career. He won its Meldola Medal and Prize two years consecutively, in 1921 and 1922. The RSC granted him its Longstaff Medal in 1951. In 1956–57, he held the society's Pedlar lectureship, and in 1961, he held its Faraday lectureship.

In 1958 Ingold was knighted, and three years later, he officially retired from University College, though he continued to contribute his work, expertise, and enthusiasm to his department. In 1969 the Sunday *Times* of London selected him as one of its 1,000 "Makers of the 20th Century." Ingold died on December 8, 1970, in London.

The Royal Society of Chemistry established the Christopher Ingold Lectureship in 1973 in honor of its former president. The lecture topic concerns the relationship between structure and reactivity in chemistry, and the RSC awards the recipient with a bronze medal and a cash prize. The 1990–91 Christopher Ingold Lecturer was Keith Ingold, the son of the scientist. University College further honored the former member of its chemistry department by naming the Christopher Ingold Laboratories in his memory.

J

Joliot-Curie, Frédéric
(1900–1958)
French
Nuclear Chemist

Frédéric Joliot-Curie shared the 1935 Nobel Prize in chemistry with his research partner and wife, IRÈNE JOLIOT-CURIE (daughter of the Nobel laureates Pierre and MARIE CURIE), for their joint discovery of artificial radioactivity. This discovery was precursor to the discovery of nuclear fission, which in turn led to the development of the atomic bomb. During World War II Joliot-Curie played a key role in deterring the Germans from building their own atomic bomb by removing vital ingredients to England before Nazi forces invaded France. He devoted the latter part of his career to the use of nuclear power for peaceful ends and opposed to the proliferation of nuclear weaponry.

Frédéric Joliot was born on March 19, 1900, in Paris, France, the sixth child born to Emilie Roederer and the merchant Henri Joliot. His parents sent him to a boarding school south of Paris, Lycée Lakanal, when he was 10 years old. Upon the death of his father, Joliot transferred to the École Supérieure de Physique et de Chimie Industrielle in Paris, where Paul Langevin introduced him to the political ideals of socialism and pacifism, which Joliot espoused the rest of his life.

In the spring of 1925 Joliot took up an assistantship to the eminent nuclear physicist Marie Curie at her Radium Institute, where he studied the chemical properties of polonium. Curie's daughter Irène was charged with the task of training the doctoral student in the use of the laboratory apparatus, and the couple soon found that they not only shared a professional affinity but also were developing a romantic attachment. Joliot married Irène Curie in 1926, and they eventually had two children, Hélène and Pierre. The couple formed a symbiotic professional relationship whereby Irène served as physicist and Frédéric as chemist; they symbolized this cooperation by adopting each other's surname professionally, signing all papers with the combined "Joliot-Curie."

Joliot-Curie wrote his dissertation on the electrochemistry of radioelements to earn his doctorate in 1930. He then received a scholarship to continue his studies on radioactivity with the Caisse Nationale des Sciences. He and his wife carried out collaborative research on atomic structure that put them on the brink of two landmark discoveries, but other scientists narrowly beat them to the punch: James Chadwick, an English physicist, discovered the neutron in 1932 (the Joliot-Curie's had misidentified it as a gamma ray), and U.S. physicist Carl David Anderson discovered the positron the next year. Undeterred, they persisted in their research, bombarding aluminum nuclei with alpha particles. As expected, the nuclei emitted neutrons during the bombardment, but to their surprise, the nuclei continued to emit positrons even after the discontinuation of the alpha bombardment.

Upon subsequent analysis the Joliot-Curie's determined that the aluminum had converted into a radionuclide of phosphorous, the first instance of elemental transmutation. More significant, this phenomenon represented the first synthesis of an artificial radioactive element. They reported their findings to the French Academy of Sciences in January 1934. They continued to pursue this line of investigation, bombarding boron, aluminum, and magnesium with alpha particles to produce isotope 13 of nitrogen, isotope 30 of phosphorus, isotope 27 of silicon, and isotope 28 of aluminum. These new elements did not occur in nature, and they decomposed spontaneously.

Recognition of the significance of this discovery was almost immediate; by 1935 the Royal Swedish Academy of Sciences had conferred the Nobel Prize in chemistry upon the duo. That same year the Collège de France appointed Joliot-Curie as a lecturer, and two years later it promoted him to a professorship. Also in 1937 the Laboratoire de Synthèse Atomique at Ivry appointed him its director. Meanwhile, he joined the Socialist Party, commencing his commitment to leftist causes, and in 1936 he joined the League for the Rights of Man. In the late 1930s he oversaw the construction of the first cyclotron in western Europe, and in 1939 he experimentally confirmed OTTO HAHN's discovery of the uranium fission phenomenon. He also collaborated with Hans Halban, Lev Kowarski, and Francis Perrin to design an atomic pile of uranium and heavy water, filing five patents for this chain reaction process in 1939 and 1940.

Anticipating the 1940 Nazi invasion of France, Joliot-Curie arranged for the deportment to England of all materials required to create radioactivity. He also ordered 6 tons of uranium oxide from the Belgian Congo (now Democratic Republic of the Congo) and almost the entire stock of heavy water from the Norsk Hydro Plant in Norway and had these raw materials for atomic fission sent directly to England. During the Nazi occupation, he actively participated in the French Resistance movement, joining the French Communist Party in 1942 and presiding over the National Front.

After World War II Joliot-Curie became the director of the Centre National de la Recherche Scientifique in 1945, and later that year he convinced President Charles de Gaulle of France to form the Atomic Energy Commission in October 1945. The following year the president appointed Joliot-Curie as the first high commissioner of the AEC; in this post Joliot-Curie oversaw the construction of the first French atomic pile, in 1948. In April 1950, however, Georges Didault removed Joliot-Curie from his commission position due to his communist sympathies and specifically his statement condemning the contribution of scientific efforts to the preparation for war against the Soviet Union.

Joliot-Curie nonetheless continued to crusade for leftist political causes, serving as president of the World Organization of the Partisans of Peace. He also continued with his commitment to the advancement of science, serving on the French National Committee for Scientific Research. When his wife, Irène, died of leukemia in March 1956, he succeeded her as director of the Radium Institute and inherited her chair of nuclear physics at the Sorbonne in Paris. That year he contracted a viral hepatitis infection, and two years later, he underwent surgery for internal hemorrhaging. Joliot-Curie died in the wake of this operation, on August 14, 1958, in Paris.

Joliot-Curie, Irène
(1897–1956)
French
Chemist, Nuclear Physicist

Irène Joliot-Curie followed in the footsteps of her parents—Pierre and MARIE CURIE, Nobel laureates for their discovery of radium and radioactivity—by sharing the 1935 Nobel Prize in chemistry with her husband, FRÉDÉRIC JOLIOT-CURIE, for synthesizing the first radioactive isotope. She inherited her mother's position as director of the Radium Institute in Paris as well as her passion for advancing both scientific and political causes internationally.

Irène Curie was born on September 12, 1897, in Paris, France. Her mother, née Marie Skłodowska, gave birth to Irène in the midst of the radium studies that garnered her parents the Nobel. After her father Pierre's tragic death in 1906 (he absentmindedly walked in front of a speeding wagon), her mother went on to receive an unprecedented second Nobel Prize (in chemistry) in 1911, for the isolation of metallic radium.

Curie's grandfather, Eugene Curie, filled the paternal role in Irène's life by teaching her botany and natural history, as well as political leftism and atheism. In addition, Curie's mother enlisted her colleagues at the Sorbonne, such as Paul Langevin and Jean Perrin, to "homeschool" her children in a teaching cooperative similar to the "floating university" education she had received from Poland's finest professors. However, the cooperative collapsed in 1911, so Irène enrolled in the Collège Sévigné, where she earned her undergraduate degree. She subsequently studied nursing at the Sorbonne, receiving a diploma before the outbreak of World War I.

Curie assisted her mother in operating X-ray equipment on the battlefields, utilizing radiographic technology to aid in the care of wounded soldiers. She returned from the war in 1918 to become her mother's assistant at the Radium Institute, founded to advance the understanding

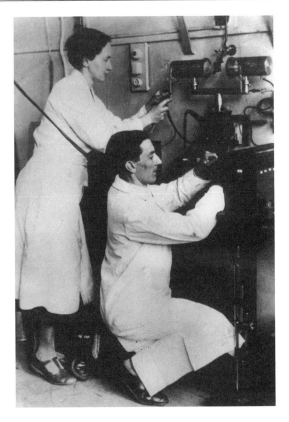

Irène Joliot-Curie with her husband, Frédéric. They were jointly awarded the 1935 Nobel Prize in chemistry for their discovery of artificial radioactivity. *(Société Française de Physique, Paris, courtesy AIP Emilio Segrè Visual Archives)*

of radioactivity for academic and medical purposes. There she worked under Fernand Holweck, the institute's chief of staff, writing her first paper—on radium, appropriately enough—in 1921. Four years later, in 1925, she submitted her dissertation on the emission of alpha particles from polonium to earn her doctorate in physics and mathematics from the Sorbonne.

That same year the Radium Institute charged Curie with the training of a young doctoral student, Frédéric Joliot, on the use of laboratory apparatus (she inherited an instinct for intuitively operating instruments from her father). The couple developed not only a professional but also a romantic relationship; they married in 1926 and from thereon, in testament to their commitment to equality, signed all their papers collaboratively as Joliot-Curie. Frédéric earned his doctorate in 1930, and two years later Irène became chief of the laboratory at the Radium Institute. The couple had two children, Hélène and Pierre.

In the early 1930s the Joliot-Curies were on the cusp of two major discoveries when other scientists beat them to it: in 1932 Englishman James Chadwick discovered the neutron, and in 1933 American Carl David Anderson discovered the positron. Nevertheless the Joliot-Curies persisted, conducting experimentation on aluminum, a light element containing 13 protons in its nucleus. When they bombarded aluminum nuclei with alpha particles, the nuclei gained two protons, thereby transforming the aluminum nuclei into phosphorus nuclei, which contain 15 protons. The Joliot-Curies were thus the first to synthesize a radioactive isotope, a discovery they announced to the French Academy of Sciences in January 1934.

That year Joliot-Curie's mother, Marie, succumbed to leukemia, a disease resulting from her long-term exposure to radiation. The next year, though, the Joliot-Curies shared the 1935 Nobel Prize in chemistry. Two years later, the Sorbonne appointed Irène as a professor, and that same year the Collège de France named Frédéric a professor of nuclear physics. Through the end of the decade, Joliot-Curie continued her fission studies, bombarding heavy elements such as uranium with alpha rays. In collaboration with Pavle Savitch, she turned to bombarding uranium nuclei with neutrons, breaking it down into other radioactive elements—a precursor to OTTO HAHN's discovery of nuclear fission, or the splitting of uranium into two atoms of equal mass.

During World War II the Joliot-Curies fought in the French Resistance against Nazi occupation before seeking refuge in Switzerland in 1944. Upon their return to Paris after the war,

the Radium Institute appointed Joliot-Curie its director, and the French government named her a commissioner of its atomic energy project. She later served on the World Peace Council, promoting the use of nuclear technology for peaceful purposes exclusively. Her affiliation with the Communist Party prevented her induction into the American Chemical Society in 1954. In 1955 she helped design the laboratory for a particle accelerator in Orsay, France. After suffering from declining health, Irène finally succumbed to the same disease that claimed her mother, leukemia, from her lifelong exposure to radiation. She died on March 17, 1956, in Paris.

⊠ Jones, Mary Ellen
(1922–1996)
American
Biochemist

Mary Ellen Jones established her reputation early in her career when, as a postdoctoral researcher, she discovered, in collaboration with Leonard Spector, carbamoyl phosphate, a precursor in the biosynthesis of arginine, urea, and pyrimidine nucleotides. She went on to become the first woman named to an endowed chair at the University of North Carolina at Chapel Hill, and the first woman appointed to chair a department in the university's medical school.

Jones was born on December 25, 1922, in La Grange, Illinois. She was one of four children born to Laura Klein and Elmer Jones. She attended Hinsdale High School, where her sophomore science teacher, Mr. Poppenhager, sparked her interest in the biological sciences. She graduated in 1940 and matriculated at the University of Chicago. While studying biochemistry there, Jones worked half time at Armour and Company as a bacteriologist in quality control. After graduating with a bachelor of science in 1944, she continued to work at Armour, which promoted her to the post of research chemist. She conducted studies with the director of the research laboratory, Paul Munson, with whom she wrote two papers on androsterone and monopalmitin, published in the *Journal of Biological Chemistry*. Jones married Munson in 1948, and together the couple had two children—Ethan V. Munson, born in 1956, and Catherine Munson, born in 1960.

In 1948 Munson accepted an assistant professorship in pharmacology at Yale University, in New Haven, Connecticut, where Jones pursued graduate studies in enzymology under Joseph Fruton. She wrote her dissertation on a characterization of the catalytic properties of cathepsin C and completed the requirements for her doctorate in a mere three years, earning her Ph.D. in 1951. That year the couple moved to Boston, where Jones secured a postdoctoral fellowship at the Massachusetts General Hospital (MGH) in the biochemistry research laboratory under Fritz Lipmann, who won the Nobel Prize in medicine or physiology two years later. Jones initially conducted research on the activation of coenzyme A.

In 1955 Jones collaborated with Leonard Spector to hypothesize that the synthesis of citrulline from ornithine requires an agent, most likely carbamoyl phosphate. After Spector crystallized carbamoyl phosphate from cyanate with lithium phosphate (a procedure he thought up while attending a performance of Claude Debussy's *Afternoon of a Faun*), Jones demonstrated that liver cell extracts converted carbamoyl phosphate to citrulline. Two years later Jones secured an assistant professorship in the graduate department of biochemistry at Brandeis University in Waltham, Massachusetts, on the strength of her nine publications while at MGH (in addition to the two papers she published while at Armour, and two more while at Yale). By 1960 she had been promoted to an associate professorship.

Jones moved to the University of North Carolina (UNC) at Chapel Hill when her hus-

Mary Ellen Jones, the first woman named to an endowed chair at the University of North Carolina at Chapel Hill and the first woman to chair a department at the university's medical school *(Courtesy News Services, The University of North Carolina at Chapel Hill)*

band was offered a chair of pharmacology in 1966. She joined UNC's department of biochemistry, and the zoology department hosted her in its laboratory, which was far inferior to the lab she was accustomed to at Brandeis. She became a full professor within two years, but soon thereafter she and Munson divorced.

In 1971 Jones moved to the University of Southern California School of Medicine as a professor of biochemistry, a post she held for the next seven years. In 1978 the UNC School of Medicine invited her to return to chair the department of biochemistry, a post she held for the next two decades. In 1980 the university named her the Kenan Professor. She retained this endowed chair until her retirement.

The 1980s brought many distinctions to Jones, as she was honored with both awards and prestigious appointments. She received the 1982 Wilbur Lucius Cross Medal from Yale University, and in 1984 the National Academy of Sciences inducted her into its fellowship. She served as president of several professional organizations in the mid-1980s: the Association of Medical School Departments of Biochemistry in 1985, the American Society for Biochemistry and Molecular Biology in 1986, and the American Association of University Professors in 1988.

Just before her planned retirement to New Mexico, in 1995, Jones was diagnosed with esophageal cancer. Despite her waning energy due to chemotherapy, she collaborated with T. W. Traut on a review article published in 1996. She eventually moved from New Mexico to Waltham, Massachusetts, where she died on August 23, 1996. On April 23 of the next year, the University of North Carolina dedicated the 11-story research center in which she had worked as the Mary Ellen Jones Building, in honor of her contributions to the university.

⊠   **Just, Ernest**
(1883–1941)
American
Chemist, Zoologist

Ernest Just established himself not simply as an outstanding African-American scientist but as a preeminent scientist regardless of his heritage. He spent the majority of his career at Howard University, a prestigious black college in Washington, D.C., and early in his tenure there the National Association for the Advancement of Colored People (NAACP) honored him with its Spingarn Medal. He also spent about two decades conducting research each summer at the Marine Biology Laboratory (MBL) in Woods Hole, Massachusetts, one of the foremost sites in the country for investigating oceanic species. His influence expanded into the international sphere during the last decade of his career, when he conducted research in Italy, Germany, and France before the advent of World War II.

Ernest Everett Just was born on August 14, 1883, in Charleston, South Carolina. His father, Charles Frazier Just, was a dock builder (like his father before him) who died when Just was only four years old. Thereafter, Just worked as a field hand to help support his younger brother and sister, while his mother, Mary Matthews, taught school. At the age of 13 he entered the South Carolina State College in Orangeburg, an all-black school. When he was 16, he went north for better schooling at Kimball Union Academy, in Meriden, New Hampshire, which he financed by working first in New York City.

As the only African American on the Kimball Union campus, Just disproved any stereotype of black inferiority: He edited the school newspaper, presided over the debating society, and completed the four-year curriculum in three to graduate as valedictorian. He matriculated to nearby Dartmouth College, in Hanover, where he similarly distinguished himself: He topped the freshman class in the study of Greek and served as the Rufus Choate scholar for the next two years. In 1907 Just graduated magna cum laude (the only student in this highest classification) with special honors in botany and history.

After teaching briefly at the M Street (now Dunbar) High School in Washington, D.C., Just joined the faculty of Howard University, a black college in the city, where his salary amounted to just $400 a year. In 1909 he commenced graduate studies in embryology under Frank Rattray Lillie at the MBL, where he spent each subsequent summer for the next two decades. In the meantime, Howard installed Just as the head of the biology department when it opened its new science building in 1910. In November 1911 he founded the black fraternal organization Omega

Psi Phi on campus, serving as its faculty adviser thereafter. Two years later the university appointed him to chair the zoology department, a post he retained the rest of his life. On June 26, 1912, he married Ethel Highwarden, a German professor at Howard, and together the couple had three children—Margaret, Highwarden, and Maribel.

In 1915 the NAACP conferred on Just its Spingarn Medal, awarded yearly to the individual who had contributed most to the progress of African-American life. Though the prize rarely went to a scientist, the NAACP honored Just (against his humble protestation) for proving that African Americans could excel in the sciences. He had yet to earn his doctorate, which he completed, on experimental embryology, the following year at the University of Chicago, submitting his dissertation, "Studies of Fertilization in *Platynereis megalops*," to earn his Ph.D. magna cum laude.

Throughout the 1920s Just served as the National Research Council's Julius Rosenwald fellow in biology, receiving a grant that supported his research with funds of $80,000 per year. He also collaborated with the likes of Lillie, his former mentor at MBL, as well as MBL president M. H. Jacobs, National Academy of Sciences president T. H. Morgan, and Princeton professor E. G. Conklin with whom he wrote the 1924 text *General Cytology*. Just edited several important journals, including *Protoplasm* (an international journal on the physical chemistry of the cell, published in Berlin), *Cytologia* (published in Japan), *Physical Zoology*, and *Biological Zoology*.

In January 1929 Just commenced a series of overseas studies, first at the Zoological Station in Naples, Italy, and the subsequent year as the first American ever invited as a visiting professor to the Kaiser Wilhelm Insitute for Biology in Berlin, Germany, where he worked in the lab of Max Hartmann. However, the rise of Nazism in 1933 forced him to flee Germany, first to Italy and then to France, where he stayed for the remainder of the 1930s. In 1939 he published his own book, *The Biology of the Cell Surface*, in which he defined the functions of the outer cytoplasm, or what he termed the *ectoplasm*. That same year he also published a slimmer and more focused volume entitled *Basic Methods for Experiments in Eggs of Marine Animals*.

Unable to flee Paris before the invasion of the Third Reich, Just was captured by the Nazis and held briefly as a prisoner of war before being released to the United States in 1940. He returned to Howard, but the progression of cancer in his digestive tract prevented him from teaching. Just died on October 27, 1941, in Washington, D.C.

K

⊠ Karrer, Paul
(1889–1971)
Russian/Swiss
Chemistry

Paul Karrer focused his research on vitamins, specifically investigating A, B$_2$, C, E, and K. He was the first scientist to elaborate the correct formula for beta-carotene, a discovery that led him to elucidate the structure of vitamin A (which he also synthesized). For this research he shared the 1937 Nobel Prize in chemistry with SIR WALTER HAWORTH, who first identified the structure of vitamin C. In addition to his valuable research, he helped disseminate ideas through his field, publishing more than 1,000 scientific papers on a wide range of fields. He also served as president of the International Congress on Pure and Applied Chemistry, thus helping define the direction his field would head into the future.

Karrer was born on April 21, 1889, in Moscow, Russia. His mother was Julie Lerch, and his father, after whom he was named, was a dentist. In 1892 the family returned to their homeland of Switzerland, where Karrer commenced his education in Wildegg. In 1908 he continued his education in Lenzburg. He pursued his doctoral studies under ALFRED WERNER at the University of Zurich, which granted him his Ph.D. in 1911.

Karrer spent the next year as an assistant at the university's Chemical Institute, before Paul Ehrlich invited Karrer to join him as a chemist in 1912 at the Georg Speyer Haus in Frankfurt am Main, Germany. Two years later he married Helena Froelich, and together the couple had two sons. In 1918 the University of Zurich beckoned him back by appointing him a reader. Within a year the university had promoted him to a professorship in chemistry and simultaneously named him director of its Chemical Institute.

Karrer focused his early research on the properties of complex metal compounds, which he also synthesized. His other focus, the study of plant pigmentation, usurped his attention in the 1920s. He particularly focused on yellow carotenoids, which make carrots orange. By 1930 he had established the structure of beta-carotene, the first vitamin component to be elucidated, and the following year he became the first to explicate the structure of a vitamin itself—in this case, vitamin A, which he later went on to synthesize. He also demonstrated the relationship between beta-carotene and vitamin A by identifying how the liver breaks the former down into the latter.

Karrer expanded his vitamin studies to include ascorbic acid, or vitamin C, and he soon verified its structure, first identified by Albert von Szent-Gyorgyi. He then proceeded to synthesize riboflavin (vitamin B$_2$) in 1935 and toco-pherol (vitamin E) in 1938. The year before he shared the 1937 Nobel Prize in chemistry with Haworth, the first scientist to synthesize a vita-min (in his case, vitamin C); the Royal Swedish Academy specifically cited Karrer's work on carotenoids, flavins, and vitamins A and B.

Besides his vitamin and carotenoidal plant pigment studies, Karrer also conducted research on coenzymes, curare (and other alkaloids), amino acids, carbohydrates, and organoarsenic compounds. In addition, he contributed to his field more generally. In 1927 he published his *Lehrbuch der Organischen Chemie (Textbook of Organic Chemistry)*, which subsequently went through 13 editions and translations into English, Italian, Spanish, French, Polish, and Japanese. His 1948 monograph on his specialty, carotenoids, also saw translation into English. He published more than 1,000 scientific papers in a variety of scientific fields besides organic chemistry, and he served his field by presiding over the 14th International Congress of Pure and Applied Chemistry, held in Zurich in 1955.

He gathered the garlands of success later in his career, as all the major national scientific societies (the French Académie des Sciences, the Royal Society of London, the American National Academy of Sciences, the Swedish Royal Academy of Sciences, the Italian National Academy, the Royal Academy of Bel-gium, the Indian Academy of Science, and the Royal Netherlands Academy of Sciences) inducted him into their fellowship. He received the Marcel Benoist and the Canizzaro prizes, as well as honorary doctorates from a bevy of pres-tigious European institutions in Basel, Breslau, Lausanne, Zurich, Paris, Lyons, Strasbourg, Sofia, Madrid, London, Turin, and Brussels, as well as in Rio de Janeiro, Brazil.

Karrer retired from his post as director of the Chemical Institute of the University of Zurich in 1959 and died a dozen years later, on June 18, 1971, in Zurich.

⊠ Kendrew, Sir John Cowdery
(1917–1997)
English
Physical Chemist, Molecular Biologist

Sir John Cowdery Kendrew discovered the struc-ture of the myoglobin molecule, the key to mus-cular oxygen storage and distribution, and constructed a model of the protein. He shared the 1962 Nobel Prize in chemistry with his collabora-tor, MAX FERDINAND PERUTZ, who not only aided Kendrew at several key junctures but also elabo-rated the structure of hemoglobin. After advanc-ing his field on the experimental and theoretical level, Kendrew devoted his career to advancing his field on more practical levels by chairing international unions and committees.

Kendrew was born on March 24, 1917, in Oxford, England. His mother, Evelyn May Gra-ham Sandberg, was an art historian who wrote (under the pen name Evelyn Sandberg Vavals) on the Italian Primitives. His father, Wilfrid George Kendrew, was a reader in climatology at the Uni-versity of Oxford. Kendrew went to the Dragon School at Oxford from 1923 to 1930, when he matriculated at Clifton College in Bristol to spend the next six years there. He finished his education on a major scholarship at Trinity College in Cam-bridge, earning first-class honors in chemistry and graduating in June 1939. Dr. E. A. Moelwyn-Hughes retained him in his lab in the physical chemistry department as a researcher in reaction kinetics, but Kendrew lasted there only about a half year.

World War II interrupted Kendrew's research; he joined the war effort on the Air Ministry Research Establishment (later known as the Telecommunication Research Establishment) working on radar. In 1940 Air Ministry scientific adviser Sir Robert Watson-Watt recruited him to conduct operational research on antisubmarine technology and bombing accuracy at the Royal Air Force (RAF) headquarters. The RAF pro-moted him from junior scientific officer to the post of scientific adviser to Lord Mountbatten's Southeast Asia Command in Ceylon and by the end of the war to the honorary rank of wing commander.

In 1946 Kendrew returned to Cambridge, where Sir Lawrence Bragg installed him alongside Max Perutz in the Cavendish Laboratory conducting X-ray crystallographic analysis on the structure of proteins. Within a year the Medical Research Council established a Research Unit for the Study of the Molecular Structure of Biological Systems (later shortened to the Unit for Molecular Biology), comprised exclusively of the Perutz-Kendrew team. Kendrew bravely took up the study of myoglobin as his doctoral research, despite the fact that the project promised to be too expensive to finish for his dissertation.

Kendrew submitted his dissertation in 1949 (without having ascertained the structure of myoglobin) and earned his Ph.D. Two years earlier he had become a fellow of Peterhouse at Cambridge, later serving as its director of studies. The next year, in 1948, he married Elizabeth Jarvie, but they dissolved their matrimonial bonds in 1956.

Kendrew commenced his protein studies on myoglobin from horse heart, which generated crystals too small for X-ray analysis. By chance Perutz happened upon a sample of sperm whale meat from Peru that, according to Perutz himself, yielded "large, sapphire-like crystals which gave beautiful X-ray diffraction diagrams" perfect for measuring the amplitudes of the diffracted rays. However, Kendrew still needed to measure their phases in order to ascertain their structure; in 1953 Perutz solved this problem by comparing the existing diffraction patterns to those generated by a crystal with heavy atoms attached to it. By 1957 Kendrew and an international team of researchers from the United States, Sweden, and Austria had built a model of the myoglobin molecule at a resolution of 6 angstroms, then within two years, at 2 angstroms, a resolution capable of generating an atomic model.

The significance of his work earned him almost immediate and universal recognition. In 1959 the *Journal of Molecular Biology* appointed him as its editor in chief, making him one of the main arbiters of his field. In 1960 the Royal Society inducted him into its fellowship, and in 1962 the British Empire named him a Companion. Nobel laurels came that same year, as he shared the prize in chemistry with his labmate Perutz, who had solved the more complex hemoglobin structure at $5^1/_2$ angstroms. Kendrew also managed to earn a doctor of science degree that year. A little over a decade later, he was knighted by Queen Elizabeth II in 1974.

Thereafter, Kendrew redirected his focus from the internal myopia of laboratory research to the outward advancement of his field. In 1964 Weizmann Institute of Science, in Israel, appointed him as its governor, and that same year the International Union of Pure and Applied Physics named him as its vice president, a post he held until 1969, when it promoted him to its presidency (he stepped down to an honorary vice presidency in 1972). In 1974 the International Council of Scientific Unions appointed him its secretary general (he stepped down to its presidency from 1980 to 1988). In 1975 the European Molecular Biology Laboratory appointed him its director general, a position he retained until 1982.

Kendrew spent the end of his career at the helm of St. John's College in Oxford, which he presided over from 1982 to 1987. A decade later he died in Cambridge, on August 23, 1997.

⊠   **King, Reatha Clark**
(1938–   )
American
Physical Chemist, Organic Chemist

Reatha Clark King, who commenced her career conducting specialized research in fluorine flame calorimetry, has worked in government, academia, and industry. She worked as a research chemist at the National Bureau of Standards

Reatha Clark King, a research chemist whose work in fluorine flame calorimetry led to the development of materials used to line rocket fuel systems. *(Courtesy Reatha Clark King)*

mother, Ola Mae, traveled in search of better-paying work. The family moved often in search of better economic opportunities, and throughout her youth Clark supplemented the family income by painting broom handles in a factory, working in cotton fields, and working as a maid after school. From the age of 12, Clark attended the Moultrie High School for Negro Youth.

Clark attended Clark College in Atlanta on a scholarship, earning her bachelor of science in chemistry and mathematics in 1958. She then pursued graduate studies at the University of Chicago, earning her master of science in 1960. She wrote her doctoral dissertation on the effects of heat on metal alloys to earn her Ph.D. in thermochemistry in 1963. The year before she had married N. Judge King, a math and science teacher. The couple had two sons, Jay and Scott.

After her graduation the Kings moved to Washington, D.C., where Reatha worked as a research chemist for the National Bureau of Standards. She followed the line of investigation commenced in her dissertation, conducting fluorine flame calorimetry at room temperature and observing the thermochemical properties of alloys with tin solution calorimetry. Using these methods she recorded the wear and tear on alloys exposed to heat and chemicals. She also invented a coiled tube that cooled hot liquids such as fuel to prevent them from exploding; this research led to the development of materials lining rocket fuel systems.

In 1968 King moved to New York City, where she secured an assistant professorship in chemistry at York College of the City University of New York. In 1970 the college promoted her to an associate professorship and appointed her associate dean of the division of natural science and mathematics. In 1974 she was again promoted, this time to a full professorship in the chemistry department and associate dean of academic affairs. During this period she also attended Columbia University to pursue a master's degree in business administration, which she earned in 1977.

before becoming a professor and associate dean at York College of the City University of New York (CUNY). She later implemented innovative educational policies as president of the Metropolitan State University in Minneapolis, Minnesota, then extended this line of work as president and executive director of the General Mills Foundation, a national philanthropic organization devoted to education, health, and the arts. She also served on several corporate boards, including Exxon and Wells Fargo.

Reatha Clark was born on April 11, 1938, in rural Pavo, Georgia. Her father, Willie Clark, was a migrant worker. When she was seven years old, Clark lived with her grandmother while her

That year she moved to the Twin Cities in Minnesota to fill the presidency of Minneapolis's Metropolitan State University, the first U.S. institution to give academic credit for life experience. She promoted the expansion of educational opportunities for women (especially single mothers), minorities, and working-class students, ushering them through undergraduate degrees and into graduate programs.

In 1988 she became president and executive director of the General Mills Foundation, which made grants to tax-exempt organizations involved in education, family life, health and nutrition, and cultural affairs. In 1989 she was named Twin Citian of the year, and in 2000 she received a Minneapolis Award. She has also received the Exceptional Black Scientist Award from the CIBA-GEIGY Corporation and the 1990 Martin Luther King Commemoration from the State University of New York at Buffalo. In 1997 she was inducted into the International Adult and Continuing Education Hall of Fame. She holds honorary degrees from 13 institutions.

King has also contributed to her community and society. In 1991 President George H. W. Bush appointed her to the U.S. Commission on National and Community Services, which she served for two years. In 1994 President Bill Clinton appointed her to the board of the Corporation of National and Community Service. She has also chaired the board of the American Council on Education. More recently she has helped organize the Hawthorne Huddle, a Minneapolis neighborhood initiative that has received national attention for its innovative solutions for crime prevention through neighborhood involvement.

Klug, Sir Aaron
(1926–)
Lithuanian/South African/English
Chemist

Aaron Klug distinguished himself as an X-ray crystallographer, using this method to help identify the structure of the tobacco mosaic virus (TMV), after which he turned his attention to spherical viruses. He collaborated on this work with ROSALIND FRANKLIN, one of the scientists responsible for the determination of the structure of deoxyribonucleic acid (DNA). He also worked on the Medical Research Council with other prominent scientists, including FRANCIS HARRY COMPTON CRICK, SIR JOHN COWDERY KENDREW, MAX FERDINAND PERUTZ, and FREDERICK SANGER. For his virus work he won the 1982 Nobel Prize in chemistry.

Klug was born on August 11, 1926, in Zelvas, Lithuania. His mother was Bella Silin; his father, Lazar Klug, was a saddler by trade, though he grew up in the cattle business. At the age of two, Klug's family moved to Durban, South Africa, where he attended Durban High School. Early in his education he read Paul de Kruif's *Microbe Hunters*, which influenced him to pursue a scientific career. He matriculated at the University of Witwatersrand in Johannesburg, where he pursued a premedical course of study while also studying physiological chemistry (now known as biochemistry), chemistry, physics, and mathematics. He graduated with a bachelor of science.

Klug then pursued graduate research in physics at the University of Cape Town, which offered him a scholarship in exchange for his experimental demonstration in laboratory classes. He conducted his own laboratory research under the X-ray crystallographer R. W. James, who taught him not only experimental techniques but also the Fourier theory. Klug earned his master of science and remained at the university, developing a formula for solving crystal structures molecularly and teaching himself quantum chemistry. While in Cape Town, he married Liebe Bobrow, a dancer who had trained in modern technique in London and later choreographed for the Cambridge Contemporary Dance Group. Together the couple had two sons, Adam (born in 1954) and David (born in 1963).

In 1949 Klug received an 1851 Exhibition Scholarship, as well as a research studentship, to attend Trinity College in Cambridge, England. He became a research student under D. R. Hartree, conducting investigations on solid-state physics to earn his doctorate. For the next year he worked under F. J. W. Roughton in the colloid science department at Cambridge, conducting studies on the state when chemical reactions and diffusion occur simultaneously, for example, when oxygen enters red blood cells. Klug simulated this situation by exposing oxygen or carbon monoxide to thin layers of blood to measure reactions.

At the end of 1953 Klug received a Nuffield Fellowship supporting his work in J. D. Bernal's laboratory at Birkbeck College in London. He conducted research on the protein ribonuclease until he was introduced to Rosalind Franklin's X-ray crystallographic photography of the tobacco mosaic virus, at which point he redirected his research. He collaborated with Kenneth Holmes and John Finch, as well as Yale researcher Donald Caspar (on leave in Cambridge from 1955 to 1956) to elaborate the structure of TMV.

Klug assumed the helm of the virus group upon Franklin's untimely death in 1958, shifting the focus from TMV to spherical viruses, with Holmes and Finch as researchers supported by a National Institutes of Health grant. In 1962, after Reuben Leberman joined the group, it established residency in the newly built Medical Research Council (MRC) Laboratory of Molecular Biology in Cambridge, which also housed the labs of Perutz, Kendrew, Crick, Sanger, and Hugh Huxley.

In 1969 the Royal Society inducted Klug into its fellowship, signaling the ascendancy of his career. At the MRC he became a joint head of the division of structural studies in 1978. International recognition of his virus work came in 1982, when the Royal Swedish Academy of Sciences granted him its Nobel Prize in chemistry for his elucidation of the three-dimensional structure of viruses, as well as his development of the crystallographic electron microscopy.

Klug's professional ascent continued; he became the director of the MRC Laboratory of Molecular Biology in 1986. Two years later he was knighted. In 1990 the Cambridge Antibody Technology (CAT) Group appointed him one of its directors, a position he retained thereafter. In 1995 he received the Order of Merit and also became the president of the Royal Society that year. In October 1996 Klug retired.

⊠ **Kuhn, Richard**
(1900–1967)
Austrian
Biochemist, Organic Chemist

Richard Kuhn is best known for his biochemical researches, particularly his discovery of eight new types of carotenoid (a yellow plant pigmentation) and his work on vitamins. His 1938 Nobel Prize in chemistry testified to the significance of these contributions, which brought about specific practical advances. His theoretical work in organic chemistry, while not as renowned, also advanced science, though not in as obvious ways. As the head of the Max Planck Institute (MPI), Kuhn exhibited keen organizational skills, as he orchestrated the coordinated efforts of his entire laboratory staff while they simultaneously examined multiple aspects of the same question from different angles. He also published prodigiously, producing more than 700 scientific papers in his career.

Kuhn was born on December 3, 1900, in Vienna, in what was then the Austro-Hungarian Empire. His mother, Angelika Rodler, was an elementary schoolteacher and his father, Richard Clemens Kuhn, was an engineer. After graduating from the gymnasium, he attended Vienna University, where he studied chemistry. He later transferred to the University of Munich, in the

Bavarian region of Germany, to study under RICHARD WILLSTÄTTER. He wrote his doctoral dissertation, entitled "Über Spezifität der Enzyme," on the specificity of enzymes, to earn his Ph.D. in 1922.

Kuhn remained at the University of Munich, which offered him a lectureship in chemistry in 1925. He only stayed on a year, though, as he took up a professorship of general and analytical chemistry at the Eidgenössische Technische Hochschule in Zurich, Switzerland. In 1928 he married Daisy Hartmann, and together the couple had six children, two sons and four daughters. In 1929 the University of Heidelberg hired him as a professor of biochemistry, and he joined its Kaiser Wilhelm Institute (KWI) for Medical Research (later called the Max Planck Institute) as one of its principals. In 1937 the institute appointed him its director, a post he retained the remainder of his career.

Kuhn focused his research on carotenoids, the ubiquitous yellow plant pigmentation, specifically investigating the double-bond structure that characterizes them. By following this line of investigation, he ended up discovering eight new types of carotenoids. He isolated a pure sample of each new carotenoid and determined the chemical composition of each as well. He announced his determination of the makeup of vitamin B_2 at the same time as PAUL KARRER, though the two worked independent of each other. Karrer won the 1937 Nobel Prize in chemistry (with SIR WALTER HAWORTH), and Kuhn received the Nobel Prize in chemistry the subsequent year, for his work on carotenoids and vitamins.

At the KWI/MPI Kuhn coordinated the research for the entire chemistry department, appointing research teams to investigate discrete elements of a common question, such as the chemical mechanism of biological resistance. He hypothesized that amino sugars in mother's milk created natural tolerance to viral and bacterial infections, so he set out to test this belief by delegating the numerous components

to different investigative teams. Adeline Gauhe and Hans Baer gathered and prepared more than 5,000 liters of mother's milk for experimentation. They then separated the milk chemically, according to molecular size, yielding several nitrogen-rich substances that promoted the growth of bifidus, an antigen. This particular substance, an oligosaccharide, also showed up in mammalian colostrum (the preliminary postpartum breast secretions) and even in human meconium (the dark intestinal substance in babies at birth), supporting Kuhn's immunoagent theory.

Kuhn even oversaw the contributions of overseas colleagues. Paul György, with whom Kuhn had collaborated on the 1930s vitamin B studies, had joined the scientific exodus from Germany to the United States but kept in contact with his former partner. After World War II Kuhn enlisted György's assistance in his bioresistance studies, asking him to test at Kuhn's lab the reactivity of the substances with bifidus as well as other intestinal bacteria and viruses. György urged his institution, the University of Pennsylvania, to woo Kuhn, who coyly accepted the entreaty until his own institution won him back with increased funding, greater autonomy, and an industrial affiliation (with BASF), topping the university's offer. Instead of sulking at Kuhn's rejection, however the University of Pennsylvania invited him to serve as a visiting professor, and the private firm that had bought the patent rights to Kuhn and György's prior collaborations hired Kuhn as a consultant.

Kuhn advanced his field more generally by serving on the senate of the Max Planck Society before becoming its vice president in 1955. He edited two of the leading chemical journals in Germany, *Liebigs Annalen der Chemie* and *Angewandten Chemie*. In 1964 the German Society of Chemists elected him its president. Unfortunately, Kuhn was diagnosed with cancer of the larynx the very next year. Two years later, on July 31, 1967, he died in Heidelberg.

L

⊠ **Langmuir, Irving**
(1881–1957)
American
Chemist

Irving Langmuir was the first nonacademic chemist to win the Nobel Prize in chemistry, which he received in 1932 for his pioneering work that established the field of surface chemistry. He spent almost his entire career in industry, conducting pure research at the General Electric Research Laboratory. In addition to inventing the high-vacuum electron tube and the gas-filled incandescent lamp, he also advanced atomic theory, developing a new vocabulary to describe his discoveries.

Langmuir was born on January 31, 1881, in Brooklyn, New York. He was the third of four sons born to Sadie Comings and Charles Langmuir, an insurance executive. Langmuir's interest in science developed early, as his brother Arthur (who became a research chemist) helped him set up a lab in his bedroom to conduct basic experiments. In 1892 the family moved to Paris, where the rigid pedagogy of the French public schools stifled him; upon their return to Philadelphia, Langmuir flourished at Chestnut Hill Academy. He attended secondary school in Brooklyn, at Pratt Institute's Manual Training High School.

Langmuir enrolled in the Columbia School of Mines, where he studied metallurgical engineering to earn his bachelor of science in 1903. He returned to Europe to pursue doctoral studies under the Nobel laureate WALTHER HERMANN NERNST, attending Germany's prestigious University of Göttingen. Langmuir focused his dissertation research on the Nernst glower, an electric lamp he used to test the reaction of gases in the presence of a hot platinum wire. He earned his Ph.D. in 1906, whereupon he took up a teaching position in chemistry at the Stevens Institute of Technology in Hoboken, New Jersey.

In 1909 Langmuir secured a summer position at the General Electric (GE) Research Laboratory in Schenectady, New York. Impressed by the young scientist's intuition and creativity, GE's director of the research laboratory, Willis R. Whitney, offered Langmuir a permanent position conducting pure research with a team of experienced experimental scientists at his command. Langmuir thrived on such freedom, investigating theoretical questions that yielded answers of great practical application. He focused initially on low-pressure chemical reactions and the emission of electrons by hot filaments in a vacuum; this work resulted in his invention of the high-vacuum electron tube in 1912 and the gas-filled incandescent lamp in 1913. In 1912 he married Marion Mersereau, and together the

121

Irving Langmuir was the first industrial scientist and the second American to win a Nobel Prize in science, which he was awarded in chemistry in 1932. *(E. F. Smith Collection, Rare Book & Manuscript Library, University of Pennsylvania)*

couple had two children, a son named Kenneth and a daughter named Barbara.

Langmuir's experimental interests ranged widely. His vacuum studies yielded the invention of a mercury-condensation vacuum pump, as well as the extension of his high-vacuum concept across a whole family of radio tubes. He was the first to isolate the element hydrogen in its atomic form and applied this discovery to the innovation of an atomic hydrogen welding torch capable of generating much higher temperatures than the oxyacetylene torch. He coined the term *plasma* to describe aggregated ionized gases that exhibited curious electrical and magnetic properties.

Langmuir contributed to the efforts in World War I by developing submarine detection devices, later redirecting these studies in collaboration with Leopold Stowkowski to improve the quality of aural recordings. After the war he concentrated his theoretical investigations on atomic theory, advancing his own concentric theory of atomic structure. He held that the preferred constitution of all atoms included an outer shell of eight orbiting electrons; the extremely stable inert gases, for example, which exhibit low reactivity rates, were thus composed, whereas atoms lacking electrons tended toward reactivity. Atoms that relinquished or accepted electrons created what he called "polar unions" (now known as *ionic bonds*), while atoms that shared electrons created what he called "nonpolar unions," *covalent bonds*. Langmuir developed a new vocabulary for his atomic theory, defining the terminology of *valence* and *isotopes*.

In collaboration with Katherine B. Blodgett, Langmuir studied how water adsorbed substances such as oils to create thin layers of surface film. They also observed the phenomenon of monolayers, or films that are a single atom or molecule thick, leading to the possibility of measuring that thickness, an especially useful device for investigating the properties of viruses and toxins. This research led to the establishment of surface chemistry, a distinct field of investigation. The Royal Swedish Academy of Sciences cited this work specifically in awarding Langmuir its 1932 Nobel Prize in chemistry.

Langmuir contributed to the cause in World War II by developing protective smoke screens and chemicals for deicing aircraft wings. The latter innovation led him to try his hand at divinity by manipulating the weather: Using dry-ice pellets and silver iodine crystals, he "seeded" clouds to promote precipitation, a line of investigation that met with some resistance. Langmuir continued to work at GE's research laboratory, which promoted him to assistant director and then to associate director, until his retirement in 1950.

Langmuir had an exceedingly productive career: He published more than 200 scientific papers; held 63 patents; presided over the American Chemical Society in 1929 and the American Association for the Advancement of Science in 1943; received 15 honorary degrees; and won 22 medals, including the Nichols Medal in 1915 and 1920, the 1921 Rumford Medal, the 1925 Cannizzaro Prize, the 1934 Franklin Medal, and the 1944 Faraday Medal. After a short illness, he died on August 16, 1957, in Woods Hole, Massachusetts.

In 1964 the General Electric Foundation (now the GE Fund) honored its enterprising employee by creating the Irving Langmuir Prize in Chemical Physics. The $10,000 prize, presented alternately by the American Chemical Society and the American Physical Society, recognizes outstanding interdisciplinary research in chemistry and physics. Other posthumous honors included the naming of Alaska's Mount Langmuir and the State University of New York at Stony Brook's Irving Langmuir College.

⊗ Lavoisier, Antoine-Laurent
(1743–1794)
French
Chemist

Antoine-Laurent Lavoisier is considered the founder of modern chemistry, primarily for his overthrow of the phlogiston theory of combustion. In his magnum opus, the 1789 text *Traité élémentaire de chimie*, he organized a list of the 55 elements known at that time (though some of these turned out not to be elements), imposing order on the field dominated by chaos up until that point.

Lavoisier was born on August 26, 1743, in Paris, France, the son of a wealthy lawyer. His mother died when he was still young, and an aunt brought him up thereafter. From 1754 until 1761, he studied astronomy, botany, chemistry, and mathematics at the Collège Marazin. He then attended the Collège des Quatre Nations, where he earned a license to practice law in 1764 in order to follow in his father's footsteps, a plan that never materialized.

Lavoisier published his first chemical investigation, on rock gypsum, in 1764. Over the next four years he helped Jean-Étienne Guettard map the geology and mineralogy of France, and in 1766 he designed a streetlamp system for Paris, earning him a gold medal from the king of France and election to the Academy of Sciences as an associate chemist in 1768. (The academy appointed him its director in 1785 and its treasurer in 1791.)

Also in 1768 Lavoisier was appointed to the Ferme Générale, a private organization that col-

Antoine-Laurent Lavoisier, whom many regard as the founder of modern chemistry *(E. F. Smith Collection, Rare Book & Manuscript Library, University of Pennsylvania)*

lected taxes for the state, as an assistant to Farmer General Baudon. In 1771, at the age of 28, he married 13-year-old Marie-Anne Pirette Paulze, daughter of a fellow tax collector, Farmer General Jacques Paulze, who bought him a title the next year. His young wife collaborated with him throughout his career, serving as his scientific assistant, note taker, and illustrator, as well as performing the vital function of translating scientific texts from English and Latin.

In 1772 Lavoisier reported to the Academy of Sciences his findings that combustion does not subtract matter, but rather adds it, as the burning of sulfur and phosphorus increases their weight due to the absorption of air. Two years later Joseph Priestley generated "dephlogisticated" air, prompting Lavoisier to theorize that oxygen (he coined the term from the Greek for "acid maker," believing that all acids contained oxygen) accounted for the weight increase in combustion, thereby overturning the phlogiston theory of combustion. He reported his findings in a series of publications, beginning in 1777 with his "Memoir on Combustion in General" and culminating with his 1783 text *Reflexions sur le phlogistique*.

In 1775 Lavoisier was appointed to the post of *régisseur des poudres*, or the commissioner of powders; in this position he devised a new method of preparing saltpeter and replaced the stocks of old powder with his new formulations. While he still held this post in 1787, Irénée du Pont apprenticed as his bookkeeper and later immigrated to the United States to found the DuPont Company, which presents the Lavoisier medals annually to its outstanding researchers. In 1785 he gained another civil post as secretary to the committee on agriculture.

Lavoisier continued to publish important texts, reporting his theory of oxygen in *Considérations générales sur la nature des acides* in 1778. His 1787 text, *Methods of Chemical Nomenclature*, established a taxonomy of chemical names, coining terms such as *sulfuric acid, sulfates,* and

sulfites. In 1789 he published his masterwork, *Traité élémentaire de chimie*, the first modern chemical textbook, which unified the field, stated the law of conservation of mass, denied existence of phlogiston, and listed 55 elements, including oxygen, nitrogen, hydrogen, phosphorus, mercury, zinc, and sulfur. He also mistakenly listed light and caloric as elements.

In 1790 Lavoisier standardized the system of weights and measurements in France, which led to the establishment of the metric system. In his civil positions Lavoisier also helped abolish the *pied forchu* tax that singled out Jews and established free schools for the poor in his tax district. Despite his progressive acts, the Reign of Terror targeted him for political persecution, forcing him to flee his home in August 1792. On December 24, 1793, he was arrested, along with 28 fellow farmers general for "plotting against the government by watering soldiers' tobacco and appropriating revenue that belonged to the state." A revolutionary tribunal condemned them all to death in May 1794. They were guillotined on May 8, 1794, and buried in a common, unmarked grave.

⊠ Le Beau, Désirée
(1907–1993)
Austro-Hungarian/American
Chemist

Désirée Le Beau advanced the field of colloid chemistry by discovering new methods of reclaiming rubber from existing sources, such as used tires, as well as innovating means of synthesizing artificial rubbers. She contributed to the war effort in World War II while as a member of the Massachusetts Institute of Technology's Department of Chemical Engineering. After the war she served as director of research of the world's largest independent reclaiming company. She filed numerous patents on rubber reclaiming processes. For her contributions to the understanding of rubber reclamation, the Society of Women Engineers (SWE) granted her its 1959 Achievement Award.

Le Beau was born on February 14, 1907, in Teschen, Austria-Hungary (now Poland). Her parents were Lucy and Philip Le Beau. She conducted her undergraduate studies at the University of Vienna, where she discovered her passion for chemistry quite by chance: Instead of attending her scheduled pharmacy lab, she accidentally attended a chemistry lab but before the end of the period realized that she had found her life's calling. After finishing her undergraduate studies, she continued with graduate studies at the University of Graz, where she earned her doctorate in chemistry, with minors in physics and mathematics, in 1931.

In 1932 Le Beau returned to Vienna, where she landed a position as a researcher at the Austro-American Rubber Works. She commenced her career in rubber studies there, remaining for three years. In 1935 she fled the encroaching Nazi regime to Paris, where she served as a consultant to the Société de Progrès Technique. However, the threat of Nazism's spread prompted her to immigrate to the United States in 1936, where she secured a position as a research chemist at Dewey & Almay Chemical Company in Massachusetts.

During World War II Le Beau made her greatest contributions to the understanding of rubber recycling in response to the great need for rubber in the war effort and in the face of shortages of raw materials. In 1940 she joined the Massachusetts Institute of Technology's Department of Chemical Engineering under the division of industrial cooperation, serving as a research associate. There she helped develop methods for reclaiming rubber from used tires and investigated means of synthesizing artificial rubbers.

After the war, in 1945, the Midwest Rubber Reclaiming Company in Illinois hired Le Beau as its director of research, placing her at the helm of the laboratory of the world's biggest independent reclaiming company of the day. In addition to overseeing and coordinating the company's research and development activities, she also published her nonproprietary findings in trade journals and books: In a 1948 edition of *Rubber Chemistry and Technology* she published "Basic Reactions Occurring during Reclaiming of Rubber I," in the October 1950 issue of *Rubber Age* she published "Reclaiming Agents for Rubber: Solvent Naphtha I," and she contributed a chapter entitled "Reclaiming of Elastomers" to Volume 7 of *Colloid Chemistry*, edited by Jerome Alexander, which appeared in 1950. That same year Pennsylvania State College appointed her as its Curie Lecturer.

On August 6, 1955, Le Beau married Henry W. Meyer. She continued to contribute to the advancement of rubber reclaiming technologies, patenting a process for producing railroad ties from reclaimed rubber in 1958. She held numerous other patents for reclaiming processes, including reclaiming techniques using amines and acids, as well as a process for producing reclaimed rubber in particulate form. The Society of Women Engineers recognized Le Beau's "significant contributions to the field of rubber reclamation" by granting her its prestigious Achievement Award in 1959.

Le Beau contributed not only research innovations and discoveries to her field but also her organizational and leadership skills. She served as the first woman chair of the American Chemical Society's division of colloid chemistry, as well as holding the first woman chair of the ACS's St. Louis section.

Le Beau died in 1993. In her honor the SWE offered companies the opportunity to sponsor the Board of Trustees Meeting and the Board of Directors Meeting in her name at the SWE's 50th anniversary national conference in Washington, D.C., in 2000. Le Beau's name thus joined other luminary women scientists, such as MARIA TELKES and DOROTHY MARTIN SIMON, in these sponsorships.

⊠   **Lee, Yuan Tseh**
(1936–   )
Chinese/American
Physical Chemist

Yuan T. Lee shared the 1986 Nobel Prize in chemistry with Dudley Herschbach and John C. Polanyi for their advancement of the field of reaction dynamics. Lee built the crossed molecular beam apparatus with which he conducted the groundbreaking experiments on reaction chemistry. Herschbach called Lee the "Mozart of physical chemistry" for his ability to combine "brilliance and attention to detail."

Yuan Tseh Lee was born on November 29, 1936, in Hsinchu, on the island of Taiwan, at that time under Japanese control. His mother, Tse-Fan Lee, was an elementary schoolteacher, and his father, Pei-Tasi Lee, was an artist and art teacher. The evacuation of his hometown to the mountains in World War II interrupted his elementary schooling, but he entered Hsinchu High School on schedule. When he graduated in 1955, his high academic standing allowed him to forgo the customary entrance examination for the National Taiwan University. There he studied chemistry under Professor Huasheng Cheng, writing his thesis on the paper electrophoresis separation of strontium and barium to earn his bachelor of science in 1959.

Lee pursued graduate studies at the National Tsinghua University, studying under Professor H. Hamaguchi. He submitted his thesis on the natural radioisotopes in the hot-spring mineral sediment Hukutolite to earn his Master of Science in 1961. The university retained him for a year as a research assistant under Professor C. H. Wong, conducting X-ray studies on the structure of tricyclopentadienyl samarium. In 1962 Lee immigrated to the United States to conduct doctoral studies at the University of California at Berkeley. On June 28, 1963, he married Bernice Chinli Wu, a friend from elementary school; they had two sons (Ted, born in 1963, and Sidney, born in 1966) and one daughter (Charlotte, born in 1969).

At Berkeley Lee studied under Professor Bruce Mahan, writing his dissertation on the chemiionization processes of electronically excited alkali atoms to earn his doctorate in 1965. He continued at Berkeley for a year and a half, conducting postdoctoral research in Mahan's lab, collaborating with Ron Gentry on measuring energy and angular distributions of ion molecule reactive scattering, and using an ion beam apparatus he had constructed expressly for this experiment.

When Harvard's Dudley Herschbach, whose research could only advance with an improved ion beam apparatus, heard of Lee's innovation, he immediately invited the instrument inventor as a postdoctoral fellow in his Cambridge, Massachusetts, laboratory. Within 10 months of his arrival in February 1967, Lee had constructed a universal crossed molecular beam apparatus with Doug McDonald, Pierre LeBreton, and, most important, machinist George Pisiello. Lee replaced Herschbach's hot wire detector, which could only sense certain reactions, with an ultrasensitive mass spectrometer, which could measure and record the products of the reaction at different points in space and time, greatly expanding the scope and usefulness of this instrument. Amazingly, he achieved this feat working on it only half of his time; he spent the other half conducting hydrogen atom–diatomic alkali molecular reactions with Robert Gordon. In late 1967 Lee and Herschbach conducted the first nonalkali experiment on chemical reactions, mapping the reaction of chlorine with two molecules of bromine to create bromium chloride and one molecule of bromium.

In October 1968 the University of Chicago hired Lee as an assistant professor of chemistry at its James Franck Institute. Within three years the university promoted him to an associate professorship, and a year and a half after that, he ascended to a full professorship. In 1974 year he returned to the University of California at Berkeley as a chemistry professor and principal investigator at the Lawrence Berkeley Laboratory. That same year he became a U.S. citizen.

In 1981 the United States Department of Energy granted him its 1981 Ernest Orlando Lawrence Memorial Award, and the University of California at Berkeley named him Miller Professor of Chemistry. Recognition showered on Lee in 1986, when he shared the Nobel Prize in chemistry with Herschbach and Polanyi, won the American Chemical Society's Peter Debye Award, and received the National Medal of Science from the National Science Foundation.

In 1994 Lee returned to his homeland of Taiwan to serve as president of the Academia Sinica, the country's most prestigious research institute. Two years later he turned down an offer to run for the Taiwanese presidency, maintaining that he could best serve his country in heading its scientific advancement and leaving the governing to others.

⊠   **Lehn, Jean-Marie**
(1939–   )
French
Organic Chemist

Jean-Marie Lehn shared the 1987 Nobel Prize in chemistry with Charles John Pedersen and Donald J. Cram for their work on supramolecular chemistry, or host-guest chemistry. All three investigated cryptates, or crown ethers, which are well suited to "hosting" bonds with multiple other molecules.

Jean-Marie Pierre Lehn was born on September 30, 1939, in Rosheim, in the Alsace region of France. He was the eldest of four sons born to Marie Salomon and Pierre Lehn, a baker and later the city's organist. Lehn attended secondary school at the Collège Freppel in nearby Obernai from 1950 to 1957, when he received *baccalauréats* in philosophy in July and in experimental sciences in September. He then commenced studies in physical, chemical, and natural sciences at the University of Strasbourg.

He narrowed his major to organic chemistry, studying under Guy Ourisson to earn his bachelor of science degree in 1960.

Lehn continued in Ourisson's laboratory as a junior member of the Centre National de la Recherche Scientifique (CNRS). He focused this doctoral research on conformational and physicochemical properties of triterpenes using the lab's nuclear magnetic resonance (NMR) spectrometer. He published his first paper in 1961, positing an additivity rule for substituent induced shifts of proton NMR signals in steroid derivatives. He earned his doctorate in June 1963.

Lehn then commenced what became an ongoing relationship with Harvard University, as he spent the next year as a visiting professor. He worked with ROBERT BURNS WOODWARD on the total synthesis of vitamin B_{12} and with Roald Hoffman on quantum mechanics. When he returned to France in 1965, he married Sylvie Lederer, and together the couple had two sons, David (born in 1966) and Mathias (born in 1969). Back at the University of Strasbourg, he focused his research on physical organic chemistry. By the next year, 1966, he was appointed as a *maître des conférences*, or assistant professor, in the chemistry department.

During this period Lehn conceived the idea of identifying the properties and structures of molecules by observing how they interact with certain molecules that are well suited to reactivity. At about the same time he was investigating the physical chemistry of biological processes, specifically how sodium and potassium ions cross nerve cell membranes to trigger electrical impulses. These studies led him to develop the term *supramolecular chemistry* also known as *host-guest chemistry*). In its more generalized form this field of research capitalizes on the fact that certain molecules recognize complementary molecules with which they can bond, if even only briefly.

Lehn followed in the footsteps of his future fellow Nobel laureate Charles Pedersen, who had already been conducting research on crown ethers, or strings or carbon and oxygen atoms that served particularly well as "hosts" for metal ions. Starting in October 1967 Lehn advanced Pedersen's studies by layering these strings into three-dimensional crown ethers, or what he called "cation cryptates." Lehn described the action of these molecules in visual terms as "lock and key" connections, with the cryptates functioning as the lock and the "guest" molecules functioning as the key. Simultaneously with Donald Cram, Lehn devised methods for synthesizing these crown ether molecules.

In early 1970 Lehn was promoted to the rank of associate professor, and in October of that same year he ascended to a full professorship. He returned to Harvard on visiting professorships in the spring semesters of 1972 and 1974 and maintained a loose affiliation with the prestigious institution until 1980. A year earlier, in 1979, the Collège de France in Paris appointed him to the chair of molecular interaction chemistry. He continued to head his laboratory in Strasbourg, while simultaneously taking up the directorship of the chemistry laboratory at the Collège de France upon the retirement of Alain Horeau in 1980.

Lehn's career has been recognized with numerous honors. The Ordre National du Mérite appointed him a chevalier in 1976, and the Légion d'Honneur appointed him to the same title in 1983, then promoting him to the rank of *officier* in 1988. He won the 1982 Paracelsus Prize, the 1983 von Humboldt Prize, and the 1989 Vermeil Medal of Paris in addition to winning the 1987 Nobel Prize in chemistry with Pedersen and Cram for their work on crown-ether cryptates and host-guest chemistry.

⊠ Leloir, Luis Federico
(1906–1987)
French–Argentine
Biochemist

Unlike many scientists who work in isolation in well-equipped laboratories, Luis Federico Leloir

brought about all his achievements while collaborating with his colleagues in a makeshift laboratory, using their ingenuity to jury-rig instruments and invent creative techniques to perform their experiments. By these means he and his team discovered the mechanism of the synthesis of sucrose. He won the 1970 Nobel Prize in chemistry for this work with sugar nucleotides.

Leloir was born on September 6, 1906, in Paris, France. His parents, Hortensia Aguirre and Federico Leloir, returned from their European vacation to their home in Buenos Aires, Argentina, when Leloir was two years old. His grandparents, emigrants from France and Spain, had amassed wealth from cattle farming, freeing his father from having to practice in the field of his training, law. After graduating from secondary school, Leloir studied medicine at the University of Buenos Aires to earn his degree in 1932.

Leloir's internship at the university hospital convinced him to apply his medical knowledge to research. He remained at the university, conducting doctoral research under the future Nobel laureate Bernardo Houssay at the Institute of Physiology. Leloir wrote his dissertation on the role of adrenaline on carbohydrate metabolism, for which the faculty awarded its annual prize for the best thesis. In 1936 Leloir conducted a year of postdoctoral research at the Biochemical Laboratory of Cambridge University in England. There he collaborated with Malcolm Dixon, N. L. Edson, and D. E. Green on enzyme studies.

When Leloir returned to Houssay's lab in Buenos Aires, he collaborated with J. M. Muñoz on the oxidation of fatty acids in the liver, a line of research that led to the discovery of the peptide angiotensin. Actually, his team (E. Braun Menéndez, J. C. Fasciolo, and A. C. Taquini) called their discovery "hypertensin" (after its hypertensive action), but a team of researchers from the American pharmaceutical company Eli Lilly made the same discovery simultaneously, called it "angiotensin," and the name stuck. In

1943 Leloir married Amelia Zuberbuhler, and together they had one daughter, Amelia.

In 1944 after the Argentine government had removed Houssay from the university faculty for political reasons, Leloir immigrated to the United States, where he landed a position as a research assistant in the lab of Carl and Gerty Cori at Washington University in St. Louis, Missouri. Within a year he moved to Columbia University's College of Physicians and Surgeons, in New York, where he worked in the Enzyme Research Laboratory under D. E. Green.

In 1945 Leloir returned to Buenos Aires to work with Houssay, who had been restored to his position at the university. Soon thereafter, though, Leloir established a private research institute, the Instituto de Investigaciones Bioquimícas, Fundación Campomar. Despite generous funding from the textile manufacturer Jaime Campomar (as well as support from Leloir's personal inheritance), the institute struggled its first years.

In a four-room suburban dwelling, Leloir and his institute colleagues constantly juryrigged solutions to technical problems stemming from insufficient funding. Leloir, for example, folded flat sheets of stainless steel like origami to conduct paper chromatography, the bathroom doubled as a darkroom for developing paper chromatography under UV light, a butcher's icebox covered with a tarp on the patio served for cold-room experiments, a toy train ran across the top shelves to collect effluent from column chromatography experiments, and the team slaughtered rabbits to collect adenosine triphosphate (ATP). The Rockefeller Foundation had donated the institute's most advanced piece of equipment, a Beckman DU spectrophotometer.

In these makeshift facilities Leloir and his colleagues (Ranwel Caputto, Carlos E. Cardini, Raúl Trucco, and Alejandro C. Paladini, among others) discovered the "missing link"—sugar nucleotides—in the conversion of carbohydrates into simple sugars. In 1953 Leloir and Enrico

Luis Federico Leloir, the first Argentine to win a Nobel Prize in chemistry, which he was awarded in 1970 (*Bernard Becker Medical Library, Washington University School of Medicine*)

Cabib discovered the mechanism for the synthesis of trehalose, and in 1955 Leloir, Cardini, and J. Chiriboga discovered the mechanism for the synthesis of sucrose.

After Campomar's death in 1957, the National Institutes of Health (NIH) in the United States stepped in for funding of Leloir's institute until rules prevented the NIH from supporting foreign ventures. In 1958 the Argentine government intervened, offering to house the institute in a former girls' school. Thereafter, the newly established Argentine National Research Council supported the institute, in part by affiliating it with the University of

Buenos Aires, which appointed Leloir as a professor in its faculty of science.

In 1970 Leloir became the first Argentine to win the Nobel Prize in chemistry (and only the third to win the Nobel in any category) for his discovery of the mechanisms of sugar nucleotides. After winning the Nobel, Leloir helped establish the Argentine Society for Biochemical Research and served as the president of the Pan-American Association of Biochemical Sciences. Leloir died on December 4, 1987.

⊠ Lewis, Gilbert Newton
(1875–1946)
American
Chemist

Gilbert Newton Lewis made significant contributions to the understanding of chemistry, though his theories were so revolutionary that they were often met with resistance, delaying their acceptance and support in the scientific community. He focused scientists' attention on free energy as a means of calculating and predicting chemical reactions. He also introduced the idea of covalent bonding, extending this notion to redefine acids and bases according to whether they give or receive an electron pair in chemical reactions. His theories languished in obscurity until championed by IRVING LANGMUIR, a more prominent scientist who recognized the simple brilliance of Lewis's bonding theory.

Lewis was born on October 23, 1875, in Weymouth, Massachusetts. His parents, Francis Wesley Lewis and Mary Burr White, home-schooled him until he went to a college preparatory school affiliated with the University of Nebraska. He matriculated into the university proper to commence his undergraduate studies but transferred to Harvard University in Cambridge, Massachusetts, in 1894. Two years later, he earned his bachelor of science in chemistry.

He remained at Harvard to conduct doctoral studies under THEODORE WILLIAM RICHARDS. He wrote his dissertation on "Some Electrochemical and Thermochemical Relations of Zinc and Cadmium Amalgams" to earn his Ph.D. in 1899. Throughout his graduate career, he simultaneously taught at Phillips Academy, in nearby Andover.

Harvard retained Lewis for a year as an instructor, after which he traveled on scholarship through Europe, working in Germany in the labs of WILHELM OSTWALD at the University of Leipzig and WALTHER HERMANN NERNST at the University of Göttingen. He then spent a year in the Philippine Islands as superintendent of weights and measures while working concurrently as a chemist for the Bureau of Science in Manila. Upon his return to the United States in 1905, he moved across the Charles River from Harvard to the Massachusetts Institute of Technology (MIT), where he joined the team of Arthur Amos Noyes as an associate professor. Within five years of becoming a full professor he was appointed the acting director of research. Throughout his tenure at MIT, he published some 30 papers on chemical dynamics and free energy.

In 1912 the University of California at Berkeley appointed Lewis the dean of its College of Chemistry, which he completely reorganized into one of the best chemistry programs anywhere. That same year he married Mary Hinckley Sheldon, and together they had three children (his two sons followed in his footsteps to become chemists).

Over the next decade Lewis collaborated with Merle Randall on free energy studies, in which he performed the experimentation necessary to prove the mathematical theorization about free energy. He identified the free energy calculations for 143 substances in his 1923 text *Thermodynamics and the Free Energy of Chemical Substances*, which he coauthored with Randall.

That same year Lewis published *Valence and the Structure of Atoms and Molecules*, in which he

Gilbert Newton Lewis had introduced the notion of covalent bonding, but his theory remained obscure until championed by Irving Langmuir, a more prominent scientist who recognized the value of Lewis's ideas. *(AIP Emilio Segré Visual Archives, E. Scott Barr Collection)*

revolutionized the scientific understanding of bases and acids by redefining them according to their ability to give or receive an electron pair in a chemical reaction (bases give, acids receive). He had proposed this theory of bonding in 1916, to the vast indifference of the scientific community. It was not until a more famous chemist, Irving Langmuir, championed his cause by supporting his theory, which included a means of notation with dots representing the sharing of electron pairs in a chemical bond, that the theory gained acceptance. It became known as the Lewis-Langmuir electron dot theory, despite the fact that Langmuir contributed nothing more than his faith in the theory to earn a spot in its title.

In 1933 Lewis abandoned the studies he had been conducting for years in order to take up isotope separation studies. He managed to separate isotopes of hydrogen to create heavy water, which contained only the deuterium isotope of hydrogen and proved vital for nuclear reaction studies. He focused his investigations in his later career on photochemistry, especially fluorescence and phosphorescence, collaborating with the noted chemist MELVIN CALVIN on a review paper. His final work returned him to his original interest in thermodynamics, this time applying these theories to study the ice ages. On March 23, 1946, he suffered a heart attack and died in his Berkeley laboratory while conducting fluorescence experiments.

⊠   **Libby, Willard Frank**
(1908–1980)
American
Chemist

Willard Frank Libby discovered the radiocarbon
dating process of identifying a dead organism's
age based on the half-life of the radioactive isotope carbon 14. For this discovery he received
the 1960 Nobel Prize in chemistry. He also discovered another radioactive dating method for
hydrogen-bearing substances with tritium. He
devoted the middle of his career to policy making
as a member of the Atomic Energy Commission.

Discoverer of the radiocarbon dating process, Willard
Frank Libby received the 1960 Nobel Prize in
chemistry. *(AIP Emilio Segré Visual Archives, Segré
Collection)*

He supported nuclear testing as a necessary
means of maintaining peace in the nuclear age, a
political stance that generated some controversy.

Libby was born on December 17, 1908, in
Grand Valley, Colorado. His parents, Ora
Edward and Eva May Libby, were farmers who
later moved Libby and his two brothers and two
sisters to an apple ranch near Sebastopol, in
northern California. In 1926 Libby entered the
University of California at Berkeley to study
mining engineering but ended up graduating
with a bachelor of science in chemistry in 1931.

Libby remained at Berkeley to conduct doctoral studies under the physical chemists GILBERT
NEWTON LEWIS and Wendell Latimer. There he
constructed the country's first Geiger-Muller
tube, a device for detecting radioactivity. He
earned his doctorate in 1933 and stayed on at
the university as an instructor in chemistry. In
1940 he married Leonor Lucinda Hickey, and
together the couple had twin daughters. In 1941
he took a one-year sabbatical to work as a
Guggenheim Fellow at Princeton University, in
New Jersey. At the outbreak of World War II he
joined the Manhattan Project, first in Chicago,
Illinois, and then at Columbia University in
New York City, where he worked with HAROLD
UREY to separate isotopes of uranium 238 by
gaseous diffusion for use in the atomic bomb.

After World War II the University of
Chicago hired Libby as a professor of chemistry
at its Institute of Nuclear Studies. There he performed his groudbreaking studies on radioactive
carbon 14. Since atmospheric carbon 14, which
is created by the decay of nitrogen that had
absorbed neutrons released by cosmic rays, oxi-
dizes into carbon dioxide, Libby theorized that
all plants must ingest trace amounts of radioactivity when they "breathe" photosynthetically.
By extension, all organisms that consume plants
must also absorb some of this radioactivity. The
long half-life of the ubiquitous carbon 14 (5,370
years, as determined by Libby's former student
Martin Kamen) thus allowed for the exact dat-

ing of almost any organism, as carbon 14 starts to decay at the organism's death.

Libby and his colleague Aristide von Grosse built a sensitive instrument for measuring carbon decay with a Geiger counter, fortified behind an 8-inch-thick iron wall that blocked out atmospheric radiation. With this apparatus they measured the carbon decay of historical artifacts with known ages, such as a wooden funerary boat from King Sesostris III of Egypt's tomb, prehistoric sloth dung from Chile, the Dead Sea Scroll wrappings, charcoal from a campfire at Stonehenge, and even an ancient sandal found in Oregon. They refined their techniques and instrumentation until they could date back 70,000 years with accuracy. Libby gathered all of his ideas on this topic into a 1952 text, *Radiocarbon Dating*.

In 1953 Libby founded Project Sunshine to collect and measure fallout from nuclear testing, and the next year he joined the Atomic Energy Commission at President Dwight D. Eisenhower's invitation. His support of nuclear armamentation led some observers to charge Libby with pandering to corporate and governmental interests, but Libby acted by his conscience, he believed in the necessity of maintaining a nuclear arsenal to keep peace. He also conducted studies that concluded nuclear fallout from testing accounts for less of our exposure to radiation than from natural sources such as cosmic rays, a finding that did not increase his popularity with environmentalists and pacifists opposed to nuclear testing.

In 1960 Libby won the Nobel Prize in chemistry for his discovery of carbon dating. The year before he had returned to the University of California, this time at its Los Angeles campus, where he served as a professor of chemistry. In 1962 the university appointed him to the directorship of its Institute of Geophysics and Planetary Physics, allowing him to return to his interest in interstellar science. (In 1946 he had discovered that cosmic rays strike hydrogen to form a radioactive isotope, tritium, which he later used to date hydrogen in the water of archaeological specimens.) In 1966 he divorced his first wife and subsequently married Leona Woods Marshall, a professor of environmental engineering at UCLA. Libby retired in 1976, and on September 8, 1980, he died of complications from pneumonia in Los Angeles.

M

⊠ **Maria the Jewess**
First Century
Egyptian
Alchemist

Maria the Jewess is considered by some to have given birth to chemistry as a field of study, as she was one of the first alchemists, a prototypical science that sought to synthesize gold through the admixture of other substances. She is best known for having invented the *kerotakis*, a kind of double boiler capable of reaching high temperatures. Although very little of her own writing survives, other writers, such as the Egyptian alchemist Zosimos, who lived about a half millennium after her, discuss her work with reverence and respect.

Maria lived in the first century A.D. in Alexandria, Egypt, at the height of its scientific renown. Very little is known of her life, making it difficult to establish even the most basic biographical facts. Some sources surmise that she could have been a Greek (or even a Syrian) who worked in Egypt. She wrote under the name Miriam the Prophetess, a moniker that has caused no end of confusion. In the Bible, Moses's sister was a prophetess named Miriam, so some commentators trace this as the source of her identification as a Jew. Zosimos referred to her as "Miriam, sister of Moses," more likely referring to her wisdom than suggesting a cultural identity.

Maria laid out her alchemical philosophy in her treatise, "Maria Practica." As the title suggests, she tended more toward the practical that the theoretical. Some commentators have likened her more to the Egyptian process engineers, who developed process chemistry for brewing beer, than to her succeeding alchemists, who concerned themselves more with alchemy's metaphysical implications than with its actual chemical transformations.

Maria's alchemical beliefs, however, were not devoid of theory: She advanced the theoretical notion that metals corresponded to the sexual identities of animals, with both male and female components. Taken a step further, she believed that chemical reactions performed the same function as sexual union—to produce something new by joining two dissimilar elements. This notion manifested itself in the practice of many early alchemists, who believed that proper chemical reactions required the complementary presence of both male and female energies. Many male alchemists enlisted the assistance of their wives in their alchemical experimentation.

Maria's practicality expressed itself in her invention of apparatuses to perform alchemical experiments. Her best-known invention was the *kerotakis*, which she used to boil mercury, arsenic, or sulfur, condensing these elements into vapors to heat copper and lead in a pan above. She believed in the spiritual essence of sulfur vapors, and she logically held that since sulfur was produced from a realgar—disulphide of arsenic—which was found in gold mines, sulfur must therefore be a constituent element of gold. Although she did not succeed in creating gold from sulfur, she did create the process for making silver sulfide, the matte black compound employed in metalwork inlays that artists call niello.

Besides the *kerotakis*, Maria also invented the three-part still, as well as the water bath (double boiler), an apparatus that can sustain a constant temperature. The French term for double boiler, *bain-marie* (Maria's bath), honors her invention of it. She also invented various other apparatuses, such as the reflux condenser. It was in her practical applications that Maria the Jewess made her most profound and lasting contributions to the science of chemistry, as her instruments are still in common use.

⊠ Martin, Archer John Porter
(1910–   )
English
Biochemist

A. J. P. Martin collaborated with RICHARD SYNGE to develop the paper chromatography method of chemical analysis. For this work they shared the 1952 Nobel Prize, and immediately after Martin went on to innovate a method for partitioning gases by chromatography.

Archer John Porter Martin was born on March 1, 1910, in London, England. He was the only son of four children born to William Archer Porter Martin, a doctor, and Lilian Kate Brown, a nurse. He received his early education at the Bedford School, graduating in 1929 to Peterhouse College of Cambridge University, which he attended on a merit scholarship. He shifted his focus from the general fields of chemistry, physics, mathematics, and engineering to specialize in biochemistry after studying under

John Burdon Sanderson Haldane. He graduated with a bachelor of science in 1932.

Cambridge retained Martin as a researcher in its physical chemistry laboratory, where he collaborated with Nora Wooster on deliquescent substance studies; they reported their results in liquefying these solid substances as they absorb air moisture in a 1932 edition of the British scientific journal *Nature*. In 1934 he received the two-year Grocers' Scholarship to support doctoral studies at Cambridge, where he commenced his collaboration with Richard Synge studying carotene, a constituent pigment in vitamin A. He conducted his dissertation research under Sir Charles Martin to earn his Ph.D. in 1936.

Martin continued his vitamin A studies over the next two years at Dunn National Laboratories, then moved in 1938 to the Wool Industries Research Association in Leeds, where he conducted amino acid studies and also investigated the chemical composure of wool felting. In 1941 he continued his collaboration with Synge, commencing their work that would lead to the discovery of paper chromatography. In the midst of these studies, on January 9, 1943, Martin married a teacher named Judith Bagenal; together they had two sons and three daughters.

By 1944 Martin and Synge had completely devised the partition chromatography technique, whereby they would place a drop of amino acid solution on porous paper, let it dry, then submerse it in a solvent that separated the amino acids, carrying them up the paper by capillary action at different rates, depending on the chemical composition. They then sprayed the paper with a reagent, such as ninhydrin, that revealed the varying colors of the "partitioned" amino acids. This simple experiment proved instrumental to a wide range of important applications, from FREDERICK SANGER's identification of the amino acid chain in insulin to MELVIN CALVIN's discovery of the mechanism of photosynthesis.

In 1946 the Boots Pure Drug Company in Nottingham hired Martin as the head of its biochemistry division in the research department. During his two-year stint there he continued his paper chromatography studies in collaboration with R. Consden and A. H. Gordon to identify lower peptides in protein mixtures. In 1948 he joined the Medical Research Council, working first at the Lister Institute of Preventive Medicine for four years before joining the National Institute for Medical Research (NIMR) in the Mars Hill section of London as the head of the physical chemistry division in 1952.

That same year, in 1952, he shared the Nobel Prize in chemistry with Synge for their innovation of the paper chromatography technique of chemical analysis. The year before, in 1951, the Swedish Medical Society had awarded him its Berzelius Gold Medal. The year after, in 1953, he commenced work with A.T. James on gas chromatography, separating chemical vapors on a porous solid by differential adsorption, and on gas-liquid chromatography, a method of separating volatile chemicals by blowing them down long tubes containing inert gas. In 1957 he stepped back to a consultant's position at the NIMR, then in 1959 he took over the helm of the Abbotsbury Laboratory, which he steered through the 1960s. In 1970 he became a consultant to the Wellcome Research Laboratories for three years, before taking up a professorship at the University of Sussex.

In 1980 Martin served as the Invited Professor of Chemistry at the École Polytechnique Fédéral in Lausanne, Switzerland. Thereafter, he returned to Cambridge, where he spent his retirement. He even indulged his interests in mountaineering and gliding.

⊠ **Massie, Samuel Proctor**
(1919–)
American
Organic Chemist

Samuel Proctor Massie distinguished himself as an outstanding chemistry professor, a fact that was duly recognized with honors. Throughout his career he also chaired the chemistry department at practically every institution with which he was affiliated. In the mid-1960s, he became the first African-American professor to be hired by the United States Naval Academy, where he remained for the rest of his academic career.

Samuel Proctor Massie Jr. was born on July 3, 1919, in Little Rock, Arkansas. Both of his parents were schoolteachers, and Massie excelled in school, graduating from high school at the age of 13. He attended Dunbar Junior College in Little Rock, graduating from there to the Agricultural Mechanical and Normal College of Arkansas (now the University of Arkansas at Pine Bluff) to study chemistry. In 1938, at the age of 18, he graduated summa cum

Samuel Proctor Massie, known for his excellence in teaching chemistry, became in 1966 the first African-American professor hired by the United States Naval Academy. *(Courtesy U.S. Department of Energy)*

laude with a Bachelor of Science. He worked as a laboratory assistant in chemistry at Fisk University in Nashvile, Tennessee, while studying for his master's degree, which he earned in 1940.

Massie returned to A.M.N. College of Arkansas as an associate professor of mathematics and acting head of the math and physics departments. In 1941 Iowa State University offered him a research associateship in chemistry while he pursued doctoral studies there in the town of Ames. World War II interrupted these studies, as he joined the Manhattan Project in 1943 to develop liquid compounds of uranium (a line of investigation that proved unsuccessful). After the war he returned to Iowa to continue his doctoral work. He submitted his dissertation, entitled "High-Molecular Weight Compounds of Nitrogen and Sulfur as Therapeutic Agents," to earn his Ph.D. in chemistry in 1946. In 1947 he married Gloria Tompkins, who became a psychology professor. Together, the couple had three sons, all of whom graduated from law school.

After receiving his doctorate Massie returned to Fisk to serve as an instructor for a year. In 1947 Langston University in Oklahoma hired him as a professor and chair of the chemistry department, posts that he held until 1953, when he served as president of the Oklahoma Academy of Sciences. Later in 1953 he returned to Fisk yet again, this time as professor and chemistry department chair. In 1954 he wrote a review article, entitled "The Chemistry of Phenothiazine," which is considered a classic in the field. He took leave of Fisk in 1956 to serve as the Sigma Xi lecturer at Swarthmore College in Pennsylvania. From 1960 to 1963 he served as an associate program director at the National Science Foundation in Arlington, Virginia. During his tenure there Howard University in Washington, D.C., hired him to chair its pharmaceutical chemistry department. In 1963 North Carolina College (now North Carolina Central University at Durham) appointed Massie as its president.

In 1966 the United States Naval Academy set a historical precedent by hiring Massie as its first African-American professor. As at the other institutions where he had worked, Massie chaired the academy's chemistry department from 1977 to 1981. He also founded the black studies program at the academy. He spent the rest of his career there, retiring to emeritus status at the end of the fall semester in 1993. However, he came out of retirement on July 1, 1994, to serve as vice president for education at the Bingwa Software Company, which develops computer software for teaching elementary students math through multiculturalism.

Massie has received numerous honors throughout his career, mostly recognizing his excellence in teaching. He won the 1961 College Chemistry Teachers Award from the Chemical Manufacturers Association. In 1980 the National Organization of Black Chemists named him Outstanding Professor, and the following year Iowa State University granted him its Distinguished Achievement Citation. In 1987 Dillard University bestowed its Henry A. Hill Award for service to the field of chemistry. In September 1988 the White House Initiative gave him its first Lifetime Achievement Award for contributions to science, technology, and community affairs. On March 1, 1995, the National Academy of Sciences unveiled a portrait of Massie in its gallery, and in 1997 *Chemical & Engineering News* named him one of its "Top Seventy-Five Distinguished Contributors to the Chemical Enterprise."

Massie's name also graces many honors. In September 1989 the Maryland State Board of Community Colleges established the Massie Science Prize in his honor. In December 1992 the National Naval Officer's Association, together with the U.S. Naval Academy's African American Alumni Association, set up the Samuel P. Massie Educational Endowment Fund to offer college tuition support to women, minorities, and low-income residents of Anne

Arundel County in Maryland. In September 1994, the U.S. Department of Energy established the Dr. Samuel P. Massie Chairs of Excellence in Environmental Engineering at the nine Historically Black Colleges and Universities (HBCUs); in 1995 a 10th chair was added at the Universidad de Turabo in Puerto Rico, a Hispanic institution.

⊠ McMillan, Edwin M.
(1907–1991)
American
Physicist

Edwin M. McMillan discovered neptunium, the first transuranium element, or element with an atomic number higher than that of uranium, which was until then thought to be the heaviest element. In collabortion with GLENN THEODORE SEABORG, he discovered plutonium, the byproduct of neptunium's radioactive decay. For these discoveries he shared the 1951 Nobel Prize in chemistry with Seaborg. He later solved a problem with nuclear accelerators by implementing phase stability, or the coordination of the electrical field to compensate for the increasing mass and decreasing acceleration particles experience in an accelerator. For this innovation he shared the 1963 Atoms for Peace Award with the Russian physicist V. I. Veksler, who simultaneously but independently developed a similar solution.

Edwin Mattison McMillan was born on September 18, 1907, in Redondo Beach, California. His mother was Anne Marie Mattison; his father, Edwin Harbaugh McMillan, was a physician. McMillan attended school in Pasadena, where his family moved when he was one year old, graduating from Pasadena High School in 1924. He matriculated at the California Institute of Technology, where he earned his bachelor of science in 1928 and his master of science the next year.

Edwin M. McMillan, who discovered neptunium, the first transuranium element *(Berkeley National Laboratory, University of California Berkeley, courtesy AIP Emilio Segré Visual Archives)*

McMillan pursued doctoral studies at Princeton University, in New Jersey. He submitted his dissertation, entitled "Electric Field Giving Uniform Deflecting Force on a Molecular Beam," to earn his Ph.D. in 1932. As a National Research fellow, he conducted two years of postdoctoral research, employing the molecular beam method to measure the magnetic moment of protons at the University of California at Berkeley. He remained there for another year as a research associate before Ernest Orlando Lawrence appointed him to the staff of the Berkeley Radiation Laboratory (later renamed the Lawrence Radiation Laboratory), where he remained the rest of his career.

Also in 1935 the University of California at Berkeley appointed McMillan to its faculty as an instructor, and the following year the university promoted him to the rank of assistant professor. In the late 1930s McMillan collaborated with Philip Abelson, bombarding the radioactive element uranium with neutrons to produce nuclear fission. In 1940 they discovered a by-product of this reaction, the transformation of the isotope uranium 238 into a new element, one atomic unit heavier. In the same spirit that designated the name uranium after the seventh planet from the Sun, McMillan and Abelson named this new element neptunium (after the eighth planet from the Sun). This represented the discovery of the first transuranium element, or elements heavier than uranium.

In 1941 McMillan collaborated with Glenn Seaborg to discover another new element, the product of the radioactive decay of neptunium. They named it plutonium, after the ninth planet from the Sun, as its atomic number was one step above its parent. That same year the university promoted him to an associate professorship. On June 7, 1941, he married Elsie Walford Blumer, daughter of the dean of the Yale University School of Medicine and sister-in-law of Lawrence. Together, the couple had three children, Ann Bradford (born in 1943), David Mattison (born in 1945), and Stephen Walker (born in 1949).

During World War II McMillan contributed to the war effort in several different capacities: from November 1940 to 1941 he worked at the Radiation Laboratory of the Massachusetts Institute of Technology, then from 1941 to 1942 he worked at the United States Navy Radio and Sound Laboratory in San Diego, and from 1942 to 1945 he worked on the Manhattan Project at Los Alamos, New Mexico. Upon his return to the Berkeley Radiation Laboratory in 1945 he sought to solve an accelerator problem: As particles accelerated, they gained mass (and thus slowed in velocity). McMillan devised an elegantly simple solution: One could change the electrical field in

the accelerator to compensate for the particles gaining mass and losing acceleration. This solution became known as *phase stability*.

In 1946 the University of California at Berkeley promoted McMillan to a full professorship. He received the 1950 Research Corporation Scientific Award, and the next year he shared the 1951 Nobel Prize in chemistry with Seaborg. The Berkeley Radiation Laboratory then promoted him to its associate director position in 1954; in 1958 it promoted him to deputy director, and later that same year he ascended to the directorship of the lab.

In 1963 McMillan shared the Atoms for Peace Award with the Russian physicist V. I. Veksler, who had devised a similar solution to the phase stability problem in accelerators simultaneously but independently from McMillan. A decade later McMillan retired. He died of complications from diabetes on September 7, 1991, in El Cerrito, California.

⊠ **Mendeleyev, Dmitri Ivanovich**
(1834–1907)
Russian
Chemist

Dmitri Mendeleyev brought order to the chaos of chemistry by asserting his periodic law, which states that the elements arrange themselves according to their atomic weight and their properties. He symbolized this arrangement in the first periodic table, which allowed for the addition of new elements as they were discovered, their positions predicted on the table. The discovery of three new elements in his lifetime, each of which accorded to his predictions, validated his theory.

Dmitri Ivanovich Mendeleyev was born on February 7, 1834, in Tobolsk, Siberia, the youngest of 17 children. His father, Ivan Pavlovitch Mendeleyev, was the director of the local gymnasium who went blind and died when Mendeleyev was a child. His mother, Maria

Dmitrievna Korniliev, was forced to manage her family's glass factory in Aremziansk to support her children. When Mendeleyev was 14 years old and attending the gymnasium that his father had run, his mother's glass factory burned down, leaving the family destitute. She moved to Moscow, intent on enrolling her youngest child in university, but officials rejected his application. She persisted, moving to St. Petersburg, where Mendeleyev passed the entrance examination to earn admission to the Pedagogical Institute on a full scholarship in 1850.

Soon thereafter Mendeleyev's mother died of tuberculosis, and he was also diagnosed with the disease in his third year at the university, forcing him to complete his last year from bed. At this time, circa 1854, he published his first paper, entitled "Chemical Analysis of a Sample from Finland." He graduated with a medal of excellence for being first in his class. In 1855 he moved to the gentler climate of Simferopol in the Crimean peninsula on the Black Sea as chief science master of the local gymnasium. Within a year his health improved vastly (there remained no trace of tuberculosis), so he moved back to St. Petersburg to complete his master's degree. He wrote his thesis on "Research and Theories on Expansion of Substances due to Heat."

In 1859 Russia's minister of public instruction dispatched Mendeleyev on a tour of Europe to gather the latest scientific information and innovations. Over the next two years he studied gas densities with Henri-Victor Regnault in Paris and spectroscopy with Gustav Robert Kirchoff in Heidelberg, Germany, as well as pursuing his own research into capillary and surface tension. He heard the Italian chemist Stanislao Cannizzaro lecture at the 1860 Chemical Conference in Karlsruhe, Germany. When he returned to St. Petersburg in 1861, he published his first book, *Organic Chemistry*, which won the Domidov Prize.

In 1863 the Technical Institute in St. Petersburg appointed Mendeleyev a professor of general chemistry. That same year his sister Olga

persuaded him to marry Feozva Nikitchna Lascheva, a woman he did not love. Together the couple had two children, a son named Volodya and a daughter named Olga. In 1866, the same year he earned his doctorate with the dissertation "On the Combinations of Water with Alcohol," he transferred to the University of St. Petersburg as a professor of chemistry. In 1868 he published *Principles of Chemistry*, which was so popular that it was translated into English in 1891 and 1897.

On March 6, 1869, Mendeleyev presented a paper entitled "The Dependence Between the Properties of the Atomic Weights of the Elements" before the Russian Chemical Society, an organization that he had helped to found the year before. In his presentation he asserted that "the

Dmitri Ivanovich Mendeleyev, who formulated the periodic table of elements *(E. F. Smith Collection, Rare Book & Manuscript Library, University of Pennsylvania)*

elements, if arranged according to their atomic weights, exhibit an apparent periodicity of properties," or his periodic law. He also outlined the periodic table, suggesting that the elements arrange themselves into an order depending on their valences and properties. Extending this logic, he declared on November 29, 1870, that the discovery of other elements could be predicted based on gaps in his periodic table, which also anticipated the properties of these elements.

In confirmation of Mendeleyev's periodic law and table, in November 1875 the French scientist Lecoq de Boisbaudran discovered gallium, an element that corresponded to Mendeleyev's predictions. The discovery of scandium in 1879 and of germanium in 1886 further validated Mendeleyev's periodic theories. In January 1882 he divorced his first wife to marry Anna Ivanova Popova, his niece's best friend. Together the couple had four children, Liubov, Ivan, and the twins Vassili and Maria.

On August 17, 1890, Mendeleyev resigned from his university position in support of student protests over unjust conditions. His immense popularity forced the Russian government to appoint him to head the Bureau of Weights and Measures in 1893. His renown, however, extended beyond his home country: The Royal Society of London had awarded him its 1882 Davy Medal and would later bestow on him its 1905 Copley Medal, the highest honor accorded by the prestigious organization. In 1906 his nomination for the Nobel Prize in chemistry fell one vote short. The next year, on January 20, 1907, he died while listening to a reading of Jules Verne's *Journey to the North Pole*.

⊠ **Moissan, Ferdinand-Frédéric-Henri**
(1852–1907)
French
Inorganic Chemist

Henri Moissan was the first chemist to isolate fluorine, one of the most volatile elements. In addition, his failed attempt to synthesize diamonds succeeded in inventing the electric-arc furnace, an instrument that proved extremely useful for many diverse applications in experimental chemistry. For these contributions Moissan won the 1906 Nobel Prize in chemistry.

Ferdinand-Frédéric-Henri Moissan was born on September 28, 1852, in Paris, France. His mother, Josephine Mitel, was a seamstress, and his father, François Ferdinand, was a clerk. Moissan was educated at the Collège de Meaux, where the family moved in 1864. In 1870 he apprenticed to a watchmaker, but soon thereafter he enlisted in the army. After defending Paris against a Prussian invasion, he apprenticed to a pharmacist in 1871.

Instead of pursuing an education in pharmacy, though, Moissan worked in Edmond Frémy's laboratory at the Musée d'Histoire Naturelle. In addition to his laboratory research, he attended lectures at the museum's School of Experimental Chemistry by E. H. Sainte-Claire Deville and Henri Debray. After a year he transferred to a laboratory at the École Practique des Hautes Études, then he ran a laboratory of his own for a short period. He finally ended up in Debray's laboratory at the Sorbonne in Paris.

Throughout this period Moissan was fulfilling the necessary coursework to earn his *baccalauréat* in 1874, followed by his *licence* (the equivalent of a bachelor's degree) in 1877, and his pharmacist's first-class degree in 1879. That year the Agronomic Institute appointed him to a junior position, and the next year he earned his doctorate with a dissertation on pyrophoric iron and its oxides. He had been teaching at the École Supérieure de Pharmacie as an assistant lecturer and a senior demonstrator; upon his receipt of his doctorate, the school named him an associate professor. In 1882 he married Marie Leonie Lugan, the daughter of the pharmacist in Meaux who had supported his pharmaceutical studies. Together the couple had one son, Louis Ferdinand Henri, who died in World War I.

Moissan commenced his studies of fluorine in 1884, following up on his mentor Frémy's attempts to isolate this, one of the most reactive elements. Moissan modified Frémy's experimental methodology, dissolving potassium acid fluoride in hydrogen fluoride to produce a mixture that remained liquid and conducted electrolytically at subzero temperatures, the two variables that had eluded his predecessor. Employing an apparatus with a platinum U-tube capped with calcium fluoride plugs and electrodes of iridium-platinum (which repelled fluorine), Moissan electrolyzed his solution at −50° Celsius to isolate gaseous fluorine in 1886.

That year the École Supérieure de Pharmacie promoted Moissan to its chair of toxicology. Over the next several years Moissan conducted a full study of fluorine's reactivity with other elements to produce compounds such as thionyl fluoride, sulfuryl fluoride, carbon tetrafluoride, and sulfur hexafluoride, among others. In 1900 he gathered together his findings to publish one of his heralded books, *Le Fluor et ses composés*. By 1903 he had isolated liquid as well as solid fluorine.

In 1892 Moissan turned his attention to a theoretical possibility: the synthesis of diamonds. He attempted this feat by constructing an electric-arc furnace, capable of generating temperatures as high as 3,500° Celsius while withstanding great pressure. The "diamonds" he created by crystallizing carbon in molten iron were not true gemstones, as the pressure generated by this method, although great, was not nearly equal to the pressures generated under the earth's surface that are necessary for diamond formation. However, the electric-arc furnace proved to be an extremely practical instrument for a wide range of experimental applications: Moissan himself used it to volatilize previously infusible compounds and create new compounds, such as silicon carbide. In 1897 he published a classic text on the new apparatus, *Le four électrique*.

In 1899 the École Supérieure de Pharmacie promoted Moissan again, this time to a professorship of inorganic chemistry, but he held this position only shortly. The very next year he moved to a chair of inorganic chemistry on the Faculty of Sciences at the University of Paris (known as the Sorbonne), which he inhabited for the remainder of his career. At this point he gathered his accumulated wisdom in his field into the five-volume *Traité de Chimie Minerale*, (*Treatise on inorganic chemistry*) published from 1904 through 1906.

Moissan garnered many awards for his work on fluorine and his invention of the electric-arc furnace: He won the 1887 Prix Lacaze from the French Academy of Sciences (which inducted him into its membership in 1891), and he received the 1896 Davy Medal from the Royal Society (which made him a foreign fellow in 1905). Most significant, he won the 1906 Nobel Prize in chemistry. Sadly, he suffered an appendicitis attack soon after returning from the awards ceremony in Stockholm, Sweden, and died on February 20, 1907, after undergoing surgery.

⊠ **Molina, Mario**
(1943–   )
Mexican/American
Chemist

Mario Molina distinguished himself while still a postdoctoral researcher studying chlorofluorocarbons (CFCs) under FRANK SHERWOOD ROWLAND. In 1974 they jointly published their CFC ozone-depletion theory, which predicted the destruction of the protective ozone layer in Earth's stratosphere if CFC emissions continued at the then-current pace. In the wake of this announcement policy makers around the globe enacted bans on CFC production, thus drastically reducing CFC ozone depletion. For this work Molina and Rowland shared the 1995

Nobel Prize in chemistry with PAUL J. CRUTZEN, who was also instrumental in CFC research.

Mario José Molina was born on March 19, 1943, in Mexico City. His mother was Leonor Henriquez; his father, Roberto Molina Pasquel, was a lawyer with a private practice who also taught at the Universidad Nacional Autónoma de México (UNAM) and later served as the Mexican ambassador to Ethiopia, Australia, and the Philippines. Encouraged by his aunt Esther, a chemist, Molina converted an unused bathroom at home into a rudimentary laboratory, where he observed paramecia and amoebas in a toy microscope. At the age of 11, he traveled to boarding school in Switzerland. He returned to Mexico to conduct his undergraduate studies at UNAM, graduating in 1965 with a degree in chemical engineering.

Molina returned to Europe to conduct graduate studies at the University of Freiburg, in Germany, where he spent two years investigating polymerization kinetics to earn the equivalent of a master's degree in 1967. He returned to an assistant professorship at UNAM, where he established a graduate program in chemical engineering. In 1968 he continued his graduate studies in the physical chemistry doctoral program at the University of California at Berkeley, where he learned how to use chemical lasers to study molecular dynamics as a member of George C. Pimentel's research group. He became proficient enough with chemical lasers to focus his own doctoral research on the distribution of internal energy produced by chemical and photochemical reactions. He earned his Ph.D. in 1972 and remained at Berkeley for a year conducting postdoctoral research. In July 1973 Molina married Luisa Y. Tan, a fellow chemist who subsequently collaborated with him extensively. Together they had one son, Felipe, born in 1977.

In the fall of 1973 Molina moved to the University of California at Irvine to continue postdoctoral studies under F. Sherwood Rowland, who presented him a list of research options. Molina chose to study the little-understood atmospheric chemistry of chlorofluorocarbons. At the time CFCs were believed to be innocuous to the environment. Within three months, however, Molina and Rowland developed the CFC ozone-depletion theory, which asserted that solar radiation breaks down CFCs in the stratosphere, freeing chlorine atoms that each destroy tens of thousands of ozone molecules, thereby depleting the ozone layer that protects Earth from harmful cosmic radiation. The pair published their findings in the June 28, 1974, issue of *Nature*, a prominent British scientific journal, and followed up with a more detailed account later in the year in *Science*, the foremost American scientific journal.

The distressing implications of the CFC ozone-depletion theory—that continued emissions of CFCs would destroy the atmosphere's protective stratospheric ozone layer—mobilized policy makers to enact preventative measures. Molina testified before the U.S. House of Representatives Subcommittee on Public Health and Environment, and in 1976 he was appointed to the National Science Foundation's Oversight Committee on Fluorocarbon Technology Assessment. The year before he had been promoted to an assistant professorship at the University of California at Irvine.

In 1982 Molina left his associate professorship at Irvine to devote himself exclusively to research as a member of the technical staff in the molecular physics and chemistry section at the Jet Propulsion Laboratory of the California Institute of Technology (CalTech). Within two years he was named senior research scientist. In 1989 he returned to teaching, joining the faculty at the Massachusetts Institute of Technology (MIT) with dual professorships in the chemistry department and the earth, atmosphere, and planetary sciences department. In 1993 MIT named him to a new chair in environmental sciences endowed by the Martin Foundation.

The vital significance of Molina's research has been recognized extensively. He won the

1987 Esselen Award of the American Chemical Society, the 1988 Newcomb-Cleveland prize from the American Association for the Advancement of Science, the 1989 NASA Medal for Exceptional Scientific Advancement, and the 1989 United Nations Environmental Programme Global 500 Award. In 1990 the Pew Charitable Trusts Scholars Program in Conservation and the Environment honored him as one of 10 environmental scientists awarded $150,000 grants. His highest honor came with his sharing of the 1995 Nobel Prize in chemistry with Rowland and Paul Crutzen, in recognition of their work on understanding the destructive atmospheric chemistry of chlorofluorocarbons.

⊠ Moore, Stanford
(1913–1982)
American
Biochemist

Stanford Moore shared the 1972 Nobel Prize in chemistry with his research partner, WILLIAM H. STEIN, and their collaborator, CHRISTIAN BOEHMER ANFINSEN, for their joint investigations on the interrelationship between the chemical structure of protein and its biological functions. Integral to the progress of Moore and Stein's research was their innovation of a technique that reduced the time needed to conduct an amino acid analysis from one full week to one full day.

Moore was born on September 4, 1913, in Chicago, Illinois, where his father, John Howard Moore, was attending the University of Chicago Law School at the time. His mother, Ruth Fowler, was a graduate of Stanford University, where she had met her husband. (Legend has it that Moore's mother named him after her alma mater.) In 1924 Moore's father joined the law faculty of Vanderbilt University in Nashville, Tennessee, where Moore attended the Peabody Demonstration School of the George Peabody College for Teachers. There, Moore's chemistry

teacher, R. O. Beauchamp, sparked his interest in the subject.

In 1931 Moore matriculated at Vanderbilt, undecided between majoring in chemistry or aeronautical engineering. Organic chemistry professor Arthur William Ingersoll swayed him toward the former, and in 1935 Moore graduated summa cum laude with a bachelor's degree in chemistry and the Founder's Medal as the most outstanding student in his class. He received stellar recommendations for a Wisconsin Alumni Research Foundation Fellowship at the University of Wisconsin at Madison, where he conducted doctoral research (in which he characterized carbohydrates as benzimidazole derivatives) under Karl Paul Link to earn his Ph.D. in organic chemistry in 1938.

Faced with a choice between accepting a four-year fellowship for study at the Harvard Medical School and immediate entry into research chemistry as an assistant in Max Bergmann's laboratory at the Rockefeller Institute for Medical Research (RIMR), Moore chose the latter hands down. He worked briefly on amino acid studies with another young biochemist, William Stein, but their collaboration had barely commenced when World War II intervened. Moore joined the Office of Scientific Research Development (OSRD) in Washington, D.C., as a junior administrative officer, coordinating academic and industrial investigations into the physiological effects of mustard gases (though these agents did not see use in the war). When the war ended, he was working in Hawaii with the Chemical Warfare Service's Operational Research Section.

Bergmann's premature death in 1944 from cancer scattered his lab, but RIMR director Herbert Gasser wisely retained Moore and Stein by offering them space in Bergmann's lab as well as modest funding to continue their amino acid studies. Following in the footsteps of ARCHER JOHN PORTER MARTIN and RICHARD SYNGE, Moore and Stein developed fractionation techniques for analyzing amino acids, transforming the British partition chromatography method into column chromatography on potato starches broken down into amino acids and peptides. This novel approach seemed promising enough to Gasser to increase the young team's budget significantly in 1949.

In 1950 Moore traveled overseas to gather know-how for his amino acid studies as well as to disseminate his knowledge to his European colleagues. E. J. Bigwood offered him the international Francqui Chair at the University of Brussels in Belgium in exchange for his consultation in the designing and building of an amino acid analysis laboratory for the school of medicine there. He then spent the spring semester of 1951 at the University of Cambridge in England, conducting research alongside FREDERICK SANGER, who was ensconced in his landmark insulin studies at the time.

Upon his return Moore and Stein set out to analyze the structure of a protein, ribonuclease, which they were familiar with from previous studies. By 1956 they had developed a new, automated method for analyzing the amino acids that took only 24 hours (as compared to the one week previously required); they published their method in 1958, only after tweaking the instrumentation perfectly, and they painstakingly documented each step for ease of replication. With this new method they endeavored to analyze the complete structure of ribonuclease, which they achieved in 1963, marking the first structural elucidation of an entire enzyme.

Just under a decade later Moore and Stein shared the 1972 Nobel Prize in chemistry with Anfinsen, with whom they had collaborated extensively on the solution of the ribonuclease structure. In 1969, after Moore had returned from a year as visiting professor of health sciences at Vanderbilt University School of Medicine, Stein suffered a crippling paralysis, but the team continued to work together, even after becoming Nobel laureates.

Moore continued to serve his field in the 1970s, presiding over the Federation of American Societies of Experimental Biology in 1970, while concurrently chairing the National Academy of Sciences' biochemistry section. He also served as a Vanderbilt trustee from 1974. His service to his field dated back to the 1950s: He had sat on the editorial board of the American Society of Biological Chemists from 1950 through 1960, serving as the society's treasurer from 1956 through 1959, as well as its president in 1966.

In the early 1980s Moore, who never married, suffered progressive nerve and muscle degeneration from amyotrophic lateral sclerosis. On August 23, 1982, he died in his New York City apartment. He endowed his estate to support biochemistry researchers. Rockefeller University hosted "A Symposium on Protein Chemistry in Tribute to Stanford Moore" on November 4, 1983.

⊠ Mulliken, Robert S.
(1896–1986)
American
Chemical Physicist

Robert S. Mulliken was one of the founders of the field of chemical physics, fusing quantum theory and the study of chemistry. In his most profound theoretical assertion he proposed the notion that molecules shared electrons. For this realization Mulliken won the 1966 Nobel Prize in chemistry.

Robert Sanderson Mulliken was born on June 7, 1896, in Newburyport, Massachusetts. His mother was Katherine W. Mulliken; his father, Samuel Parsons Mulliken, was a professor of organic chemistry at the Massachusetts Institute of Technology (MIT). As a youngster, Mulliken proofread galleys of his father's textbooks, habituating him to deciphering the Latinate names of chemicals. He attended Newburyport

For his bold assertion that molecules share electrons, Robert S. Mulliken won the 1966 Nobel Prize in chemistry. *(Harris & Ewing, courtesy AIP Emilio Segré Visual Archives)*

High School and delivered the salutatory address, which he titled "The Electron: What It Is and What It Does." He won the school's Wheelwright Scholarship to MIT, matriculating in 1913.

After briefly toying with the idea of majoring in chemical engineering, Mulliken graduated from MIT in 1917 with a bachelor of science in chemistry. He contributed to the war effort as a junior chemical engineer for the Department of the Interior's Bureau of Mines in Washington, D.C., conducting research on poisonous gases at American University, though he demonstrated a degree of experimental inepti-

tude when he spilled some mustard gas. After World War I he pursued doctoral studies under W. D. Harkins and Robert A. Millikan at the University of Chicago, concentrating on developing methods for separating mercury isotopes from one another.

After receiving his Ph.D. in 1921, Mulliken remained at the University of Chicago to conduct postdoctoral studies as a National Research Council Fellow, improving on his "isotope factory" for separating mercury isotopes. In 1923 he applied to extend his fellowship for two more years, but he had to switch institutions and fields of concentration, so he transferred to Harvard University in Cambridge, Massachusetts, where he studied the diatomic band spectra of boron nitrite, predicting his own discovery of isotopes of boron that had previously escaped notice.

It was on his European tours of 1925 and 1927 that Mulliken realized the necessity of fusing physics with chemistry, taking his cue from Friedrich Hund, a German chemist who applied Werner Karl Heisenberg's uncertainty principle to chemical investigations and theorization. In his classic 1928 paper Mulliken expressed the radical notion that when atoms combine to form molecules, their individual electrons cease acting atomically (as if they still belonged to each atom exclusively) and start acting molecularly (as if the electrons actually belonged to the whole compound).

Upon his return in 1926 from his first continental tour, Mulliken landed an assistant professorship in the physics department of the Washington Square College of New York University. In 1928 he returned to the University of Chicago as an associate professor of physics. The next year, on the day before Christmas, he married Mary Helen von Noé, and together the couple had two daughters, Lucia Maria and Valerie Noé.

Mulliken stayed on at the University of Chicago for the rest of his career, gaining a full professorship in 1931. He returned to Europe on Guggenheim Fellowships in 1930 and 1932 and again from 1952 to 1954, when he served as a Fulbright Scholar at Oxford University and a visiting fellow at St. John's College of Oxford University. In between these trips he served during World War II as the director of editorial work and information for the Plutonium Project, an extension of the Manhattan Project.

In 1956 the University of Chicago named Mulliken the Ernest de Witt Burton Distinguished Service Professor of Physics and Chemistry, a chair he filled until 1961, when he stepped down to joint appointments in the physics and chemistry departments. Prestigious U.S. universities, such as Cornell and Yale, invited him to disseminate his wisdom, the former as the Baker Lecturer in 1960 and the latter as the Silliman Lecturer in 1965.

From 1964 on he spent every winter at Florida State University in Tallahassee as a Distinguished Research Professor of Chemical Physics at its Institute of Molecular Biophysics.

In the 1960s Mulliken received increasing recognition for his contribution to science in the fusion of chemistry with physics. He won the American Chemical Society's 1963 Peter Debye Award and its 1965 Willard Gibbs Medal, before receiving the 1966 Nobel Prize in chemistry. The year before, he published a memoir of sorts, an article entitled "Molecular Scientists and Molecular Science: Some Reminiscences," which appeared in the November 15, 1965, issue of *The Journal of Chemical Physics*. He died on October 31, 1986, in Arlington, Virginia.

N

⊠ **Natta, Giulio**
(1903–1979)
Italian
Chemist

Giulio Natta shared the 1963 Nobel Prize in chemistry with KARL ZIEGLER, his close friend and collaborator until Natta disregarded their information-sharing agreement when he discovered polypropylene using the same catalytic process Ziegler had used to discover polyethylene. Polypropylene, an isotactic (or symmetric) polymer, proved to mold objects stronger, spin fibers lighter, and resist heat better than polyethylene, as well as resolving into thin, clear films. During his career Natta published more than 500 scientific papers and received almost as many patents.

Natta was born on February 26, 1903, in Imperia, Italy. His mother was Elena Crespi, and his father was Francesco Natta, a lawyer and judge. He attended primary and secondary school in Genoa, some 60 miles northeast of Imperia. He entered the University of Genoa at the age of 16 to study mathematics but transferred to the Milan Polytechnic Institute in 1921 to study chemical engineering. In 1924, at the age of 21, he earned his doctoral degree.

The Polytechnic Institute retained Natta as an instructor, promoting him to assistant professor of general chemistry within a year and to full professor two years after that, in 1927. In 1933 the University of Pavia hired Natta as a professor of general chemistry and director of its chemical institute. During this period he conducted research in inorganic chemistry, analyzing substances by X ray and through electron diffraction. These studies yielded his discovery of a catalyst for methanol synthesis.

In 1935 Natta moved to the University of Rome, which had appointed him to chair its Department of Physical Chemistry. That same year he married Rosita Beati, a professor of literature at the University of Milan, and together the couple had two children, a daughter named Franca and a son named Giuseppe. In 1937 Natta moved his family to Turin, where he held a professorship at the Turin Polytechnic Institute and the directorship of its Institute of Industrial Chemistry. The next year he moved to a professorship of chemistry and directorship of the Industrial Chemical Research Center.

In this early part of his career Natta pioneered the use of certain catalysts in several important industrial synthetic applications. He used methyl alcohol (methanol) to synthesize formaldehyde (methanal); propylene (propene) and carbon monoxide to synthesize propionaldehyde (butanal); and acetylene (ethyne), carbon monoxide, and synthetic gas to synthesize succinic acid (butandioic acid).

Natta spent World War II conducting polymer studies to investigate new ways to synthesize rubber. Even after the war ended he continued his polymer studies, no longer supported by Benito Mussolini's fascist government but by the Montecatini Company. This huge chemical interest allowed Natta to invite Karl Ziegler down to its facilities in Milan. Natta had seen the director of the Max Planck Institute of Coal Research lecture in 1952 in Frankfurt on growth (Aufbau) reactions eliciting large molecules from the petroleum by-product ethylene. Ever the pragmatist, Natta had immediately foreseen the practical implications of this process as applied to polymer chemistry, so he secured an agreement with Ziegler to share all of their research mutually, and Montecatini secured Italian rights to Ziegler's discoveries.

However, Natta applied Ziegler's catalytic polymerization process to propylene (which was much cheaper than ethylene, Ziegler's catalyst) and neglected to inform his collaborator of his discovery: On March 11, 1954, he synthesized linear polypropylene, an even stronger and more versatile polymer than linear polyethylene, which Ziegler had just patented. Only after Natta had filed for a patent on the substance did he inform Ziegler of his discovery, which the German chemist considered bald disregard for their information-sharing agreement, and refused to speak to the Italian chemist for years thereafter.

Besides its practical applications, polypropylene also represented the discovery of coordination polymerization, Natta realized. Whereas most types of polymerization distance the catalyst from the growth of the polymer chain, coordination polymerization places a monomer directly adjacent to the solid surface of the catalyst to control the geometry of the chain reaction generating the polymer. Natta also realized upon X-ray crystallographic examination of the crystalline structure of polypropylene that the chain reaction was very ordered, creating a symmetrical, or what he coined as *isotactic*, structure.

In 1959 Natta developed Parkinson's disease, which crippled him. By the time they shared the 1963 Nobel Prize in chemistry, Natta and Ziegler had reconciled their differences well enough to share the stage. Natta retired in the early 1970s, and in the late 1970s he had surgery performed on his broken femur; complications from the operation killed him on May 2, 1979, in Bergamo, Italy.

⊠ **Nernst, Walther Hermann**
(1864–1941)
Prussian
Physical Chemist

Walther Nernst won the 1920 Nobel Prize in chemistry "in recognition of his work in thermochemistry." Nernst revolutionized this field with his 1906 assertion of the third law of thermodynamics, which demonstrated the impossibility of actually reaching absolute zero, while simultaneously allowing for the calculation of the maximum work a process is capable of generating based on its rate of entropy as it nears absolute zero. Nernst contributed other significant advances to thermochemistry, as well as to the administration of his field by helping found and directing two institutes, the Institute for Physical Chemistry and Electrochemistry at the University of Göttingen and the Institute for Experimental Physics at the University of Berlin.

Walther Hermann Nernst was born on June 25, 1864, in Briesen, West Prussia, Germany (now Wabrzezno, near Toruń, Poland). His father, Gustav Nernst, was a district judge. He attended gymnasium at Graudenz (now Grudziądz, Poland), then continued his education in traditional European fashion by attending a succession of universities, studying physics and mathematics in Zurich, Switzerland; Berlin, Germany; and Graz, Austria. At the latter institution he conducted research under Ludwig Boltzmann and Albert von Ettinghausen, before pursuing his doctorate under Friedrich Kohlrausch at Würzburg, in Germany, where he wrote his dissertation on the electromotive forces produced by magnetism in heated metal plates. He received his Ph.D. from the University of Würzburg in 1887.

Nernst then joined the distinguished faculty at Leipzig University, working under WILHELM OSTWALD in the physical chemistry department, which also included such luminaries as SVANTE AUGUST ARRHENIUS and Jacobus Van't Hoff. While there he formulated the Nernst equation in 1888, which calculated the interdependence between the chemical properties of a battery cell and its voltage, and the Nernst distribution law in 1891, which applied to the concentration of soluble substances in liquid. In 1892 he married Emma Lohmeyer, and together the couple had two sons (both of whom were killed in World War I) and three daughters.

In 1894 the Universities of Munich and Berlin offered him chairs in physics, but he instead accepted the invitation from the University of Göttingen, which offered him a chair in physical chemistry. While at Göttingen, he founded the Institute for Physical Chemistry and Electrochemistry, which he subsequently headed as its director. In 1905 the University of Berlin again wooed him, successfully this time, to fill its chair of chemistry, and later a chair of physics.

In 1906 Nernst formulated his famous third law of thermodynamics, also known as the Nernst heat theorem: As the temperature approaches absolute zero, the entropy change in a reaction between perfect crystalline substances also approaches zero. In essence this law expounded the impossibility of actually reaching absolute zero, or 0° on the Kelvin scale. Previous attempts to quantify the effects of entropy on chemical reactions had not taken temperature into account. In 1914 Nernst and Max Planck persuaded Albert Einstein to visit Berlin.

In 1918 Nernst proposed his theory of photoinduced atomic chain reaction, which held that quantum reactions could sustain themselves internally for long periods after the initial induction. In 1920, after 16 years of opposition by Arrhenius, the Royal Swedish Academy of Sciences finally recognized the significance of Nernst's thermochemical work by granting him the Nobel Prize in chemistry. In 1924 the University of Berlin founded a new Institute for Experimental Physics and named Nernst its director; he retained this position throughout the remainder of his career.

During his career Nernst published several significant texts. *Theoretische Chemie vom Standpunkte der Avogadro'schen Regel und der Thermodynamik* (*Theoretical Chemistry from the Standpoint of Avogadro's Rule and Thermodynamics*), which was first published in 1893, went through 10 editions by 1921 (the fifth English edition appeared in 1923). *Einführung in die mathematische Behandlung der Naturwissenschaften* (*Introduction to the Mathematical Study of the Natural Sciences*) similarly saw 10 editions by 1923. His monograph, *Die theoretischen und experimentellen Grundlagen des neuen Wärmesatzes*, was first published in 1918 and reprinted in 1923; an English edition, entitled *The New Heat Theorem*, appeared in 1926.

In 1930 Nernst collaborated with the Bechstein and Siemens companies to produce an electronic piano by replacing the sounding board with pickups to capture the natural acoustic sound for amplification and manipulation. Three years later he retired from his positions in Berlin. He died on November 18, 1941, in Berlin. Humboldt University in Berlin honored his memory by subdividing its chemistry department into the Walther Nernst Division and the Emil Fischer Division, housing the former in the Walther Nernst Building.

⊠ Newlands, John Alexander Reina
(1837–1898)
English
Chemist

John Alexander Reina Newlands was one of the first scientists to notice a periodicity of similarity in the chemical properties of elements, prompting him to organize the known elements at the time into a periodic table. His organizing principle rested on the fact that every eighth element (in ascending order according to atomic weight) exhibited similar characteristics, much as every eighth musical note corresponds, leading Newlands to label his theory the law of octaves. Other scientists dismissed (and even outright ridiculed) his theory; four years after he proposed his law of octaves, though, the Russian chemist DMITRI IVANOVICH MENDELEYEV proposed a similar periodic table, which gained acceptance when his predictions of the atomic weights and characteristics of then-undiscovered elements proved true. The periodic table is thus commonly attributed to Mendeleyev.

Newlands was born on November 26, 1837, in Southwark, London, England. He was the second son born to Maria Reina, who was of Italian ancestry. His father, a Presbyterian minister, schooled him at home until he entered the Royal College of Chemistry in London in 1856, where he studied for a year under August Hofmann. He then served as assistant to the chemist of the Royal Agricultural Society, J. T. Way. In 1860 he volunteered in Giuseppe Garibaldi's Italian unification campaign, then returned to Way's laboratory in London, where he remained until 1864, when he set up an independent laboratory as an analytical chemist.

Between February 7, 1863, and March 9, 1866, the journal *Chemical News* published five accounts of Newlands's developing theory of a periodic table. His first communication, in the form of a letter to the editor, established a connection between the properties of elements and their atomic weights, or "equivalents" (as was the case with many of his contemporaries, Newlands employed these two terms interchangeably, obfuscating an argument that would have benefited from more clarity in terminology). Although Newlands did not acknowledge it, the German chemist Johann Döbereiner had initi-

ated this reasoning in 1817, when he proposed his law of triads and grouped together three elements with similar chemical properties, with the atomic weight of the middle element being the mean atomic weight of the other two elements.

In subsequent communications Newlands defended and advanced his theory, pointing out the periodicity of eight in July 1864 and provisionally calling his theory the "law of octaves" a little more than a year later, in August 1865. On March 1, 1866, Newlands presented a paper entitled "The Law of Octaves, and the Causes of Numerical Relations among the Atomic Weights" to the Chemical Society. His theory met with opposition verging on ridicule when G. F. Foster, a professor of physics at University College, sarcastically inquired if he had tried to arrange the elements according to their first initials. Foster also advanced a more serious objection, censuring Newlands's separation of manganese from chromium and iron from nickel and cobalt.

Although Newlands acknowledged the likelihood that newly discovered elements would disrupt the organization of his periodic table, he did not allow for these additions in his schematic. Mendeleyev wisely left gaps in his proposed periodic table and even predicted the atomic weights and characteristics of as-yet undiscovered elements. The subsequent discovery of those elements vindicated Mendeleyev and forever associated his name with the periodic table, despite the fact that Newlands had already predicted, for example, the existence of germanium (with its atomic weight being the mean between silicon and tin).

In 1868 Newlands abandoned his independent practice in analytical chemistry to become the chief chemist at James Duncan's sugar refinery. Despite the fact that Newlands and Duncan collaborated to introduce a new system for cleaning sugar, as well as several improvements in processing techniques, foreign competition drove Duncan's company out of business. Newlands reestablished his analytical chemistry practice with his brother, B. E. R. Newlands. The brothers collaborated with C. G. W. Lock to revise his treatise on sugar growing and refining.

Despite Newlands's claim of priority over Mendeleyev in the proposal of the periodic table, the Royal Society attributed the accomplishment to the Russian chemist and awarded him its Davy Medal in 1882. Five years later, the Royal Society finally acknowledged Newlands's role in the establishment of the periodic table by awarding him its 1887 Davy Medal. Newlands died in London a little more than a decade later, on July 29, 1898.

⊠ Nobel, Alfred
(1833–1896)
Swedish
Industrial Chemist

Alfred Nobel invented dynamite and amassed a huge fortune on this and other inventions. He posthumously put this wealth to good use by endowing in his will a foundation that would award annual prizes to the people who made the greatest contributions to humankind in one of several areas: physics, chemistry, physiology or medicine, literature, and peace. Established in 1901, the Nobel prizes quickly became (and continue to be) the most prestigious awards in the world.

Alfred Bernhard Nobel was born on October 21, 1833, in Stockholm, Sweden. His father, Immanuel Nobel, was an inventor and engineer who went bankrupt when barges carrying bridge and building construction supplies were lost. While his father traveled to Finland and Russia to recoup financially, his mother, Andrietta Ahnsell, established and ran a grocery store to support her family. Nobel attended St. Jakob's Higher Apologist School in Stockholm until 1842, when his family moved to St. Petersburg, where his father had amassed some wealth sell-

ing the naval mines he invented (wooden casks filled with gunpowder submerged just beneath the water's surface) to the Russian military.

Nobel continued his education in St. Petersburg under private tutors, learning four foreign languages (Russian, French, English, and German) fluently. His father disapproved of his love of literature and poetry, instead encouraging his innate interest in chemistry and physics. In 1850 his father sent him on a two-year continental and transatlantic chemical engineering tour, studying in the United States, Sweden, Germany, and France. In Paris, his favorite city, he worked in T. J. Pelouze's chemistry laboratory. There he met the Italian chemist Ascanio Sobrero, who had invented nitroglycerine three years earlier by mixing glycerin with sulfuric and nitric acid to produce a highly volatile and explosive liquid (it produces 1,200 times its volume in gas upon explosion).

Upon his return to St. Petersburg in 1852, Nobel worked in his family's business, which provided munitions to the Russian military. At the end of the Crimean War in 1856, however, the munitions market collapsed and Nobel's father again went bankrupt, prompting him to move the family back to Sweden in 1859. There Nobel and his father worked independently on explosives studies. In 1862 the elder Nobel devised a means of producing nitroglycerine industrially, and the next year Nobel himself invented a mercury fulminate detonator. Tragedy struck in 1864, when the family's nitroglycerine factory exploded, killing Nobel's brother Emil and several others.

Undeterred, Nobel continued to experiment with nitroglycerine, though Stockholm officials forced him to conduct these studies outside city limits, on a barge anchored in Lake Mälaren. Within three years he had devised a means of stabilizing nitroglycerine by adding the silica keiselguhr, a porous powder that turned the liquid into a paste that could be shaped. In 1867 he patented this invention (in Sweden,

Britain, and the United States) as "dynamite," a product that proved extremely successful commercially. Over the course of his lifetime, he established manufacturing plants and laboratories in 90 different locations in 20 different countries.

The volatility of "guhr dynamite," as it was called, prompted Nobel to continue researching, striving to improve upon his invention. By 1875 he had created blasting gelatin, or gelignite, which added a colloidal solution of gun cotton (cotton fiber nitrated with a mixture of nitric and sulfuric acids) to nitroglycerin, making it more stable as well as more explosive. He also improved on blasting powder by mixing nitrocellulose as well as camphor and other additives with nitroglycerin to create ballistite, which was nearly smokeless, in 1887. He continued to conduct research on explosives, as well as other commercial applications, such as synthetic rubber and leather and artificial silk. For these and other inventions, he held 355 patents. He also conducted less market-oriented research in electrochemistry, optics, biology, and physiology at his laboratories in Hamburg, Germany; Ardeer, Scotland; Karlskoga, Sweden; Paris and Sevran, France; and San Remo, Italy.

When Nobel died on December 10, 1896, in San Remo, Italy, his relatives were surprised to discover that he had willed his $9 million fortune to endow a foundation for awarding annual prizes in physics, chemistry, physiology or medicine, literature, and peace to "those who, during the preceding year, shall have conferred the greatest benefit on mankind." The responsibility for establishing the Nobel Foundation fell upon the engineers Ragnar Sohlman and Rudolf Lilljequist. The first Nobel prizes were awarded in 1901. In 1905 the Nobel Peace Prize went to Countess Bertha Kinsky, a peace activist and author of the classic book *Lay Down Your Arms,* who had worked briefly as Nobel's assistant. It was she who probably planted the seed of the idea for establishing prizes that honored great

contributions to humankind. Nobel was honored in return in 1958, when element number 102 was named nobelium after him.

⊠   **Norrish, Ronald G. W.**
(1897–1978)
English
Physical Chemist

Ronald G. W. Norrish shared the 1967 Nobel Prize in chemistry with his collaborator, SIR GEORGE PORTER, for their development of the flash photolysis method of analyzing reaction kinetics by eliciting a dissociative reaction with a burst of light, then examining the return to equilibrium by shining successive flashes of light at different intervals. MANFRED EIGEN also shared the prize for his development of a similar process, the relaxation technique, which disturbed the equilibrium not with light, but chemically. Norrish is credited with the establishment of reaction kinetics as a distinct specialization in the field of physical chemistry.

Ronald George Wreyford Norrish was born on November 9, 1897, in Cambridge, England, to Amy and Herbert Norrish. He attended the Perse Grammar School before winning a scholarship in natural sciences at Emmanuel College of Cambridge University in 1915. However, his studies were interrupted by World War I, in which he served in the Royal Field Artillery in France as a lieutenant. The Germans captured him and held him for a year as a prisoner of war. In 1919 he returned from the war to Cambridge, where he completed his bachelor of science in chemistry in 1921.

Norrish remained at Cambridge to conduct doctoral studies on the photochemistry of potassium-permanganate solution reactions under E. K. Rideal. In 1924 he earned his Ph.D. in chemistry, and Cambridge's Emmanuel College named him a fellow. The next year the college appointed him a demonstrator in chemistry. In

1926 he married Anne Smith, a lecturer at the University of Wales, and the couple eventually had twin daughters. In 1928 he published a paper on the photochemistry of glyoxal in which he observed that light breaks down aldehydes and ketones into either stable molecules or free radicals (unstable molecules with unpaired electrons). These two types of reactions became known as Norrish type I and type II reactions.

By 1930 Cambridge named Norrish the Humphrey Owen Jones Lecturer in Physical Chemistry. In 1937 he ascended to a professorship of physical chemistry and the directorship of the physical chemistry department, positions that he retained through the remainder of his career. In the 1930s Norrish focused his research on the kinetics of polymerization (the synthesis of large molecules), identifying the *gel effect,* a term he coined to describe the viscosity that occurs in the last stages of free-radical polymerization.

Through the mid-1930s Norrish also studied the correlation between physical phenomena, such as phosphorescence and spectral character, and photodecomposition. Other early photochemistry studies include his collaboration with M. Ritchie, in which they investigated the reaction to light flashes of hydrogen and chlorine. In addition, he studied hydrocarbon combustion, discovering that formaldehyde is formed as a necessary intermediary step in the process of combusting methane and ethylene.

During World War II Norrish chaired the Incendiary Projectiles Committee, while conducting studies on gun-flash suppression. After the war he collaborated with George Porter, a former student, to develop the flash photolysis process of eliciting rapid dissociation of a compound into free radicals by means of a photoflash, or burst of light. They subsequently shone weaker flashes of light to illuminate the resulting chemical reaction for analysis at different intervals, thereby allowing them to study the entire kinetic process microsecond-by-microsecond. They later refined this spectrographic process until they could analyze intervals as short as a thousandth of a millionth of a second. They coauthored several papers reporting their findings, including the 1949 *Nature* article entitled "Chemical Reactions Produced by Very High Light Intensities" and the 1954 *Discussions of the Faraday Society* paper entitled "The Application of Flash Techniques to the Study of Fast Reactions."

In 1958 Norrish received both the Liversidge Medal from the Chemical Society as well as the Davy Medal from the Royal Society. In 1964 he won the Bernard Lewis Gold Medal from the Combustion Institute, and the following year he retired from active research to senior fellow status at Emmanuel College. In 1967 the Royal Swedish Academy of Sciences conferred its Nobel Prize in chemistry on Norrish, his laboratory partner George Porter, and Manfred Eigen. About a decade later he died in Cambridge, on June 7, 1978.

⊠ **Northrop, John Howard**
(1891–1987)
American
Biochemist

John Howard Northrop shared the 1946 Nobel Prize in chemistry with WENDELL MEREDITH STANLEY and JAMES BATCHELLER SUMNER for their collaborative work on enzymes, first purifying and isolating pepsin, trypsin, and chymopepsin, then proving that enzymes are in fact proteins.

Northrop was born on July 5, 1891, in Yonkers, New York. Before he was born, his father, John Isaiah Northrop, died in a fire at his laboratory at Columbia University, in New York City, where he taught zoology. His mother, Alice Belle Rich, supported the family after her husband died by returning to teaching botany at Hunter College, also in New York City. Northrop entered Columbia University in 1908, graduating in 1912 with a bachelor of science in biochemistry. He remained at Columbia to earn his master's degree in chemistry in 1913 and his Ph.D. in 1915.

Northrop spent the next year as the W. B. Cutting Traveling Fellow in Jacques Loeb's laboratory at the Rockefeller Institute for Medical Research (RIMR), commencing a career-long affiliation with what was later known as Rockefeller University. He first studied the fruit fly *Drosophila*, concluding that heat, not light or energy expenditure, determined its life span. At the conclusion of his fellowship the RIMR appointed him to its staff, but soon thereafter he joined the war effort. In recognition of his development of a fermentation process for acetone, a chemical component of explosives, the United States Army commissioned him as a captain in its Chemical Warfare Service. He spent the rest of World War I in Terre Haute, Indiana, supervising an acetone production plant. In 1917 he married Louise Walker; the couple had one son, John, who later became an oceanographer, and one daughter, Alice, who married the Nobel laureate Frederick C. Robbins.

Also in 1917 the RIMR promoted Northrop to the rank of associate and then to the rank of associate member in 1920. The year before he had commenced the line of research that would later gain him notoriety—on the digestive enzymes pepsin and trypsin. Over the next five years he also conducted simultaneous studies on the vision of the *Limulus* crab, on the chemical composition of gelatin (with Moses Kunitz), and on bacterial suspensions (with Paul De Kruif). In 1924 Northrop became a full member of the RIMR at the same time that he transferred to its animal pathology department in Princeton, New Jersey, where he focused on his enzyme studies.

In 1929 Northrop collaborated with M. L. Anson in innovating the diffusion cell, a surprisingly simple means of isolating substances. Using this method he isolated pepsin in pure crystalline form and soon thereafter did the same with trypsin and chymotrypsin. By 1931 he had

collaborated with Kunitz to test the purity of all three enzymes by the phase-rule solubility method, which determined that they were indeed proteins, as James Sumner had hypothesized. He and Kunitz, along with their colleague Roger Herriott, gathered their findings in the 1939 text *Crystalline Enzymes*.

In the 1930s Northrop shifted his focus to bacteriophages, importing his diffusion cell method first to isolate the tobacco mosaic virus (with Wendell Stanley) and then by 1938 to purify and isolate nucleoprotein, in the process proving that nucleic acid is the active agent of bacteriophages. In response he produced a crystalline antibody, the first of its kind, for diphtheria. He went on to create a similar antibody for pneumococcus after World War II.

Northrop remained at RIMR during World War II, working in conjunction with the United States Office of Scientific Research and Development (OSRD) to develop the Northrop titrator, a device for detecting the concentration of airborne mustard gas at a distance as well as a battery-operated version for use in the field. After the war he shared the 1946 Nobel Prize in chemistry with Stanley and Sumner, in recognition of their joint work in purifying and isolating enzymes, and proving they were indeed proteins. Northrop remained at RIMR until it closed its Princeton laboratories in 1949, whereupon he moved to the University of California at Berkeley. He remained a visiting professor (first of bacteriology, then of biophysics) throughout his tenure at Berkeley, however, maintaining his primary affiliation with Rockefeller. He retired to emeritus status there in 1961.

Besides the Nobel Northrop received Columbia University's 1931 Stevens prize, the 1936 Chandler Medal, the 1939 Daniel Giraud Elliot Medal, a Certificate of Merit from the U.S. government in 1948, and the 1961 Alexander Hamilton Medal. He served as the editor of the Rockefeller Institute's *Journal of General Physiology* for an amazing 62 years, from 1924 until his death. He had retired in 1959 from Berkeley to Wickenburg, Arizona, where he died on May 27, 1987.

⊠   **Nyholm, Sir Ronald Sydney**
(1917–1971)
Australian
Inorganic Chemist

Ronald Sydney Nyholm advanced the study of coordination compounds by creating complexes of transition metals in stable valence states. Capitalizing on the use of diarsine as a ligand, he reversed the instability of many coordination compounds. He continued to research coordination compounds throughout his career, later using X-ray crystallography and nuclear magnetic resonance (NMR) spectography techniques. He devoted his late career to improving education by reorganizing the science curricula in British schools. In recognition of his scientific contributions, he was knighted.

Nyholm was born on January 29, 1917, in Broken Hill, New South Wales, Australia, the fourth of six children. His father was a railroad employee who had emigrated from Finland to Australia. In 1934 he received a scholarship to study natural sciences at Sydney University, where he first encountered coordination chemistry while studying under George Burrows. Together student and teacher conducted research on the reactions between ferric chloride and simple arsines. After graduating he worked briefly as a research chemist at the Ever Ready Battery Company near Sydney.

In 1940 the Sydney Technical College hired Nyholm as a lecturer, later promoting him to senior lecturer. There he collaborated with F. P. J. Dwyer, who was researching coordination compounds of rhodium using arsines as ligands. In 1947 Nyholm received an ICI Fellowship to study at University College London (UCL) under SIR CHRISTOPHER INGOLD. When his fel-

lowship ended the university retained Nyholm as a lecturer. In 1948 he commenced using the arsenic compound diarsine (ortho-phenylendimenthylarsine, first synthesized by Chatt and Mann in 1937) as a ligand, first creating complexes with palladium chloride. Using diarsine, which he abbreviated as "diars," he proceeded to prepare stable compounds of transition metals in valence states that had previously been unstable.

In 1951 Nyholm returned to Australia, where the New South Wales University of Technology in Sydney had appointed him associate professor of inorganic chemistry. He remained in his homeland for only four years before returning to University College London as a professor of chemistry. Also in 1955 he published what became a classic paper, "The Renaissance of Inorganic Chemistry." He continued his coordination compounds research, and in an effort to elucidate a comprehensive understanding of coordination compounds, he determined their structures and properties using diverse experimental methods. He determined oxidation states by employed potentiometric titrations; he examined the charged species containing the complexes by electrical conductivity measurements; and later, utilizing X-ray crystallography and nuclear magnetic resonance spectography, he determined that the magnetic moment of coordination compounds achieves the closest connection between electronic structure, chemical structure, and stereochemistry.

In 1963 UCL promoted Nyholm to chair the Department of Chemistry. With his academic and scientific careers firmly established, he turned some of his attention to the state of education. Along these lines he presided over the Association of Science Education in 1967 and served on the Science Research Council from 1967 until 1971. Under these auspices he induced the Nuffield Foundation to establish the Science Teaching Project. He chaired the Chemistry Consultative Committee that updated the chemistry syllabi in British schools and supported the integration of physical, inorganic, and organic chemistry, stressing their complementariness over their distinctness.

Nyholm served his field extensively. He was president of the Royal Society of New South Wales in 1954. In 1966 he acted as president of the chemistry section of the British Association for the Advancement of Science and in 1968 as president of the Chemical Society of the United Kingdom. He delivered the Tilden Lecture to the Chemical Society of London in 1960, and in 1968, he served as the society's Liversidge Lecturer.

Nyholm's career was recognized with many honors: The Chemical Society of London granted him its Corday Morgan Medal and Prize in 1953, the Royal Australian Chemical Institute awarded him its H. G. Smith Medal in 1955, the Royal Society of New South Wales gave him its Royal Medal in 1959, the Italian Chemical Society honored him with its Gold Medal in 1968, and the University of Bologna presented him its Sigillum Magnum in 1969. In 1967 he was knighted. His life was cut short on December 4, 1971, when he was killed in a car accident near Cambridge.

O

Ochoa, Severo
(1905–1993)
Spanish/American
Biochemist

Severo Ochoa focused his research on the linking of nucleotide units in ribonucleic acid (RNA) and deoxyribonucleic acid (DNA). In 1955 he synthesized an artifical RNA, a discovery that led to his receipt of the 1959 Nobel Prize in physiology or medicine, which he shared with Arthur Kornberg, who had synthesized DNA.

Severo Ochoa de Albornoz was born on September 24, 1905, in Luarca, Spain, the youngest of seven children born to Carmen de Albornoz. His father, Severo Ochoa, was a lawyer and a businessman. He graduated from secondary school at the Instituto de Bachillerato de Málaga in 1921, though some sources cite this as the date of his graduation from Málaga College, where he earned his bachelor's degree.

Inspired by the biological writings of the neurologist Ramón y Cajal, Ochoa pursued medical studies at the University of Madrid, where he worked as an assistant to Juan Negrín. He spent the summer of 1927 visiting the University of Glasgow, working under D. Noel Paton. He returned to the University of Madrid to earn his medical degree with honors in 1929. The Spanish Council of Scientific Research sup-

ported his travel to the Kaiser Wilhelm Institute in Heidelberg, Germany, where he conducted research on the physiology and biochemistry of muscle under Otto Meyerhoff.

In 1931 Ochoa returned to the University of Madrid, which appointed him to a lectureship in physiology. That year he married Carmen García Cobián, and the following year he again traveled abroad, conducting enzymological research at the National Institute for Medical Research in London under H. W. Dudley. In 1934 he returned again to Madrid, where he had retained his lectureship in physiology and added a lectureship in biochemistry. He later ascended to the head of the physiology department of the Institute for Medical Research in Madrid.

In 1936 Ochoa returned to Heidelberg as Meyerhoff's guest research assistant, investigating the enzymatic processes of glycolysis and fermentation. The next year he traveled again to England as the Ray Lankester Investigator at the Plymouth Marine Biological Laboratory. In 1938 he moved to Oxford University, which had appointed him demonstrator and Nuffield Research Assistant. His studies, conducted under R. A. Peters, focused on the biological function of vitamin B_1. He also studied the enzymatic processes of biological oxidation.

Ochoa crossed the Atlantic in 1941 to reach St. Louis, Missouri, where he worked as an

Severo Ochoa, the first scientist to synthesize ribonucleic acid (RNA) *(Bernard Beker Medical Library, Washington University School of Medicine)*

and prompted a flurry of activity throughout the scientific community, including Kornberg's discovery of an enzyme that elicited the formation of a synthetic DNA. In 1956 Ochoa became a citizen of the United States of America.

Ochoa's other research centered on the metabolism of carbohydrates and fatty acids, the utilization of carbon dioxide in the biochemical pathways of the human body, oxidative phosphorylation, polynucleotide phosphorylase, the reductive carboxylation of ketoglutaric and pyruvic acids, the photochemical reduction of pyridine nucleotides in photosynthesis, and the condensing enzyme of the Krebs citric acid cycle.

Ochoa and Kornberg shared the 1959 Nobel Prize in physiology or medicine for their independent syntheses of RNA and DNA. That year Ochoa also won the Medal of the Société de Chimie Biologique and the Medal of New York University, and earlier that decade he had won the 1951 Neuberg Medal in Biochemistry. He also served as the president of the International Union of Biochemistry. In 1975 Ochoa retired from New York University and took up a position at the Roche Institute of Molecular Biology, where he remained for the rest of his career. Ochoa died on November 1, 1993, in Madrid.

instructor in pharmacology at the Washington University School of Medicine and a research associate studying enzymology in the laboratory of Carl and GERTY CORI. The next year he moved to the New York University School of Medicine, serving first as a research associate in medicine, then climbing the academic ladder to assistant professor in 1945 and professor of pharmacology in 1946. In 1954 the university appointed him to a professorship and the chair of the Department of Biochemistry.

In 1955 Ochoa used a bacterial enzyme as a catalyst to string together a series of nucleotides into a nucleic acid that simulated RNA. This was the first artificial synthesis of a nucleic acid

Onsager, Lars
(1903–1976)
Norwegian/American
Chemist

Lars Onsager was ahead of his time, proposing chemical theories that the scientific community ignored for decades. In 1931 he first asserted his theory of reversible processes; however, it was not until the late 1940s that the theory found general acceptance. Onsager finally received the acknowledgment he deserved when he won the 1968 Nobel Prize in chemistry.

Onsager was born on November 27, 1903, in Christiania (now Oslo), Norway. His mother was Ingrid Kirkeby; his father, Erling Onsager, practiced law before the Norwegian Supreme Court. His mother and private tutors home-schooled him at first, but eventually he attended the Frogner School, where he kept himself abreast of the professional chemical journals from his freshman year on. He skipped a grade to graduate a year early, in 1920; he then matriculated at the Norges Tekniski Høgskole in Trondheim to study chemical engineering. He passed his qualifying exams in 1925.

That year Onsager traveled through Europe to Zurich, Switzerland, where he reported to PETER DEBYE that the Debye-Hückel theory (a mathematical formula proposed by the Dutch chemist and his German collaborator, Erich Hückel) was flawed. What's more, he supplied the solution to the problem: Debye and Hückel had assumed that one particular ion moved through solution in a straight line, while all the other ions followed the laws of Brownian motion; Onsager simply applied Brownian motion laws to *all* of the ions, thereby arriving at the correct calculations as predicted by experimentation. Debye offered the brilliant young scientist a research assistantship at the Eidgenössische Technische Hochschule immediately.

In 1928 Onsager immigrated to the United States, where Johns Hopkins University in Baltimore, Maryland, hired him as an associate in chemistry. Onsager retained this position only one semester, though, as his lectures on introductory chemistry defied the understanding of even his best students. Charles A. Krauss, head of the chemistry department at Brown University in Providence, Rhode Island, appointed Onsager to a position more appropriate to his mathematical bent—teaching statistical mechanics. However, his students still found his explanations esoteric and inaccessible, earning his class the title of "Sadistical Mechanics" (later students dubbed his course "Advanced Norwegian I and II"). Brown retained Onsager for five

years, more on the strength of his independent research than for his pedagogical skills.

At Brown, Onsager submitted a dissertation advancing his theory of reversible processes. Although it is now highly regarded and known as Onsager's law of reciprocal relations, the theory was ignored when he first published it in 1931; it wasn't until the late 1940s that it earned the respect it deserved. He demonstrated that his theory was thermodynamically equivalent to Hermann von Helmholtz's principle of least dissipation. Onsager generalized his theory to apply to such diverse situations that it earned the informal title of the "fourth law of thermodynamics."

Onsager's appointment at Brown lapsed in 1933, but Yale University in New Haven, Connecticut, picked him up as its Sterling and Gibbs Fellow. He filled this position after spending the summer in Europe visiting the Austrian electrochemist H. Falkenhagen. Onsager and Falkenhagen's sister, Margarethe Arledter, fell in love while he was there and married on September 7, 1933. They eventually had four children—three sons (Erling Frederick, Hans Tanberg, and Christian Carl) and one daughter, Inger Marie. Onsager's promotion to assistant professor the next year required him to earn his doctorate, so he submitted an esoteric paper he had published, "Solutions to the Mathieu equation of period 4π and certain related functions." The obscure mathematical formulations completely baffled both the chemistry and physics faculties, but they trusted the assessment of the mathematics department's chair regarding the significance of Onsager's research. Onsager earned his Ph.D. in chemistry in 1935.

By the next year Onsager had solved yet another anomaly in a Debye theory: He modified his former mentor's dipole theory of dielectrics to account for all situations, even cases uncovered by Debye's theory, such as liquids with high dielectric constants. Debye's refusal to publish the paper in *Physikalische*

Zeitschrift stunned Onsager and effectively banished the correction to obscurity for the next decade until Debye finally accepted Onsager's theory. By then, 1945, World War II had ended, and Onsager had become a citizen of the United States. (Incidentally, his Norwegian citizenship prevented him from contributing to the war effort, though he surely would have proven invaluable to the Manhattan Project racing to build an atomic bomb.)

In 1949 Onsager explained the superfluidity of liquid helium through statistical analysis (Princeton's Richard Feynman arrived at the same conclusion independently two years later). In 1951 Onsager served as a Fulbright scholar conducting research at the Cavendish Laboratory in Cambridge, England, where he advanced a theory of diamagnetism of metals.

Not until 1968 did the Royal Swedish Academy of Sciences acknowledge the significance of his reciprocal relations theory by awarding him the Nobel Prize in chemistry. A few years later, in 1972, when Onsager was up for retirement to emeritus status at Yale, Onsager instead continued his biophysics and radiation chemistry investigations as a Distinguished University Professor at the Center for Theoretical Studies at the University of Miami. Four years later, on October 5, 1976, he died of an apparent heart attack.

⊠   Osborn, Mary Jane
(1927–   )
American
Biochemist

As a doctoral student in the late 1950s, Mary Jane Osborn became the first scientist to demonstrate the chemotherapeutic effects of methotrexate, which was subsequently used to combat cancers, particularly leukemia. She also conducted important research on the biosynthesis of the bacterial toxin lipopolysaccharide, which

controls immunological action. Later in her career she contributed to her field by serving on many important committees, councils, and boards, presiding over many of them.

Osborn was born Mary Jane Merton in 1927. Her parents encouraged her aspirations beyond the boundaries then limiting women; when, at age 10, she announced she wanted to become a nurse, her parents asked her if she might want to become a doctor instead. She enrolled in the University of California at Berkeley as a premed student, but ended up preferring laboratory research to clinical practice, so she majored in physiology to earn her bachelor's degree in 1948. Two years after her graduation she married Ralph K. Osborn.

Osborn pursued graduate studies in biochemistry at the University of Washington, investigating the functions of vitamins and enzymes activated by folic acid for her dissertation. In 1957 she identified the action of methotrexate, which blocks the physiological effects of folic acid; this discovery later led to the use of methotrexate in the chemotherapeutic treatment of cancers, specifically leukemia. She earned her Ph.D. in 1958.

She conducted postdoctoral research on the biosynthesis of lipopolysaccharide, a complex polysaccharide molecule found in bacteria such as salmonella, shigella, and the cholera bacillus. She continued to study the biogenesis of bacterial membranes throughout her career. In 1961 Osborn joined the New York University School of Medicine as an instructor, and the school promoted her to an assistant professorship the very next year. In 1963 she moved to a similar position as assistant professor of molecular biology at the Albert Einstein College of Medicine in the Bronx, New York. She was promoted to the rank of associate professor in 1966. In 1968 the University of Connecticut (UConn) Health Center in Farmington appointed her to a professorship of microbiology, a position she retained for the remainder of her career.

After establishing herself at UConn, Osborn devoted much of her time to her field, serving on committees and councils and in administrative positions of various professional organizations. From 1974 to 1978 she served on the advisory council of the Max Planck Institute of Immunobiology. In 1975 she served for a year as chair of the division of biological chemistry of the American Chemical Society. In 1978 she was elected to the National Academy of Sciences, a prestigious honor, and in 1980 she became head of the Department of Microbiology at the University of Connecticut Health Center.

That same year Osborn joined the National Science Board, the body that advises the National Science Foundation, and she remained on the board until 1986. She also served on the president's committee on the National Medal of Sciences from 1981 through 1982. From 1982 to 1983 she presided over the Federation of American Societies for Experimental Biology, then from 1983 to 1985 she chaired the board of scientific advisers for the Roche Institute for Molecular Biology. She was inducted into the fellowship of the American Academy of Arts and Sciences, serving on its council from 1988 through 1992. From 1992 to 1994 Osborn chaired the advisory council of the division of research grants of the National Institutes of Health, a group that she had joined in 1989.

As chair of the Committee on Space Biology and Medicine of the Space Studies Board, Osborn testified before the Subcommittee on Space and Aeronautics of the U.S. House of Representatives Committee on Science in the year 2000. Her testimony pleaded for budgetary support for continuing research on the effects of space travel on the human body and mind, for the sake of astronauts' safety. She also praised the National Aeronautical and Space Administration's research projects involving protein crystal growth in space, as well as its cell science program. She urged the representatives for continued budgetary support for hardware, such as

X-ray crystallographers, that are necessary for cutting-edge research. In the end she thanked the subcommittee for its budgeting of financial support for this and related NASA programs.

⊠ Ostwald, Wilhelm
(1853–1932)
Latvian/German
Physical Chemist

Wilhelm Ostwald is considered one of the founders of the discipline of physical chemistry, the study of the chemical dynamics of physical phenomena. He promoted this new field by publishing several defining textbooks, founding both a journal that chronicled advances in the field and a society of practitioners in the field, and organizing one of the first academic departments of physical chemistry. He also oversaw the construction of an institute of physical chemistry. For his research in the field, specifically his measurement of chemical reactions and his experiments in catalysis, he won the 1909 Nobel Prize in chemistry.

Friedrich Wilhelm Ostwald was born on September 2, 1853, in Riga, Latvia. His mother was Elisabith Leuckel; his father, Gottfried Wilhelm Ostwald, was a master cooper and a painter. His parents inculcated in him an appreciation for the arts, which he retained throughout his life, but his first passion was chemistry—at the age of 11, he made his own fireworks. He attended secondary school at Riga's *Realgymnasium*, graduating in 1872 to Dorpat University (now the State University of Tartu in Estonia) to study chemistry and physics, the two disciplines that he would fuse together in his career.

Ostwald earned his bachelor's degree in 1875 and stayed on at Dorpat University's Physics Institute as an assistant to Arthur von Oettingen and then at the Chemistry Laboratory assisting Carl Schmidt. He investigated the electrochemical and chemical dynamics of the law of mass action of water, specifically focusing on chemical affinity. He eventually collated affinity tables for 12 acids. For his master's thesis, he studied the density by volume of substances submerged in water to earn his degree in 1876. While conducting research for his doctoral dissertation on optical refraction, he took up an unpaid lectureship at the university in 1877. The next year he earned his Ph.D. In 1880 he married Helene von Reyher, and together the couple had two daughters and three sons.

In 1881 the Riga Polytechnic University appointed Ostwald as a professor of chemistry. He continued his chemical affinity investigations, measuring chemical reaction rates, specifically the velocity at which acids split esters into alcohol and organic acids. In 1885 he started writing what became *Lehrbuch der Allgemeinen Chemie* (*Textbook of General Chemistry*), published in 1887. That year, Leipzig University appointed him to the first chair of physical chemistry in Germany, thus validating the burgeoning discipline. He further established the field by founding the journal *Zeitschrift für Physikalische Chemie*, editing 100 volumes from 1887 through 1922.

Ostwald's research led to the establishment of the dilution law, which related electrolytic dissociation to conductivity. He followed this up with research into catalysis, or the process by which a chemical induces or quickens reactions between other chemicals without its being transformed. Although his catalytic theory later proved to be incorrect, his research advanced the knowledge in the field of physical chemistry.

Leipzig University charged Ostwald with the task of organizing the new Department of Physical Chemistry. He enlisted Jacobus Van't Hoff, a promising young physical chemist who went on to win one of the first Nobel Prizes. In 1889 he further defined this new field with the publication of *Grundriss der Allgemeinen Chemie* (*Outline of General Chemistry*). In 1894 he founded the Deutsche Electrochemische Gesellschaft (German Electrochemical Society), which in 1902 grew into the German Bunsen Society for Applied Physical Chemistry. In 1898 Leipzig established a physical chemistry institute, where Ostwald developed a process for converting ammonia into nitric acid by burning it in the presence of platinum three years later. In 1899 the king of Saxony made him a Geheimrat (privy councilor).

In the 20th century Ostwald expanded his studies to include natural philosophy (in 1902 he founded the journal *Annalen der Naturalphilosophie*) as well as color theory (he published several texts and founded a periodical). He also expanded his influence across the Atlantic, promoting physical chemistry in the United States by serving as the first Exchange Professor at Harvard University in 1904 to 1905. In 1906 he retired to Energie, his suburban estate, where he continued to pursue an ever-diversifying list of interests: He wrote several texts on monism, he studied the phenomenon of genius, and he supported the implementation of a global language, Esperanto (he also invented his own language, Ido).

In 1909 Ostwald won the Nobel Prize in chemistry; he used his prize money to establish The Bridge, an "international institute for the organizing of intellectual work." The organization, cofounded in Munich by Karl Wilhelm Buehrer and Adolf Saager, published its manifesto, "The Organizing of Intellectual Work," in June 1911. Ostwald died from uremia at his estate outside Leipzig on April 3, 1932.

P

⊠ **Pasteur, Louis**
(1822–1895)
French
Chemist, Microbiologist

Louis Pasteur is considered one of the most important scientists in the history of the field. His work led to the establishment of the new fields of stereochemistry and microbiology. He developed the process of pasteurization, which has become indispensable in the preparation of safe foods. Later in his career he theorized the idea of attenuating microbial diseases as a means of immunization, and he developed the first rabies vaccination. Most significant, though, he combined many of the elements of these investigations to propose the germ theory of disease, which revolutionized the practice of medicine.

Pasteur was born on December 27, 1822, in Dôle, in the Jura region of France. His mother was Jeanne-Etiennette Roqui, and his father, a tanner, was Jean-Joseph Pasteur. Pasteur had two older siblings—a brother who died within months of his birth and a sister named Jeanne-Antoinne (called Virginie)—and two younger sisters—Joséphine (who died of consumption at 25) and Emilie (who died of epilepsy at 26). The family moved to Arbois, where he attended the local primary school, though his father largely educated him at home. At the age of 16, he traveled to Paris to continue his education at the Collège Saint-Louis, but the homesick youngster soon returned to his family.

In 1839 Pasteur entered the Royal Collège of Franche-Comté at Besançon, where he fashioned himself as a portraitist (an artistic perspective informed all his lifework) while conducting general studies in *philosophie*. He earned his bachelor of letters degree at Besançon on August 29, 1840, and prepared to take his *baccalauréat* examination the next year. He failed in his first attempt, requiring him to stay at Besançon for a second year of advanced mathematics studies. He passed his second *baccalauréat* examination (with a mediocre grade in chemistry) to earn his Bachelor of Science degree in 1842. The next year he commenced doctoral studies at the École Normale Supérieure in Paris, earning his doctorate in 1847.

That year Pasteur was appointed professor of physics by the Dijon Lycée, where he commenced his first important studies, on the molecular asymmetry of tartaric acid: He discovered that the two sides of the molecule rotate plane-polarized light in opposite directions and that living microorganisms can absorb one side but not its antipode. He presented these findings to the Paris Academy of Sciences in 1848, sowing the seeds that would establish a new field of study—stereochemistry.

In 1849 the University of Strasbourg appointed Pasteur as a chemistry professor; that same year he married Marie Laurent, the university rector's daughter. The couple had five children together, three of whom passed away in childhood. In 1857 the École Normale Supérieure named Pasteur director of scientific studies, a post he relinquished a decade later to relieve himself from administrative duties. In 1863 the school created a chair for him that combined chemistry, physics, geology, and the fine arts. Also in that year Lille University appointed him dean of its new science faculty.

Pasteur conducted his next round of significant research at Lille. At the request of a distiller named Bigo, he investigated fermentation, discovering differentiation between yeast cells in good wine (which were spherical globules) and in wine that had gone bad (which were elongated molecules). He thus ascertained that different types of fermentation result from different types of yeast. Pasteur also discovered fermentation to be an anaerobic phenomenon. In an attempt to prevent wines from spoiling, he devised a means of killing yeast cells after they had caused fermentation by gently heating the wine to 55° Celsius. This process came to be known as *pasteurization* and was applied to many other foods to prevent spoilage.

Other important studies included Pasteur's conclusive disproof of the theory of spontaneous generation, as well as his identification of a parasite infecting silkworms from 1865 to 1868. He extended these studies to develop the germ theory of disease, which held that microbial germs spread disease and that sterile conditions could reduce the spread of disease. This theory proved of extreme significance to the medical field.

In 1867 Emperor Napoléon III granted Pasteur a pension sufficient to support his scientific studies without requiring him to perform bureaucratic duties. The next year he suffered a stroke that partially paralyzed him, though he recovered well enough to continue his work. Over the next two decades his work helped establish the field of microbiology; in the latter decade of this period alone, he discovered three bacteria that caused illness: staphylococcus, streptococcus, and pneumococcus. In 1881 he turned his attention to creating vaccines by heating pathogenic microorganisms to attenuate their virulence enough to trigger a response from the immune system without eliciting the disease itself. He applied this method successfully to anthrax bacili in sheep, as well as creating a vaccine for fowl to combat chicken cholera.

The crowning achievement of Pasteur's already-distinguished career came when he developed a vaccine against rabies. He started this work in 1882, and by July 6, 1885, he saved the youngster Joseph Meister from rabies. He presented his findings to the French Academy of Sciences on March 1, 1886; as a result the Pasteur Institute for rabies and infectious disease studies was approved by the French president in 1887, and the next year it opened. Pasteur remained there as director for the rest of his career, expanding the institute's influence by opening satellites internationally (the first, in Saigon, Vietnam, opened in 1891). Pasteur died in Paris on September 28, 1895.

⊠   **Pauling, Linus Carl**
(1901–1994)
American
Chemist

Linus Pauling holds the unprecedented distinction of having won two unshared Nobel prizes: the first in his primary field of study, chemistry, in 1954; the second, the 1963 Nobel Peace Prize, for his subsequent work fighting the dangers of nuclear testing, the threat of nuclear war, and the insanity of nuclear proliferation. Pauling was essentially apolitical before winning his first Nobel, focusing all of his attention on advancing the understanding of chemical bonding by applying quantum mechanics and X-ray crystallography to his studies. Later his renown as a Nobel laureate spurred him to apply his fame toward productive ends, namely, the end of war.

Linus Carl Pauling was born on February 28, 1901, in Portland, Oregon. He was the first of three children born to Lucy Isabelle ("Belle") Darling and Herman Henry William Pauling, a druggist in constant search of establishing a successful business that forever eluded him. He moved the family throughout Oregon to establish new pharmacies, first in Oswego in 1903, then in Salem in 1904, in Condon in 1905, and finally back in Portland in 1909, before he died of a perforated ulcer in 1910.

Despite the fact that Pauling failed to graduate from Washington High School in Portland in 1917, as he had not filled a compulsory American history credit, he nevertheless matriculated at Oregon State Agricultural College (now the College of Agricultural Sciences at Oregon State University). At the end of his sophomore year he was suffering financial shortcomings, but the college offered him a $100-a-month job teaching quantitative analysis so that Pauling did not have to drop out. He graduated with a degree in chemical engineering in 1922 and married his classmate Ava Helen Miller on June 17, 1923. Together the couple had four children, Linus Jr. (1925), Peter Jeffress (1931), Linda Helen (1932), and Edward Crellin (1937).

Pauling conducted doctoral studies on X-ray crystallography under Roscoe Gilley Dickinson at the California Institute of Technology (Caltech), publishing his first paper, "The Crystal Structure of Molybdenite," in the *Journal of the American Chemical Society* (*JACS*) in 1923. He published six more papers before graduating summa cum laude with a doctorate in chemistry in 1925. He then traveled through Europe on a postdoctoral fellowship, conducting research in the laboratories of several prominent scientists working on atomic and molecular structure: Arnold Sommerfeld in Munich, Germany; Niels Bohr in Copenhagen, Denmark; Erwin Schrödinger in Zurich,

Switzerland; and William Henry Bragg in London, England.

Pauling returned to Caltech in 1927, taking up an assistant professorship in theoretical chemistry. He continued his X-ray crystallographic studies, while also turning his attention to quantum mechanics at an atomic and molecular level. In 1929 Caltech promoted Pauling to an associate professorship and the next year to a full professorship. In the summer of 1930 Pauling again toured Europe, visiting Bragg and Sommerfeld again, as well as Herman Mark in Ludwigshafen am Rhein, Germany. He brought back the technical understanding necessary to conduct electron diffraction experiments, skills he taught his student L. O. Brockway, and the pair used this technique to analyze the molecular structure of more than 225 substances over the years.

The April 6, 1931, issue of the *JACS* published a classic paper by Pauling, "The Nature of the Chemical Bond: Applications of Results

Linus Carl Pauling, the only person ever to win two Nobel prizes in different disciplines, shown picketing the White House as part of a mass demonstration opposing the resumption of U.S. atmospheric nuclear tests. *(NARA, courtesy AIP Emilio Segrè Visual Archives)*

Obtained from the Quantum Mechanics and from a Theory of Paramagnetic Susceptibility to the Structure of Molecules." Pauling introduced the application of quantum mechanics to the study of molecular structure; he used X-ray and electron diffraction as well as thermochemical techniques to analyze interatomic distances and bond angles. In this particular paper (six more followed in the same vein over the next two years), Pauling described the quantum mechanics behind the electron-pair bonding of two atoms, each donating one unpaired electron to form a complete bond between them. Later in 1931 Pauling was named the most noteworthy pure scientist under 30 years of age by the American Chemical Society in awarding him its Langmuir Prize.

In 1935 Pauling published his *Introduction to Quantum Mechanics*, and the next year the Gates and Crellin Laboratories appointed him its director, a post he retained the next two decades. At the end of the 1930s he summarized his understanding of chemical bonding in his classic text, *The Nature of the Chemical Bond, and the Structure of Molecules and Crystals*. In it he distinguished between ionic bonding (in which atoms give or receive electrons) and covalent bonding (in which atoms share electrons), as well as explaining the symmetry found in most carbon compounds. The text essentially established the study of chemical bonding and earned Pauling the 1954 Nobel Prize in chemistry.

In the 1940s Pauling turned his attention to biological applications, investigating hemoglobin and the structures of amino acids and polypeptides. The former studies, performed in conjunction with his graduate student Charles Coryell, revealed that hemoglobin (the oxygen-bearing substance in our bloodstream) transforms structurally upon oxygenation. The latter studies, conducted in collaboration with Robert Corey, yielded models that revealed the structure of these proteins to be helical, or twisted around a central axis like a winding staircase; FRANCIS HARRY COMPTON CRICK and James Watson adapted this technique to construct their model of deoxyribonucleic acid (DNA). At the end of the 1940s Pauling published two textbooks: *General Chemistry* (1948) and *College Chemistry* (1950).

After winning his first Nobel Prize, Pauling used his influence as a laureate to promote the cause of peace, specifically by opposing the testing of nuclear weapons. He gathered 11,021 signatures from scientists around the world who opposed nuclear weapons testing, and he delivered this petition to the United Nations. He explained such actions in his 1958 text *No More War!* These contributions to the international antiwar movement were the basis for his receipt of the 1963 Nobel Peace Prize.

In 1964 Pauling shifted from chemical research to studying political policy at the Center for the Study of Democratic Institutions in Santa Barbara, California. After four years there he returned to science with a professorship at the University of California, San Diego, where he remained for two years. In 1969 he moved to Stanford University, in Palo Alto, California, where he taught until his compulsory retirement in 1974.

Instead of resting and relaxing, after retirement, though, Pauling helped found the Institute of Orthomolecular Medicine in Menlo Park, California (now the Linus Pauling Institute of Science and Medicine). Throughout this latter period of his career, he devoted himself to vitamin C studies, asserting its importance in preventing diseases such as cancer. At the ripe age of 93, Pauling succumbed to cancer at his Big Sur, California, ranch on August 19, 1994.

⊠   **Pennington, Mary Engle**
(1872–1952)
American
Chemist

Mary Engle Pennington was sometimes called the Ice Lady, in reference to her expertise in refrigeration and frozen foods. She designed the first refrigerated boxcars for railroad transportation of perishables, while also developing the standards governing railway refrigeration until the 1940s. Earlier in her career she had implemented standards for milk inspection that were subsequently adopted by health boards throughout the United States.

Pennington was born on October 8, 1872, in Nashville, Tennessee. Her mother was Sarah B. Molony, and her father was Henry Pennington, a label manufacturer who piqued his daughter's interest in food and botany by gardening together. She graduated from high school in 1890 and then attended the Towne Scientific School (now the School of Engineering and Applied Science) at the University of Pennsylvania in Philadelphia to study chemistry and biology. The university did not grant bachelor's degrees to women, only certificates of proficiency, which she earned in 1892; but in an ironic twist, she continued with doctoral work in chemistry (with minors in zoology and botany) and obtained her Ph.D. in 1895 through a bureaucratic loophole allowing women to earn doctorates in "extraordinary cases."

Pennington remained at the University of Pennsylvania for two years (from 1895 to 1897) on postdoctoral fellowships in chemical botany; she next spent a postdoctoral year (1897–98) at Yale University, in New Haven, Connecticut, conducting research in physiological chemistry. She then returned to Philadelphia to serve as a lecturer and director of the chemical laboratory at the Women's Medical College of Pennsylvania until 1906. Simultaneously she worked as a researcher at the University of Pennsylvania until 1901, when she established her own laboratory, the Philadelphia Chemical Laboratory, to provide bacteriological analyses for physicians.

In 1904 the Philadelphia health department hired Pennington as director of its bacteriological laboratory. She focused her efforts at this post on reducing bacterial contamination of dairy products, especially ice cream. During her three-year tenure there, she simultaneously performed some consulting work for the Bureau of Chemistry of the United States Department of Agriculture (USDA). The head of this bureau, Harvey W. Wiley, was also a family friend who encouraged her to apply for a newly created government position overseeing compliance with the Pure Food and Drug Act of 1906. Aware of the inherent sexism she would encounter, she veiled her sex by applying as "M. E. Pennington." Her qualifications earned her a position in 1907 as a bacteriological chemist, much to the consternation of the Food Research Laboratory, which did not expect a woman to fill the post.

Within a year, however, the excellence of Pennington's work revealed that her sex was immaterial to her performance; she was appointed as head of the lab. Over the next decade she spearheaded multiple advancements in the storage and transportation of perishable foods: She developed milk inspection standards that were later adopted by many health departments throughout the United States, designed egg packaging that reduced breakage, and improved methods for freezing and storing fish safely. During this period she also invented refrigerated train cars, designing insulation systems to retain the cold from ice beds. President Herbert Hoover conferred on her the Notable Service Medal for her work on railway refrigerator cars during World War I.

In 1919 Pennington left her position as a civil servant for a post in industry as the director of research and development at the American Balsa Company, an insulating materials manufacturer. She remained there only three years, though, before launching her own consulting firm in New York City in 1922. Her firm specialized in dispensing information and advice on the processing and storage of frozen foods, both at the industrial and the household level. In 1933, she coauthored the book *Eggs*.

In 1940 the American Chemical Society presented Pennington the Garvan Medal, awarded annually to an outstanding woman chemist in the United States. She served as vice president of the American Institute of Refrigeration and was the first female member of the American Society of Refrigerating Engineers. She was also the first woman inducted into the Poultry Historical Society's Hall of Fame in recognition of her development of slaughtering methods that extended the freshness of poultry. Pennington continued to work until the end of her life; she died in New York City on December 27, 1952.

⊠   **Perutz, Max Ferdinand**
(1914–   )
Austrian
Biochemist

Max Perutz helped establish the field of structural biochemistry with his elucidation of the structure of hemoglobin, the complex protein that carries oxygen from the lungs attached to blood cells for delivery to bodily tissue. He later constructed a model of the structure of hemoglobin. For this work he shared the 1962 Nobel Prize in chemistry with his collaborator, SIR JOHN COWDERY KENDREW, who elaborated the structure of myoglobin, hemoglobin's muscle-bound relative.

Max Ferdinand Perutz was born on May 19, 1914, in Vienna, Austria-Hungary. Both his mother, Dely Goldschmidt, and his father, Hugo Perutz, hailed from families that had amassed wealth by introducing mechanical spinning and weaving into Austrian textile manufacturing. He attended grammar school at the Theresianum school (named after the empress Maria Theresa), where his interest in chemistry diverted him from his parentally ordained pursuit of legal studies to enter the family business.

In 1932 Perutz entered Vienna University, where he studied organic chemistry; upon hearing of the biochemical work of Sir. F. G. Hopkins at Cambridge University during a lecture by

F. von Wessely, Perutz determined to pursue his doctoral studies there. With the financial assistance of his father, he secured a position as a research student under J. D. Bernal at Cambridge's Cavendish Laboratory in September 1936. However, the Nazi Party usurped the family business, bankrupting and expatriating his family.

The Rockefeller Foundation came to Perutz's rescue, granting him a research assistantship under Sir Lawrence Bragg from January 1, 1939, through the end of World War II, which allowed him to stay at Cambridge. This period proved to be intensely eventful for Perutz: He earned his Ph.D. from Cambridge in 1940; as an Austrian, Perutz was interned in Canada briefly during the war; and in 1942 he married Gisela Peiser. The couple eventually had two children, Vivien (born in 1944) and Robin (born in 1949). After the war ended in 1945 he received an Imperial Chemical Industries Research Fellowship. Then in October 1947 the Medical Research Council (MRC) established the Unit for Molecular Biology, consisting solely of Perutz and his colleague, Kendrew, in the Cavendish Laboratory.

Perutz's work on hemoglobin—an interest aroused by a conversation with F. Haurowitz in Prague, Czechoslovakia, in September 1937—commenced in 1938, when he collaborated with Bernal and I. Fankuchen to take X-ray crystallographic photographs of horse hemoglobin provided by G. S. Adair. Publication of this work in the journal *Nature* captured the attention of D. Keilin, a Cambridge professor of biology and parasitology, who recognized the project's joint demands of X-ray crystallography facilities (amply provided by the Cavendish Laboratory) and biochemical facilities, which Perutz and his colleagues lacked, so Keilin donated space in his Molteno Institute laboratory.

Perutz continued his hemoglobin studies while Kendrew studied its less-complex derivative, myoglobin, or muscular hemoglobin. Perutz contributed to Kendrew's work by securing a sperm whale sample from Peru, which generated larger crystals than the horse samples Kendrew was using. Also Perutz suggested a means of comparing Kendrew's myoglobin readings to those of similar molecules with heavy atoms (such as mercury) attached for differentiation, a technique that he later employed in his own hemoglobin studies. By 1953 Perutz's studies had identified the structure of hemoglobin, and within the next six years Perutz had constructed a complete model of hemoglobin, consisting of four separate polypeptide chains forming a tetrameric structure. The Royal Swedish Academy of Sciences recognized the joint accomplishments of Perutz and Kendrew by conferring the 1962 Nobel Prize in chemistry on the pair. That year the MRC founded a new Laboratory for Molecular Biology, installing Perutz at its helm. He retained this post until his 1979 retirement.

Perutz began receiving recognition for his work early in his career. The Royal Society inducted him into its fellowship in 1954 and subsequently awarded him its Copley Medal and Royal Medal. He also won the Ehrenzeichen für Wissenschaft und Kunst. The year after becoming a Nobel laureate, he became a Companion of the British Empire. In 1975 he was named a Companion of Honor, and in 1988 he was appointed to the Order of Merit. In 1997 Rockefeller University awarded him its Lewis Thomas Prize, established in 1993 to honor lucid writing in the fields of science. Perutz's 1989 collection of essays, *Is Science Necessary?* established his reputation for clear, concise explications of abstruse concepts, and the later book *I Wish I'd Made You Angry Earlier* cemented his reputation for writing excellence.

Perutz's name also graces several institutions, actively honoring his work by promoting continued scientific discovery. The Research Institute of Molecular Pathology of Vienna named the Max Perutz Library in honor of its hometown hero. The Liverpool John Moores University honored him with the naming of the Max Perutz building, a wing of the School of Biomolecular Science designed in 1970 to house nuclear studies.

⊠ Polanyi, Michael
(1891–1976)
Hungarian/English
Physical Chemist, Philosopher

Michael Polanyi, who first trained as a medical doctor, shifted his specialty to physical chemistry when he proposed his theory on the thermodynamics of adsorption in his doctoral dissertation. He continued to contribute to physical chemistry, specifically reaction kinetics, until the late 1930s, when he again shifted his focus toward philosophy and economics. However, he did not fully abandon science, as his philosophical treatises promoted the necessity of intellectual freedom in the advancement of scientific research.

Polanyi was born on March 11 or 12, 1891, in Budapest, Hungary, then part of the Austro-Hungarian Empire. His father was a businessman and engineer who lost his wealth before World War I on railroad investments. His Russian mother, Cecile, wrote a fashion column in Budapest's German-language newspaper and hosted a coterie for Hungarian literati. As a youth, Polanyi joined his brother Karl, later a noted economist, in founding the Galilei Circle, a group promoting Hungarian nationalism and independence. In 1909 Polanyi enrolled in Budapest University to study medicine, publishing his first scientific paper, "Chemistry of Hydrocephalic Liquid," the following year. He earned his medical degree in 1913.

Polanyi studied chemistry briefly under Georg Bredig at the Technische Hochschule in Karlsruhe, Germany, before serving as a medical officer in the Austro-Hungarian army in World War I. During his own hospitalization he wrote a dissertation on the thermodynamics of adsorption, advancing his theory that an attractive force exerted by solids created multiple layers of molecules of gases on the solid's surface. He published this theory in 1916, and the chemistry department of Budapest University accepted it as a dissertation, awarding him a Ph.D. in 1917.

After the Great War Polanyi returned to Karlsruhe briefly, then moved to the newly established Institute of Fiber Chemistry of the Kaiser Wilhelm Institute in Berlin, where he conducted an X-ray diffraction study of cellulose fibers. In 1923 FRITZ HABER invited him to join the Kaiser Wilhelm Institute for Physical Chemistry and Electrochemistry, where his studies included the growth of crystals on metals, measuring their shear and rupture strengths by utilizing an apparatus he developed. During his decade-long tenure there he corresponded and collaborated with the most prominent scientists of the time, including the Polish chemists Georg Bredig and Kasimir Fajans, as well as the German radiochemist OTTO HAHN. During this time he also married Magda Kemenev, a Roman Catholic woman (Polanyi was Jewish), and together they started a family.

When the National Socialist (Nazi) Party ascended to power in Germany in 1933 and enacted anti-Semitic laws, Polanyi decided to immigrate to England, where the University of Manchester created a chair in physical chemistry expressly for him. There he collaborated with Eyring in extending a line of study commenced with Haber in Berlin, using the angular momenta of the colliding particles to advance the theories of rates of association and dissociation. Existing collision theories in the field of reaction kinetics addressed only molecules with specific critical energies. Polanyi and Eyring applied quantum mechanics to extend collision theories to other molecules, plotting variations in their potential energy (based on the distance between reacting nuclei) to predict peaks that represented the activation energy of a reaction.

By the late 1930s Polanyi had begun his shift away from a strictly scientific focus, as reported by the future Nobel laureate MELVIN CALVIN, who conducted postdoctoral work under him: "Toward the end of my stay [in Manchester], in 1937, it got so it became difficult often for me to talk with [Polanyi] because he was thinking in terms of economics and philosophy, and I couldn't understand his language." Polanyi joined the Society for Freedom in Science, but it was not until a decade later that this shift away from pure science reached its fruition, when the University of Manchester created a chair in social studies for him in 1948.

In 1951 Polanyi was prevented from accepting a Rockefeller Foundation grant funding a position on the University of Chicago's Committee on Social Thought, when the U.S. State Department blocked his immigrant visa under the McCarran Act due to his involvement in the Galilei Circle as a youngster in Budapest. Ironically Polanyi's economic philosophy opposed Soviet communism on the grounds that it infringed on individual freedom. In 1958 Polanyi published what is considered his magnum opus, *Personal Knowledge*, which synthesized his philosophy of science with his philosophies of politics, economics, culture, and life. That year he also retired from Manchester to a position as senior research fellow at Merton College of Oxford University. Polanyi continued to publish philosophical treatises over the next several decades until his death in a Northampton hospital on February 22, 1976.

⊠ **Porter, Sir George**
(1920–   )
English
Chemist

Sir George Porter shared the 1967 Nobel Prize in chemistry with his mentor, RONALD G.W. NORRISH, with whom he had developed flash photolysis, and with MANFRED EIGEN, who developed a similar "relaxation" technique for analyzing very rapid chemical reactions. Whereas Eigen's method chemically disrupted equilibrium in order to study the return to equilibrium, Porter and Norrish's method did so with short bursts of light energy.

Porter was born on December 6, 1920, in Stainforth, in West Yorkshire, England. His parents were Alice Ann Roebuck and John Smith Porter. He attended Thorne Grammar School, and in 1938 he entered the University of Leeds as the Ackroyd Scholar. He studied chemistry under M. G. Evans, specializing in physical chemistry and chemical kinetics. After receiving his bachelor of science in 1941, he joined the Royal Naval Volunteer Reserve Special Branch as an officer, applying the knowledge he had gained in a special course on radio physics and electronics at Leeds. As a radar specialist his training included pulse techniques, methods that he later returned to and advanced.

After World War II Porter pursued graduate studies at Emmanuel College of Cambridge University as a postgraduate research student of Ronald G. W. Norrish. His research topic concerned free radicals generated by gaseous photochemical reactions, which he studied by flow techniques. Seeking to improve his methodology, Porter replaced flow techniques with pulse techniques. In the summer of 1947 he constructed an apparatus that elicited chemical reactions by disrupting chemical equilibrium with a very brief pulse of light energy, immediately followed by a weaker pulse of light to illuminate the reactive parts, namely, the free radicals, for study. This technique became known as flash photolysis. In 1949 he married Stella Brooke, and together the couple had two sons, John Brooke and Christopher.

In 1949 Porter earned his doctorate from Cambridge University, which retained him as a demonstrator in chemistry. By 1952 Emmanuel College had inducted him as a fellow, and Cambridge appointed him as assistant director of research in its Department of Physical Chemistry. Throughout this time he and Norrish collaborated on flash photolysis studies of combustion and of gaseous free radicals. Their partnership ended in 1954, when Porter left for an assistant director of research position at the British Rayon Research Association. He applied flash photolysis methods, recording reactions as short as 1 millisecond, to investigate dye fading and phototendering of fabrics during his brief tenure in industry.

In 1955 Porter returned to academia as a professor of physical chemistry at Sheffield University in South Yorkshire. He continued to extend his flash photolysis technique into new applications, developing a new method of stabilizing free radicals called matrix isolation. He utilized his flash photolysis technique on both animal and plant, investigating reactions between oxygen and hemoglobin in the former, while examining chlorophyll properties and photosynthesis processes in the latter. In 1963 Sheffield named him Firth Professor and appointed him to head the chemistry department.

Porter moved to the Royal Institution in London, where he was named Fullerian Professor of Chemistry (succeeding Sir Lawrence Bragg) and director of the Davy Faraday Research Laboratory in 1966. The next year he shared the 1967 Nobel Prize in chemistry with Norrish and Eigen for their contributions to the study of very fast chemical reactions. Honors continued to grace Porter thereafter: He received the Royal Society's 1971 Davy Medal, in 1972 he was knighted, and in 1978 he won both the Robertson Prize of the American National Academy of Sciences and the Rumford Medal of the Royal Society (which had inducted him into its fellowship in 1960). He also held a visiting professorship at University College London from 1967 and an honorary professorship at the University of Kent at Canterbury.

In 1990 Porter took over the chair of the Center for Photomolecular Sciences at Imperial College of London, and he continued with active research throughout the 1990s, publishing a chapter in a 1994 collection on *Femtosecond Chemistry*, a lecture he delivered in Pune, India, in 1995; a journal article titled "Chlorine—An Introduction," which appeared in a 1996 issue of *Pure & Applied Chemistry*; and a book in 1997, *Chemistry in Microtime*. He also extended his influence into other fields, using television as a medium for advancing interest in and study of the sciences.

⊠   **Pregl, Fritz**
(1869–1930)
Austro-Hungarian
Analytical Chemist

Fritz Pregl won the 1923 Nobel Prize in chemistry for his improvement of existing equipment and techniques for the study of microchemistry. His improvements reduced the size of samples necessary for experimentation, as well as the time necessary to take measurements. He also universalized the equipment, thus standardizing their readings for easier comparison of results throughout the world.

Pregl was born on September 3, 1869, in Laibach, Austria-Hungary (now Ljubljana, the capital of Slovenia). He was the only child born to Friderike Schlacker and Raimund Pregl, a bank treasurer who died when Pregl was young. When Pregl graduated from the local gymnasium in 1887, he and his mother moved to Graz, where he enrolled at the University of Graz to study medicine. He studied chemistry thoroughly under Skraup and histology and physiology under Alexander Rollett, who appointed him as an assistant in his laboratory. Pregl earned his medical degree in 1893 and remained at Graz in Rollett's lab.

After becoming a doctor, Pregl established a practice in ophthalmology. The University of Graz also appointed him an assistant lecturer in histology and physiology. In the lab he focused his research on human physiology, specifically investigating the chemistry of bile and urine. He received a university lectureship in 1899 as a result of his studies of chloric acid in bile and the relative concentrations of carbon and nitrogen in human urine.

In 1904 Pregl toured Germany, studying with Gustav von Hüfner in Tübingen, EMIL FISCHER (the 1902 Nobel laureate in chemistry) in Berlin, and WILHELM OSTWALD (the 1909 Nobel laureate in chemistry) in Leipzig. The next year he returned to Graz to work as an assistant in K. B. Hofmann's medical and chemical laboratory at the University of Graz's Medico-Chemical Institute. In 1907 he added to his duties forensic chemist for the province of Styria, which contained Graz.

While investigating albuminous bodies and bile acids, Pregl realized the inefficiency of traditional methods that required him to process several tons of raw bile, just to prepare enough material for experimentation. He therefore sought to minimize the amounts necessary for analysis. Thus his transition into analytical chemistry occurred due to his interest in improving on the existing techniques and apparatuses for conducting research on organic material.

In 1910 he became a professor of medical chemistry at the University of Innsbruck, where he continued to conduct his microanalysis studies. He improved the microbalance developed by W. H. Kuhlman, increasing its sensitivity from 10ths of milligrams to 100ths of milligrams. Similarly he minimized the size of the equipment for the microanalysis of combustion, while also adding a universal filling to the instrument's combustion tube (besides reducing the scale, Pregl sought to standardize his equipment so that results could be compared globally). These improvements reduced the necessary sample size from grams to milligrams, as well as reducing the amount of time necessary to conduct the experiment. He unveiled his improvement publicly to the German Chemical Society in Berlin in 1911 and to a scientific congress in Vienna in 1913.

That year Pregl returned as a full professor to the University of Graz, where he opened his laboratories at the Medico-Chemical Institute to scientists from around the world, visiting to learn firsthand about his microanalytical techniques and instruments. In 1916 the university named him dean of its medical faculty, and in 1920 he was promoted to vice-chancellor of the university. In 1917 he published a comprehensive text on his microanalytical techniques and instruments, *Die quantitative organische Mikroanalyse*, which subsequently went through a second edition in 1923, a third in 1930 (revised and expanded), and several posthumous editions (edited by H. Roth), until its seventh printing in 1958. It also appeared in French and English editions.

The Royal Swedish Academy of Sciences awarded Pregl its 1923 Nobel Prize in chemistry not for an original invention, but rather for his improvements of existing microanalytical instrumentation and techniques. A bachelor throughout his life, Pregl established an endowment for the Vienna Academy of Sciences to award a yearly stipend to fund research in microanalytical chemistry by an Austrian. He died in Graz after a short illness on December 13, 1930.

⊠ Prelog, Vladimir
(1906–1998)
Bosnian/Swiss
Organic Chemist

Vladimir Prelog shared the 1975 Nobel Prize in chemistry with SIR JOHN WARCUP CORNFORTH for their independent work on stereochemistry. Along with Robert Cahn and SIR CHRISTOPHER INGOLD, Prelog shared the initial of his surname in the CIP system for abbreviating stereochemical shapes and configurations that now enjoys almost universal use.

Prelog was born on July 23, 1906, in Sarajevo, Bosnia, then a province of the Austro-Hungarian Empire. His mother was Mara Cettolo; his father, Milan Prelog, was a high school and university history professor. At the age of eight, in 1914, Prelog witnessed the assassination of Archduke Franz Ferdinand, the event that set World War I in motion. The next year his parents separated and sent him to Zagreb, Croatia, where an unmarried aunt raised him while he attended gymnasium. At the age of 15, he published his first scientific paper in the journal *Chemiker Zeitung*, through the auspices of Ivan Kuria, a chemistry teacher. Prelog graduated from gymnasium in 1924.

That year Prelog matriculated at the Czech Institute of Technology in Prague, where the lecturer Rudolf Lukes mentored his chemistry studies. He wrote his dissertation on the structure of rhamnoconvolvulinoic acid, a natural sugar derivative, under the supervision of Emil Votoček, a prominent sugar chemist, to earn his doctorate in 1929. Unable to secure an academic position in the depressed economy, he worked for the next five years in the home laboratory of Gothard J. Driza, synthesizing the antimalarial drug quinine, as well as the chemicals used in tear gas and in hairdressing solutions. His quinine studies, which yielded his determination of the structure of quinine alkaloids, took on added relevance after he contracted malaria while serving a stint in the Royal Yugoslav Navy in 1932. The next year he married Kamila Vitek, who gave birth in 1949 to their one son, Jan.

In 1935 Prelog accepted a lectureship at the University of Zagreb, which saddled him with a full professor's workload but paid him only a laborer's wage. He secured a second job synthesizing chemicals (such as sulfanilamide) for the upstart pharmaceutical company Kastel. Nazi occupation in 1941 prompted Prelog to seek a way to leave the country, which materialized in the form of invitations to lecture in Germany from RICHARD KUHN and in Switzerland from LEOPOLD RUŽIČKA, who had just won the 1939

Nobel Prize in chemistry. Prelog and his wife stayed in Zurich, where he served as an assistant in the Organic Chemistry Laboratory of the Federal Institute of Technology (Eidgenössische Technische Hochschule, or ETH).

In 1942 the ETH promoted Prelog, and three years later he disproved SIR ROBERT ROBINSON's formulas for strychnine alkaloids while conducting studies on the structure and synthesis of solanine, a potato-based chemical. In 1947 he became an associate professor. He spent the years 1950 and 1951 traveling and lecturing in the United States, where he received job offers from Harvard University and the Hoffmann-LaRoche Company. The ETH retained him by creating a professorship *ad personam* for him in 1952. Then in 1957 he succeeded Ružička as head of the laboratory. Two years later he obtained his Swiss citizenship.

Prelog's most significant work focused on stereochemistry, or the study of shapes and configurations of molecules. Working in collaboration with Cahn and Ingold, he developed a system using single letters to symbolize specific stereochemical characteristics. He then went on to develop new methods of synthesizing stereoisomers. The Royal Swedish Academy of Sciences recognized his contributions to stereochemistry in awarding him the 1975 Nobel Prize in chemistry.

Throughout his tenure at the ETH Prelog worked in conjunction with the Swiss pharmaceutical conglomerate CIBA (later CIBA-Geigy), which appointed him to its board of directors from 1960 to 1978. In this capacity he helped found the Woodward Institute in Basel specifically to support the research of the Nobel laureate ROBERT BURNS WOODWARD.

Prelog retired in 1976, though he continued to conduct research on the chromatographic separation of complex chemical compounds as a "postdoctoral student" (hence the title of his 1991 autobiography, *My 132 Semesters of Chemistry Studies*). He published about 400 scientific papers in his career and lectured at more than 150 places worldwide (and more than once at many of these places). Besides winning the Nobel, he also won the Werner medal in 1945 and the Marcel Benoist award in 1965. Prelog died on January 7, 1998, in Zurich.

⊠ **Priestley, Joseph**
(1733–1804)
English
Chemist

Joseph Priestley was considered by French naturalist Georges Cuvier to be the father of modern chemistry. He was the first to prepare many gases, most notably oxygen. SIR HUMPHRY DAVY remarked of Priestley, "No single person ever discovered so many new and curious substances." Priestley also distinguished himself as an educator, writer, theologian, and political theorist. He wrote approximately 150 books, the majority of them on nonscientific topics; however, his skill as a practical experimenter earned him lasting renown for his scientific achievements.

Priestley was born on March 13, 1733, in Birstal Fieldhead, near Leeds, in Yorkshire, England. He was the oldest son of Jonas Priestley, a handloom weaver. His mother, Mary Swift, died in 1739. Thereafter he lived with his aunt, Sarah Priestley Keighley, until he was 19. In 1752 he entered the Nonconformist Academy at Daventry, where he studied to become a preacher despite an inherited speech impediment. Largely an autodidact, he skipped the first year and a half of the curriculum. In 1755 he graduated from Daventry to assume a ministry at Needham Market in Suffolk and then at Nantwich in Cheshire.

After his ministry at Nantwich proved unsuccessful, Priestley opened a school so successful that Warrington Academy invited him to serve as a tutor of languages and belles lettres in 1761. The next year he married Mary Wilkinson, daughter and sister to a triad of renowned ironsmiths (Priestley taught one of the brothers at Nantwich and Warrington). The couple eventually had three sons and one daughter. In 1764 Priestley was ordained after the University of Edinburgh granted him a Doctor of Laws.

In 1765 Priestley embarked on his scientific career at the behest of Benjamin Franklin and John Canton, who urged him to write a history of electricity. Not content to simply chronicle these events, he experimentally verified all the results he reported, discovering some new results in the process. These feats earned him induction into the fellowship of the Royal Society the following year, though the *History of Electricity* was not published until 1767.

That same year Priestley took up the ministry at Mill-Hill Chapel in Leeds. During his tenure there he produced some 30 books, among them the *Essay on the First Principles of Government* (published in 1768) that purportedly inspired Thomas Jefferson to write the Declaration of Independence and admittedly inspired Jeremy Bentham's famous formulation of "the greatest happiness of the greatest number." Priestley also produced scientific tracts on *Perspective*, on *Vision, Light and Colours*, and on the *History of Optics*.

In 1773 William Petty, or Lord Shelburne, secured Priestley's services—officially as a librarian, unofficially as an intellectual mentor. In 1774 Priestley accompanied Shelburne on a continental tour, during which Priestley met many famous scientists in Paris. That same year Priestley became the first scientist to prepare oxygen when he heated mercuric oxide. Then, using a similar method of collecting gases over mercury instead of water, he proceeded to discover a host of other gases, including ammonia, sulphur dioxide, carbon monoxide, hydrochloric acid, nitric oxide, and hydrogen sulfide. He discussed these discoveries in his three-volume work *Experiments and Observations on Different Kinds of Air*. Ironically these were not his most famous discoveries of this period; working with "fixed air" (carbon dioxide) produced by the

brewing process, he came up with the idea of dissolving this bubbling gas in water, thus discovering soda water, which became an instant hit throughout Europe.

In 1780 Priestley departed from Shelburne's services for the ministry of the New Meeting House in Birmingham, though Shelburne kept his word to provide an annuity until the scientist's death. Priestley soon ensconced himself in the Lunar Society, a group of prestigious intellectuals who supported the theologian's scientific experimentation, both financially and socially. However, his increasingly polemical theological and political statements earned him violent opposition, as both his church and his residential laboratory were destroyed by arsonists. Reluctantly Priestley immigrated to the United States in 1794.

The University of Pennsylvania offered Priestley a chair of chemistry, which he turned down in favor of settling in Northumberland, Pennsylvania, a planned community of English political exiles that never materialized as such, though Priestley remained there for the last decade of his life nonetheless. Priestley died in Northumberland on February 6, 1804. His house was transformed into a museum honoring his lifework.

⊠ **Prigogine, Ilya**
(1917–)
Russian/Belgian
Theoretical Chemist

Ilya Prigogine's citation for his 1977 Nobel Prize in chemistry characterized him as the "poet of thermodynamics," as he applied these physical laws to biological systems to demonstrate how order staves off the propensity toward chaos by sustaining states of disequilibrium. His theory of dissipative structures explains how ordered systems interact creatively with their environs, exchanging energy back and forth. His work on irreversible processes supports the "arrow of time" notion, which demonstrates how phenomena proceed forward without recourse to a past change.

Prigogine was born on January 25, 1917, in Moscow, Russia, just months before the Russian Revolution. His father, Roman Prigogine, was a chemical engineer at the Moscow Polytechnic. His mother remembered that Prigogine could read musical scores before printed words. The family emigrated from Bolshevik Russia in 1921, migrating through Germany over the next eight years, finally settling in Brussels, Belgium, in 1929.

The "poet of thermodynamics," Ilya Prigogine explored the relationship between ordered systems, such as biological systems, and the second law of thermodynamics, which holds that physical systems naturally tend toward entropy. *(Courtesy Ilya Prigogine Center for Studies in Statistical Mechanics and Complex Systems, The University of Texas at Austin)*

Prigogine graduated from secondary school in the classical section of Ixelles Athenaeum and followed his brother Alexander to the Université Libre de Bruxelles to study chemistry under Théophile De Donder, who peaked his interest in theoretical thermodynamics, and Jean Timmermans, who grounded his thermodynamic interest in experimentalism. In 1939 he earned his *licencié* in chemistry and physics. He remained at the university after earning his doctorate in chemistry in 1941, ascending to a professorship by 1947. Two years later he became a Belgian national.

In 1947 Prigogine published his *Introduction to Thermodynamics of Irreversible Processes* (translated into English in 1954), and over the next two decades he developed this theory of dissipative structures. Starting from the second law of thermodynamics, which holds that physical systems naturally tend toward entropy (disorder), he questioned how ordered systems, such as biological systems, sustained themselves in the face of a tendency toward disorder. He ascertained that ordered structures thrive best at a remove from equilibrium and that they feed on energy from their surroundings to compensate for their natural dissipation of energy (movement toward the equilibrium of entropy). He developed mathematical models to confirm his theoretical notions.

In 1959 Prigogine was appointed director of the International Solvay Institutes of Physics and Chemistry in Brussels, a position he has retained throughout his career together with his chair at the Université Libre de Bruxelles. On February 25, 1961, he married Marina Prokopowicz. He has two children, Yves Prigogine (born July 5, 1945) and Pascal Prigogine (born February 6, 1970).

From 1961 to 1966 he served as a chemistry professor at the Enrico Fermi Institute for Nuclear Studies and the Institute for the Study of Metals at the University of Chicago. During his tenure there he published *Nonequilibrium Statistical Mechanics*, in 1962. In 1967 he founded the Center for Statistical Mechanics (now the Ilya

Prigogine Center for Studies in Statistical Mechanics and Complex Systems) of the University of Texas at Austin Department of Physics, where he has served as the Regental Professor since 1977 and as the Ashbel Smith Professor of Physics and Chemical Engineering since 1984.

In 1977 Prigogine won the Nobel Prize in chemistry for his development of the theory of dissipative structures. That year he published *Self-Organization in Non-Equilibrium Systems: From Dissipative Structures to Order through Fluctuations* with G. Nicolis. He has since published a steady stream of books: *From Being to Becoming: Time and Complexity in the Physical Sciences* in 1980, *Order Out of Chaos* (with I. Stengers) in 1983, *Exploring Complexity* (with G. Nicolis) in 1989, *The End of Certainty, Time, Chaos and the New Laws of Nature* (with I. Stengers) in 1997, and *Modern Thermodynamics: From Heat Engines to Dissipative Structures* (with D. Kondepudi) in 1998.

Besides the Nobel Prize, Prigogine has received some of the world's most prestigious honors, including the 1969 Svante Arrhenius Golden Medal of the Royal Swedish Academy of Sciences; the 1976 Rumford Gold Medal of the Royal Society of London; the 1979 Descartes Medal of the Université Descartes in Paris, as well as appointment as a commander of the French Legion of Honor in 1988 and the 1991 Medaille d'Or; and the Gold and Silver Medals of Japan's Imperial Order of the Rising Sun in 1991. On July 21, 1989, the king of Belgium bestowed hereditary nobility on him in granting him the personal title of viscount. Prigogine also holds 48 honorary degrees and belongs to 63 professional organizations.

⊠ **Proust, Joseph-Louis**
(1754–1826)
French
Chemist

Joseph-Louis Proust proposed the law of constant composition, which states that compounds maintain constant proportions in their composition. His position came under attack by the chemist Claude-Louis Berthollet, who relied on theoretical arguments to advance the idea that compounds could vary widely in their composition. However, Proust's perfectionism as an analytical chemist prevailed in the end; his experimental results were simply irrefutable. Although Proust did not recognize it himself, his law of constant composition was a precursor to JOHN DALTON's atomic theory.

Proust was born on September 26, 1754, in Angers, France. He was the second son born to Rosalie Sartre and Joseph Proust, an apothecary. His godparents supervised his early education, until he was sent to the local Oratorian *collège*. After that, he apprenticed in pharmacy under his father, who planned to pass his apothecary on to Proust; however, Proust chose instead to travel to Paris around 1774 to continue his pharmaceutical education. He also studied chemistry under Hilaire-Martin Rouelle.

In 1776 the Salpêtrière hospital appointed Proust as its chief pharmacist; during this period he published his first papers. In 1778 he traveled to Spain to take up a chair in chemistry at the Real Seminario Patriótico Vascongado at Vergara, newly established by the Real Sociedad Económica Vascongada de Amigos del País, a society that itself had been established only recently, in 1765. Proust returned to France within two years, arriving back on native soil in June 1780.

The following year, in 1781, Pilâtre de Rozier established a *musée*, where Proust taught chemistry until his ineffective teaching style prompted his ouster in 1784. The next year the Spanish government invited him to return, and in 1786 he obliged, teaching for two years in Madrid before moving to Segovia in 1788 to take up a chemistry professorship at the Royal Artillery College. Over the next decade he also worked for the Spanish government, conducting geological surveys and mineralogical analyses.

On June 30, 1798, he married Anne Rose Châtelain Daubigné, a French national.

While in Segovia Proust formulated his law of definite proportions, first stating this idea in a 1794 paper, "Recherches sur le bleu de Prusse." His theory (which seemingly appeared out of nowhere, as none of his previous work hinted at it) generalized that most metals form two oxides at constant proportions (a minimum and a maximum), each of which could spawn a series of compounds. Over the next dozen years he developed his theory while defending it against Berthollet, who attacked it from a theoretical perspective, maintaining that the composition of compounds could vary widely from Proust's constants.

In April 1799 Proust returned to Madrid to conduct research in a newly established chemistry laboratory that became renowned for its well-appointed facilities. During this period he continued to conduct experiments in support of his law of constant composition. Berthollet mustered his best argument for his theoretical position in his 1803 text *Essai de statique chimique*. Proust answered in an 1804 paper, "Sur les oxidations métalliques," in which he not only exposed inconsistencies in Berthollet's argument but also asserted that Berthollet mistook oxide mixtures between Proust's minimum and maximum to be pure oxides.

In 1806 Proust returned to France, dividing his time between Craonne and his family farm in the Loire valley. Despite his success as a chemist in Spain, he lived in reduced circumstances in France. Ironically Proust could have secured his financial future had he accepted the Napoleonic government's offer of 100,000 francs in 1810 to help establish a factory for producing grape sugar (by a process that Proust had discovered in 1799). It is unclear why this project did not reach fruition.

Both Proust and his wife became ill in 1810, and she died in 1817. The year before he was elected to the Institut de France, succeeding Guyton de Morveau. In 1819 the Legion of Honor appointed him a chevalier. In 1820 Louis XVIII granted Proust a pension, at which point he returned to his birthplace of Angers to take over the family pharmacy that had been passed along to his brother, Joachim, who was then ailing. Proust died in Angers on July 5, 1826.

R

⊠ **Ramart-Lucas, Pauline**
(1880–1953)
French
Chemist

Pauline Ramart-Lucas followed in the footsteps of MARIE CURIE to become the Sorbonne's second woman professor. Throughout her career, she published more than 200 scientific papers and advised some 50 doctoral theses and graduate memoirs, particularly by women students. In recognition of her accomplishments she received the 1928 Ellen Swallow Richards Research Prize from the American Association of University Women. That same year the French Legion of Honor named her a chevalier and in the year of her death promoted her to commander.

Pauline Ramart-Lucas was born on November 22, 1880, in Paris, France. Despite her family's poverty, she determined to pursue an academic career. Toward this end she took evening classes while arranging artificial flowers in the Sorbonne district during the days; in addition, a pharmacist sparked her interest in chemistry while instructing her in English. After earning her secondary-school diploma, she matriculated at the Sorbonne, where she worked in Albin Haller's organic chemistry laboratory. While still an undergraduate, she published her first academic paper, an article cowritten with Haller and M. Bauer and entitled "Action de l'amidure de sodium sur cétones aromatiques et mixtes," which appeared in a 1908 issue of the *Bulletin de la Société Chimique de France*. She obtained her *licence,* or the equivalent of a bachelor's degree, in 1909.

Ramart-Lucas continued to work in Haller's lab, conducting research toward her doctorate. She also published independent works, such as "Déshydration du pseudobutyldiphénylcarbinols," which appeared in a 1911 edition of the *Comptes rendus l'Académie des Sciences*. After she earned her doctorate in 1913 she served in the field in World War I, working with radiology (as did Marie Curie). After the war she returned to Haller's lab, where she was appointed his laboratory manager. In 1917 she coauthored another professional paper with him, "Synthèses au moyen de l'amidure de sodium," which appeared in the *Annales de chimie*. She also wrote more independent papers, including "Transpositions moléculaires dans la série du pseudobutyl-diphénylcarbinol," which she published in 1923 in the *Comptes rendus l'Académie des Sciences*.

Up until this point Ramart-Lucas had focused her research on how dehydration transformed the molecular structure of alcohols. After 1925, when the Sorbonne promoted her to a lectureship, she focused her studies on the structure, chemical reactivity, and ultraviolet absorption spectrum of organic compounds. In 1927 she published two papers subsequently in the *Comptes rendus l'Académie des Sciences:* "Sur le mécanisme des transpositions moléculaires I" and "Sur le mécanisme des transpositions moléculaires II." The following year she began to earn recognition for her accomplishments, as the American Association of University Women awarded her its 1928 Ellen Swallow Richards Research Prize and the French Legion of Honor named her a chevalier. Perhaps her highest honor, though, came in 1930, when the University of Paris promoted her to a professorship, an extremely rare distinction for a woman at the time.

In 1931 Ramart-Lucas collaborated with M. Hoch to write a paper, "Stabilité comparée des isomers éthyléniques etnsynthèses par l'ultraviolet," which they published in the journal where so much of her work had appeared, *Comptes rendus l'Académie des Sciences.* In 1934 she copublished "Sur la structure des arylamides maloniques" in *Bulletin de la Société Chimique.* In 1936 she contributed a long summary chapter on molecular structure and absorption spectra to FRANÇOIS-AUGUSTE-VICTOR GRIGNARD's multivolume organic chemistry handbook.

In 1938 the French Legion of Honor promoted Ramart-Lucas to the rank of officer. In 1944 she diversified her contributions when the Consultative Assembly appointed her vice president of its educational section. She also served on the council of the Palais de la Découverte, France's national science museum, and the École de Physique et de Chimie. In 1953 the French Legion of Honor raised her to commander. Ramart-Lucas, who never married, died, on March 13, 1953.

⊠   **Ramsay, Sir William**
(1852–1916)
Scottish
Chemist

Sir William Ramsay discovered argon, an element that had not even been predicted; in fact,

the discovery opened a whole new column on the periodic table, which Ramsay helped to fill in with his 1898 discovery of three more new elements—krypton, neon, and xenon. As well, he discovered helium on Earth; it had previously only been identified as a constituent of the Sun. For this work he won the 1904 Nobel Prize in chemistry.

Ramsay was born on October 2, 1852, at Queen's Crescent, in Glasgow, Scotland. His mother, Catharine Robertson, hailed from a family of physicians, and his father, after whom he was named, was a civil engineer. Ramsay's paternal grandfather founded the Glasgow Chemical Society, though Ramsay himself exhibited little predilection toward chemistry as a youngster. After graduating from Glasgow Academy in 1866, at the young age of 14, he matriculated at the University of Glasgow to study liberal arts. Two years later he apprenticed to the local chemist, Robert Tatlock, at the Glasgow city analyst's lab from 1869 to 1870.

That year Ramsay traveled to Germany to conduct doctoral studies in organic chemistry under Rudolf Fittig at the University of Tübingen. After he submitted his dissertation on toluic and nitrotoluic acids to earn his doctorate in 1873, he returned to Glasgow to become a research assistant at Anderson's College (later called the Royal Technical College), conducting research on quinine- and cinchonine-related compounds.

In 1880 the newly established University College of Bristol (later renamed Bristol University) appointed Ramsay a professor of chemistry, and in 1881 he became principal of the college. In August 1881 he married Margaret Buchanan, and together they had two children. While at Bristol he collaborated with his assistant, Sydney Young, on the relationship between molecular weight and physical properties in liquids.

In 1887 Ramsay inherited Alexander Williamson's chair of chemistry at University College London. As his first order of business he reorganized the school's outdated laboratory, while also conducting research on diketones—the metallic compounds of ethylene (ethene)—and the atomic weight of boron. The Royal Society recognized the significance of these studies by inducting him into its fellowship in 1888, though none of his early studies distinguished him as a chemist who would impact the history of science.

In a September 1892 article in *Nature*, Lord Rayleigh (Robert Strutt) reported a discrepancy he had discovered between the density of airborne nitrogen as compared to chemically derived nitrogen. Ramsay and his assistant, Percy Williams, burned magnesium metal in a sample of air containing nitrogen, to form magnesium nitride. As expected, some residue of an unknown inert gas remained. Although they lost this sample in a laboratory accident, they repeated the experiment to produce another sample for spectroscopic analysis, which revealed this to be a new element. Ramsay named this inert gas *argon*, after the Greek for "lazy," *argos*.

Ramsay's career took off from this point on, as his discovery of argon opened up a whole new column of unknown elements on the periodic table. The American chemist William F. Hillebrand had discovered another inert gas upon heating uranium minerals. In 1895 Ramsay sought to identify a sample of this gas Hillebrand had generated in 1895 by heating the mineral clevite in sulfuric acid. William Crookes performed a spectrographic analysis of Ramsay's sample, revealing the substance to be helium, a gas known to exist in the Sun but until then undetected on Earth. Ramsay expounded on his discovery, relating helium to argon, in the 1896 book *The Gases of the Atmosphere*.

Ramsay then concentrated his efforts on identifying the other unknown elements, first generating a new periodic table with blanks where these new elements should fit. In 1898 Ramsay collaborated with Morris Travers to find these elements by means of fractional distillation of air. Within four months, Ramsay announced their discovery of krypton (on June 6), neon (on June 16), and xenon (on September 8).

In 1901 Ramsay collaborated with Robert Whytlaw-Gray to identify radon, a by-product of radioactive decay, and they determined its atomic weight. The next year Ramsay was knighted. In 1903 he collaborated with FREDERICK SODDY to prove that the radioactive decay of radium created helium. In 1904 the Royal Swedish Academy of Sciences awarded Ramsay the Nobel Prize in chemistry for his discovery of these previously unknown elements. Prior to the Nobel he had won the 1895 Davy Medal from the Royal Society and the 1903 August Wilhelm von Hofmann Medal from the German Chemical Society.

In 1912 Ramsay retired to High Wycombe, in Buckinghamshire, where he converted stables into a makeshift laboratory. During World War I he served on numerous committees, but his deteriorating health prevented him from contributing much chemical research. He died at his home in High Wycombe on July 23, 1916.

⊠ **Richards, Ellen Henrietta Swallow**
(1842–1911)
American
Chemist

Ellen Swallow Richards is considered the mother of euthenics, which is the study of environmental conditions, ecology, and home economics as a means of improving human well-being. She was the first woman ever to enroll at the Massachusetts Institute of Technology (MIT), where she remained the rest of her career, and the first woman in the United States to earn a Bachelor of Science in chemistry.

Ellen Henrietta Swallow was born on December 3, 1842, in Dunstable, Massachusetts. Her mother, Fanny Gould Taylor, was a teacher, as was her father, Peter Swallow. She helped her

Now recognized as the woman who founded ecology, Ellen Henrietta Swallow Richards was a 19th-century pioneer in the field of environmental and sanitary engineering. *(Courtesy The Massachusetts Institute of Technology Museum)*

parents tend their farm and keep their grocery store, and in return they educated their only daughter at home until her intellectual appetite outstripped their ability to feed it. At that point the family moved to the nearby town of Westford, home to Westford Academy, which she attended. After graduating in 1863, she taught in Littleton, Massachusetts, where she worked various other jobs (nursing, tutoring, storekeeping, cooking, and housecleaning) to save money for furthering her education.

In September 1868, at the age of 26, Swallow enrolled at Vassar, a women's college in Poughkeepsie, New York, where she studied astronomy under Maria Mitchell and applied chemistry under Charles Farrar. She completed its four-year curriculum in two years to become a member of

Vassar's first graduating class, earning her bachelor of arts in 1870. Seeking further education in chemistry, in January 1871 she became the first woman to enroll at the Massachusetts Institute of Technology (MIT) in Cambridge. MIT dubbed her studies the "Swallow experiment," considering her a kind of guinea pig for testing the educability of women. In 1873 she submitted a thesis entitled "Notes on Some Sulpharsenites and Sulphantimonites from Colorado" to become the first woman in the United States to earn a bachelor of science in chemistry.

That same year Swallow isolated a new metal, vanadium, from iron ore and composed a thesis on this research to earn a master of arts from Vassar College. She was granted an Artium Omnium Magistra, teacher of all arts, by MIT, which retained her as an untitled and unsalaried instructor and laboratory assistant. She continued her studies, although MIT had voted to disallow women from receiving doctorates; however, her achievements convinced the institute to vote, on May 11, 1876, in favor of allowing women to officially enroll at MIT. The year before she had married Robert Hallowell Richards, the head of MIT's Department of Mining Engineering, on June 4, 1875.

Ellen Swallow Richards appealed to the Women's Education Association of Boston to fund a women's laboratory at MIT, which was established in 1876. Serving as an assistant to lab director John M. Ordway, she instructed a small group of women in chemical analysis, industrial chemistry, mineralogy, and biology. MIT closed the women's laboratory in 1883 (allowing all its students to matriculate into the regular curriculum), but the next year it opened a new laboratory for the study of sanitary chemistry. As the lab's principal instructor, Richards taught air, water, and sewage analysis.

In 1881 Richards collaborated with Marion Talbot to form the Association of Collegiate Alumnae (now the American Association of University Women, or AAUW). She published her first book, *The Chemistry of Cooking and Cleaning: A Manual for House-keepers,* in 1882, following up with *Food Materials and Their Adulterations* in 1885 and *Home Sanitation* in 1887. That year she also commenced a two-year survey of Massachusetts inland waters, analyzing some 40,000 water samples. Her study raised concern over the management of the environment, exposing more human damage than had been suspected. She fused these concerns with the more practical concerns of home economics, or the management of the household, to found a discipline that she named *ecology,* though the title met with resistance (Melvill Dewey refused to include ecology in his classification system).

In Boston in 1890 Richards founded the New England Kitchen to teach women about proper food hygiene. Nine years later she organized a conference on home economics at Lake Placid, New York; nine years after that, in 1908, she helped establish the American Home Economics Association, serving as its president for its first two years. She even underwrote the publication of the association's *Journal of Home Economics.* She published almost a dozen books in the 20th century, including new editions of *Food Materials and Their Adulterations* in 1906 and *The Chemistry of Cooking and Cleaning* in 1910, as well as *Euthenics: Science of Controllable Environment* in 1910.

In 1910 Richards finally became a doctor when Smith College for women, in Northampton, Massachusetts, granted her an honorary doctorate. She died of heart disease in Jamaica Plain, Massachusetts, the following year, on March 30, 1911.

⊠ **Richards, Theodore William**
   (1868–1928)
   American
   Chemist

Theodore Richards defined anew the periodic table by redetermining atomic weight measurements for dozens of elements with greater accuracy than had ever been achieved before. He exploded the notion that all atomic weights are whole numbers, and his atomic weight measurements confirmed the existence of radioactive isotopes. For this work he received the 1914 Nobel Prize in chemistry, becoming the first American chemist to receive this honor.

Theodore William Richards was born on January 31, 1868, in Germantown, Pennsylvania, the fifth of six children. His father, William Trist Richards, was a seascape painter, and his mother, Anna Matlack Richards, was a poet and writer who homeschooled her son. In the summer of his sixth year Richards met his future mentor, the Harvard chemist Josiah Cooke, and at age 13 he began attending chemistry lectures at the University of Pennsylvania in Philadelphia. The next year, in 1882, he entered the sophomore class at Haverford College, located near Philadelphia, to study astronomy, but his poor eyesight prompted him to study a science closer to home, namely, chemistry. He graduated as class valedictorian with a baccalaureate in chemistry in 1885.

Richards matriculated at Harvard University in Cambridge, Massachusetts, to study chemistry under Josiah Cooke, and within a year he had earned a second bachelor's degree summa cum laude. He then embarked on doctoral research on the atomic weight of hydrogen, investigating the ratio of hydrogen to oxygen in water. Through painstakingly precise measurement he discovered not the expected whole-number weight ratio of 16:1 but a fractional weight of 15.869, thereby blowing a hole in the existing atomic weight theory. For this work he earned his doctorate in 1888, at the young age of 20.

Richards then embarked on a one-year traveling fellowship of Europe, visiting the laboratories of such prominent chemists as Viktor Meyer and Lord Rayleigh. Upon his return to Harvard in 1889, he took up a teaching assistantship in analytical chemistry. Within two years Harvard promoted him to instructor and then to assistant professor in 1894.

Upon Professor Josiah Clarke's death in 1895, Richards assumed his duties teaching physical chemistry. In preparation for this responsibility he toured Europe again, visiting the laboratories of the Nobel laureates WILHELM OSTWALD and WALTHER HERMANN NERNST. In 1896 he married Miriam Stuart Thayer, the daughter of Harvard Divinity School professor Joseph H. Thayer. Together they had three children: William, who studied chemistry under his father at Harvard before teaching at Princeton University; Greenough, who became an architect; and Grace, who married James B. Conant, a former student of Richards's who eventually became president of Harvard.

In 1901 the University of Göttingen in Germany offered Richards a professorship that he turned down to assume a full professorship at Harvard. The next year he dropped his analytical chemistry courses, and in 1903 he assumed the chairmanship of the chemistry department. He retained this post until 1911, after which he took up the directorship of Harvard's Wolcott Gibbs Memorial Laboratory, where he remained for the rest of his career.

Richards distinguished himself as an exceedingly precise and thorough experimentalist. He re-created the experiments performed by Jean-Servais Stas in the 1860s that had established the "standard" atomic weights, discovering several inaccuracies in Stas's findings resulting from impurities contained in his samples. In response Richards developed a method for maintaining the purity of the precipitates he worked on by sealing the samples from the contaminant's moisture and exposure. He also invented the nephelometer, a device that compared the turbidity of two solutions as a means of determining the final titration of silver nitrate.

Richards essentially revised the periodic table by determining the correct atomic weights of 25 elements, and his coworkers corrected the atomic weight measurements for 40 more elements. He also demonstrated a constancy in elements' atomic weights, despite their geographic (or cosmic) origin. In 1913 he measured a difference between the atomic weights of ordinary lead and lead resulting from the radioactive decay of uranium, thus confirming FREDERICK SODDY's assertion of the existence of radioactive isotopes, or molecules that differed from their parent element due to radioactivity. For these achievements the Royal Swedish Academy of Sciences awarded Richards the 1914 Nobel Prize in chemistry.

Richards contributed to his field in the latter part of his career by presiding over several prominent professional organizations: In 1914 he assumed the presidency of the American Chemical Society; in 1917, of the American Association for the Advancement of Science; and in 1920, of the American Academy of Arts and Sciences. He continued to work up until his death on April 2, 1928, in Cambridge, Massachusetts.

⊠ **Robinson, Sir Robert**
(1886–1975)
English
Organic Chemist

Sir Robert Robinson is considered one of the founders of modern organic chemistry, in recognition of his wide-ranging research in the field. His most significant studies focused on alkaloids, or naturally nitrogen-rich compounds, which include such substances as the opioid morphine and the poison strychnine. For his alkaloid studies he received the 1947 Nobel Prize in chemistry. A believer in pure science, he spurned the idea of conducting research with a specific pragmatic end in mind, though several of his discoveries did find commercial applications: His steroid and sex hormone research led to the development of oral contraception and treatments for female infertility.

Robinson was born on September 13, 1886, in Bufford, near Chesterfield, England. His mother was Jane Davenport, and his father, the inventor William Bradbury Robinson, ran a medical-products business that had been in the family for almost a century. When Robinson was three, the family (which included eight half siblings from his father's first marriage and eventually four younger siblings) moved to New Brampton, where Robinson attended the Fulneck School, a private institution run by the Moravian Church. In 1902 he matriculated at Manchester University to study chemistry (in preparation for entering the family business) under William H. Perkin Jr. He graduated with high honors in 1905.

Instead of entering the family business, Robinson continued his work in chemistry: He spent the next five years working in Perkin's private laboratory, while concurrently conducting doctoral research to earn his Ph.D. from Manchester University in 1912. That year he married Gertrude Maude Walsh, and together the couple had two children, a son and a daughter. He then traveled to Australia to assume the chair in organic chemistry at Sydney University. He held similar positions at the universities of Liverpool (1915–20), St. Andrews (1920–22), and Manchester, his alma mater (1922–29), before settling down for two and a half decades at Oxford University, where he inherited the Waynflete chair in chemistry from his mentor, Perkin.

Robinson commenced his organic chemistry studies investigating plant materials, specifically the components of brazilwood: brazilin, a colorless material, and brazilein, its red oxidation by-product. His dyestuff studies extended to anthocyanins, or red and blue flower pigments, and anthoxanthins, or yellow and brown pigments, finding correlations between their structure and their color. Along this same line of investigation he succeeded in synthesizing callistephin chloride in 1928.

Robinson's alkaloid studies focused on the biochemistry of this class of substances. In 1925 he elucidated the structure of morphine, an opioid with psychoactive properties. Later he elucidated the structure of strychnine, and by 1946 he had discovered a means of synthesizing this poison as well as brucine. He also devised means of synthesizing other alkaloids: papaverine, hydrastine, narcotine, and tropinone. The Royal Swedish Academy of Sciences cited these alkaloid studies in awarding him the 1947 Nobel Prize in chemistry.

During his Oxford years Robinson conducted research on steroids and sex hormones that had far-reaching implications. He discovered that certain synthetic steroids could mimic and produce the same physical effects as their naturally occurring estrogenic sex hormones, which led to the application of stilbestrol, hexestrol, and dienestrol in the contraceptive pill and female infertility treatments. Other of his scientific advances found practical applications; for example, he introduced the use of curly arrows to symbolize electron transfer.

Robinson contributed to the effort during World War II by studying the properties of antibiotic penicillin, elucidating its structure as well. He also conducted research on antimalarial drugs and explosives. After the war he turned his attention to geochemical studies, specifically addressing the question of the composition and origin of petroleum; he hypothesized that petroleum might originate from amino acids present on Earth before our life-form existed here.

Robinson served his profession by presiding over the British Association for the Advancement of Science, the Society of the Chemical Industry, and the Royal Society. This last organization honored him repeatedly, awarding him its Longstaff, Faraday, Davy, Royal, and Copley medals. The British Empire honored him with knighthood in 1939. He retired in 1955, the year after his first wife died; in 1957 he married Stearn Hillstrom. That same year he collaborated with ROBERT BURNS WOODWARD to found the organic chemistry journal *Tetrahedron*. He died in Great Missenden on February 8, 1975.

⊠ Rowland, Frank Sherwood
(1927–)
American
Atmospheric Chemist

F. Sherwood Rowland shared the 1995 Nobel Prize in chemistry with his colleague, MARIO MOLINA, as well as PAUL J. CRUTZEN, for their roles in identifying the erosion of the protective ozone layer by chlorofluorocarbons (CFCs), which the scientific community had assumed until that point to be inert in the atmosphere. Their informing the world of this phenomenon elicited an extremely swift response from the international community, and within a quarter century of their initial discoveries, CFC production ceased.

Frank Sherwood Rowland was born on June 28, 1927, in Delaware, Ohio, the middle son of three children. Rowland's family had moved there the year before, when his father, Sidney A. Rowland, was appointed chairman of the mathematics department at Ohio Wesleyan University. His mother, Margaret Drake, taught Latin. Rowland's parents encouraged his scientific curiosity and study, and he developed his interest in atmospheric science by volunteering at the local weather station, measuring temperature highs and lows as well as the amount of precipitation. He attended an accelerated public school system that promoted him through its ranks to graduate from high school at the age of 16.

Ineligible for the draft during World War II, Rowland entered college at Ohio Wesleyan University in 1943, studying year-round for two years before he enlisted in the navy to train radar operators. After the war ended (he saw no action), Rowland returned to Ohio Wesleyan, electing to finish his remaining studies in two instead of just one year. He graduated in 1948 with a bachelor's degree in chemistry, physics, and mathematics. He immediately matriculated at the University of Chicago to pursue graduate studies under some of the most prominent scientists of the day: the chemist WILLARD FRANK LIBBY, the physical chemists HAROLD UREY and Edward Teller, the inorganic chemist Henry Taube, and nuclear physicists Maria Goeppert Mayer and Enrico Fermi.

Rowland earned his master's degree in radiochemistry in 1951, then focused his dissertation on the chemical state of radioactive bromine produced in the cyclotron to earn his doctorate the following year. On June 7, 1952, he married a fellow graduate student, Joan E. Lundberg, and the couple embarked for Princeton University, in New Jersey, where Rowland had secured an instructorship in chemistry. There, their first child, Ingrid Drake, was born in the summer of 1953; a son, Jeffrey Sherwood, joined the family in the summer of 1955, born on Long Island during one of Rowland's summers conducting research in the chemistry department of the Brookhaven National Laboratory.

Rowland remained at Princeton until 1956, when the University of Kansas hired him as an assistant professor. Then in 1964 the University of California at Irvine appointed him a full professor and first chairman of its chemistry department. He chaired the department until 1970, when he stepped down to pursue a new direction in research. That same year he chanced upon that direction: After attending the International Atomic Energy Agency meeting on the environmental applications of radioactivity in Salzburg, Austria, he shared a train compartment with an Atomic Energy Commission (AEC) official who introduced Rowland to the notion of studying the atmosphere at the chemical level, an approach the scientist immediately espoused.

The AEC representative invited Rowland to his 1972 chemistry-meteorology workshop in Fort Lauderdale, Florida, the second in a series that featured the speaker James Lovelock, who suggested the utility of following the atmospheric movement of an extremely stable molecule—chlorofluorocarbon—to track wind patterns. The idea fascinated Rowland, who wondered what happened when this molecule eventually destabilized, as it must through solar photochemistry. With financial support from the AEC, Rowland commenced a study of the atmospheric chemistry of CFCs, enlisting the services of his new postdoctoral researcher, Mario Molina.

Within three months, Rowland and Molina determined that CFCs drift into the stratospheric ozone layer, some 8 to 30 miles above Earth, where ultraviolet radiation breaks down the bond, releasing an atom of chlorine. This free radical readily combines with ozone molecules, destroying their radiation-blocking properties. What's more, a single chlorine atom recombines with thousands of ozone molecules. That year the United States alone produced 400,000 tons of CFCs, most of which was released into the atmosphere. Rowland and Molina announced these initial observations in a landmark paper that appeared in the June 28, 1974, issue of the prominent British scientific journal *Nature*, creating an almost immediate sensation.

The National Academy of Sciences confirmed Rowland and Molina's findings in 1976; their testimony in legislative hearings convinced the Environmental Protection Agency (EPA) to ban production of aerosol CFCs in October 1978. They continued to study CFC ozone depletion, as did a spawning generation of atmospheric chemists who realized the dire implications of such diminution. In late 1984 Joe Farman, a scientist of the British Antarctic Survey, discovered what the CFC ozone-depletion theory inherently prophesied: a "hole" in the ozone above Antarctica, as revealed by satellite pictures. Atmospheric chemist SUSAN SOLOMON led two Antarctic expeditions, in fall 1986 and summer 1987, gathering data that confirmed the disappearance of the ozone layer above the South Pole. Rowland had collaborated with Solomon (as well as Rolando R. Garcia and Donald J. Wuebbles) on the paper that announced her hypothesis explaining the polar concentration of atmospheric destruction, "On Depletion of Antarctic Ozone," published in the June 19, 1986, edition of *Nature*.

In 1987 the industrial nations of the world adopted the Montreal Protocol to phase out CFC production before the end of the century, and 1992 amendments foreshortened this time frame. The University of California at Irvine recognized Rowland's accomplishments by naming him to endowed chairs: from 1985 to 1989 he served as the Daniel G. Aldrich Jr. Professor of Chemistry; from 1989 to 1994, as the Donald Bren Professor of Chemistry; and from 1994 on, as the Donald Bren Research Professor of Chemistry. Also since 1994 he has served as the foreign secretary of the United States National Academy of Sciences. In 1995 Rowland received global recognition for his contributions to the welfare of humankind with the awarding of the Nobel Prize in chemistry, which he shared with Molina and Paul Crutzen.

⊠   **Rutherford, Ernest, Lord**
(1871–1937)
New Zealander/English
Physicist

Ernest, Lord Rutherford is considered one of the founders of nuclear physics, as he discovered many of the key characteristics of nuclear radioactivity, including the fact that radioactivity actually transmutes elements into other elements. This discovery earned him the 1908 Nobel Prize in chemistry, though he continued to make significant discoveries concerning radioactivity thereafter. He proposed a theory of atomic structure that stood up to experimental verification and conformed to laws of quantum physics. He later induced an atomic transformation of

Lord Rutherford, who confirmed that the atom consists of a dense nucleus orbited by electrons *(E. F. Smith Collection, Rare Book & Manuscript Library, University of Pennsylvania)*

Rutherford's invention of an electromagnetic wave detector (a result of his research on the effects of electromagnetic waves on the magnetism of lead) earned him yet another scholarship, endowed by proceeds from the Great Exhibition of 1851, to study at Trinity College of Cambridge University in England. Beginning in 1895 he served as the first research student at the Cavendish Laboratory, working under the noted physicist J. J. Thomson. In 1896, only weeks after Wilhelm Röntgen had discovered X rays, Rutherford commenced experimentation on the phenomenon, discovering that it broke gas molecules into negative and positive ions. In 1897 Rutherford shifted his focus to studying radioactivity, discovered the previous year by ANTOINE-HENRI BECQUEREL. By wrapping uranium in layers of aluminum foil, he ascertained that radioactive emissions came in at least two distinct types, the less-penetrating alpha rays and the more-penetrating beta rays.

In 1898 he immigrated to Montreal, Canada, to become the second MacDonald Professor of Physics at McGill University, home of one of the best-equipped laboratories in the world. He continued his radioactivity studies, now investigating thorium, which upon decay produced a radioactive gas. He realized that the radioactivity of this by-product, which he called "emanation," decreased at a geometric rate, thus discovering what is now called a radioactive substance's *half-life,* or the length of time it takes for half of the substance to decay. In 1900 he married Mary Newton, whom he had met in New Zealand; the couple had one daughter, Eileen, who later married Ralph Fowler, one of Rutherford's lab assistants.

In 1901 Rutherford commenced a collaboration with FREDERICK SODDY in which they isolated a distinct component of thorium (with its own, much shorter half-life), which they called thorium X. They realized that thorium X decayed more rapidly because the radioactivity consumed itself up, whereas the remaining thorium, which had seemed inert upon the removal of

nitrogen into hydrogen and oxygen, the first example of artificial radioactive transmutation.

Rutherford was born on August 30, 1871, near the village of Spring Grove, on South Island, New Zealand. He was the fourth of 12 children born to the schoolteacher Martha Thompson and the wheelwright and flax farmer James Rutherford. He attended secondary school on a scholarship at Nelson College and then earned a scholarship to Canterbury College in Christchurch, New Zealand. He earned a bachelor of arts in 1892 and continued there, studying mathematics and mathematical physics, to earn a master of arts with honors in 1893 and a bachelor of science the next year.

thorium X, regained its radioactivity. By 1903 Rutherford and Soddy were able to assert that radioactivity amounts to atomic transformation, in other words, that an unstable substance emits alpha, beta, and gamma rays in the process of decaying into an element of lower atomic number. This announcement, unprecedented as it was, might have met more resistance had it not been expressed so eloquently by Rutherford.

Both Yale and Columbia Universities in the United States wooed Rutherford with generous offers, but he remained at McGill. In 1903 the Royal Society inducted him into its fellowship and the following year granted him its Rumford Medal. In 1905 he published *Radio-Activity,* which saw a second edition within a year. Then in 1906 Sir Arthur Schuster offered that he would only resign his chair of physics at Manchester University in England if Rutherford filled it, which he did the next year. In 1908 Rutherford received the Nobel Prize in chemistry for his explanation of radioactivity's atomic nature.

Rutherford did not rest on his laurels after receiving the Nobel; quite the opposite he immediately commenced on work of even greater significance, as it turned out. Previously, at McGill in 1903 he had proven alpha particles to be positively charged, and now at Manchester in 1908 he proved that alpha particles are nuclei of helium atoms. That same year he collaborated with Hans Geiger to devise a method of counting alpha particles with precision, a technique that Geiger would develop into his famous Geiger counter.

Based on alpha-particle scattering experimentation performed by Geiger and Ernest Marsden, Rutherford proved his theory that atoms consist of tiny tightly bound nuclei surrounded by orbiting electrons. In 1911 he announced his atomic theory that the atom consists mostly of empty space filled by electrons in constant motion. The experimental proof provided by Geiger and Marsden was complemented by Niels Bohr's confirmation that the theory conformed to laws of quantum physics.

In 1914 Rutherford diffracted gamma rays through a crystal to discover that they consisted of electromagnetic waves. World War I interrupted his research on radioactivity, as he contributed to the war effort by serving on the Board of Invention and Research of the British navy, developing techniques for detecting German U-boats at sea. After the war he discovered that alpha-particle bombardment split nitrogen atoms into hydrogen and oxygen nuclei; this was the first evidence of the artificial transmutation of one element into another element. Experimental verification of this theory came in 1925.

Rutherford succeeded Thomson as director of the Cavendish Laboratory in 1919. Administrative responsibilities diverted some of his attention from active research, and he also served as the president of the British Association for the Advancement of Science from 1925 through 1930. He further exerted his influence by presiding over the Academic Assistance Council to harbor Jews escaping from Nazi Germany. He did not, however, abandon his scientific research altogether: In 1934 he collaborated with Marcus Oliphant and Paul Hartbeck to produce the first instance of nuclear fusion by bombarding deuterium with deuterons to produce tritium.

Rutherford was knighted in 1914, and in 1925 King George V bestowed on him the Order of Merit, Britain's highest civilian honor. In 1931 he ascended to the title of peer, thus transforming his name from Sir Rutherford to Lord Rutherford of Nelson. Rutherford died in Cambridge on October 19, 1937, from complications after surgery on a strangulated hernia.

Ružička, Leopold
(1887–1976)
Croatian/Swiss
Chemist

Leopold Ružička discovered that the musk perfumes muscone and civetone contained carbon rings with more than eight carbon atoms, a phenomenon scientists had previously believed to be impossibly unstable. He also discovered the molecular structure of the male sex hormones testosterone and androsterone, earning him the 1939 Nobel Prize in chemistry. Ružička and his corecipient, ADOLF BUTENANDT, who discovered several female sex hormones, had to wait until after World War II to receive their laurels.

Leopold Stephen Ružička was born on September 13, 1887, in Vukovar, Croatia, then part of the Austro-Hungarian Empire. He was the first of two sons born to Ljubica Sever. His father, Stjepan Ružička, was a cooper who died when Ružička was only four years old. At that point his mother moved back to her birthplace, Osijek, where Ružička attended primary school and then gymnasium.

In 1906 Ružička matriculated at the Technische Hochschule in Karlsruhe, Germany, where he completed all his laboratory courses within a year and three-quarters. He then commenced his doctoral studies on ketenes under HERMANN STAUDINGER. After only two years of dissertation research, Ružička received his doctorate. He remained in Karlsruhe as Staudinger's assistant, conducting research on pyrethrins, byproducts of chrysanthemums toxic to insects and other cold-blooded animals that thus functioned as insecticides.

When the Eidgenössische Technische Hochschule, or ETH where Ružička had considered attending university), in Zurich, Switzerland, hired Staudinger to replace RICHARD MARTIN WILLSTÄTTER as professor in September 1912, he brought Ružička with him as an assistant. Also in 1912 Ružička married Anna Housmann. By 1916 he had established himself independent of Staudinger, with financial support from the German perfume company Haarman & Reimer of Holzminden. Ružička conducted research on the Tiemann formula for drone, resulting in the synthesis of fenchone; he also extended the so-called Wagner rearrange-ment. In 1917 he became a citizen of Switzerland.

The next year, in 1918, he completed his habilitation work to become a *privatdozent*, a lecturing position that carried no salary. At that time it was understood that industry supported the work of academics, which in return conducted research of potential benefit to industry, So in 1918 the firm of Ciba, Basle, took over funding of his research into the preparation of compounds related to quinine. This line of investigation yielded the first synthesis of beta-collidine and linalool, as well as a partial synthesis of pinene.

In 1921 Ružička commenced work under funding from Chuit, Naef & Firmenich, a Geneva-based perfume company. Ružička's team accomplished much under this contract: It synthesized nerolidol and fernesol, established the structure of jasmone, and corrected the formula for Tiemann's irone. Most significant, the team identified the structures of muscone (derived from male musk deer) and civetone (from male and female civets). This represented a major breakthrough in organic chemistry, as the discovery of substances with 15 and 17 carbon rings, respectively, disproved the supposition that carbon rings larger than eight would be too unstable. The team proceeded to prepare all of the alicyclic ketones, with up to 30 carbon rings.

In October 1926 Ružička moved from Switzerland to the Netherlands to take up a chair in organic chemistry at the University of Utrecht; however, he only remained there three years. The ETH invited him back to Zurich to take up the directorship in 1929; he readily accepted the offer. The next year the Ciba perfume company reinstituted its support for Ružička's research. His team proceeded to discover the molecular structure of the male sex hormones testosterone and androsterone, which it then synthesized. In 1937 the U.S. Rockefeller Foundation added its financial support to Ružička's research, allowing him free reign to direct that research in the direction he chose. He decided to investigate tripertenes and steroids.

In 1939 Ružička shared the Nobel Prize in chemistry with Adolf Butenandt. During World War II he founded the Swiss-Yugoslav Relief Society and also aided in the flight of several Jewish scientists from Nazi persecution, including VLADIMIR PRELOG, who succeeded Ružička as a Nobel laureate and as the director of the ETH upon his 1957 retirement. In 1950 Ružička divorced his first wife, and the next year he married Gertrud Acklin. He died in Zurich on September 26, 1976.

S

Sabatier, Paul
(1854–1941)
French
Chemist

Paul Sabatier shared the 1912 Nobel Prize in chemistry with FRANÇOIS-AUGUSTE-VICTOR GRIGNARD. Sabatier's work established the catalytic hydrogenation of gaseous hydrocarbons. He went on to assert a theory on the action of catalysts called "chemisorption," or the formation of unstable intermediate compounds between the catalyst and the reactants. This theory transformed the study of organic chemistry.

Sabatier was born on November 5, 1854, in Carcassone, France. His uncle taught at the school he attended, and his older sister tutored him in Latin and mathematics. He transferred to the Toulouse Lycée when his uncle moved there. After graduating he matriculated in 1874 at the École Normale Supérieur (turning down acceptance into the equally prestigious École Polytechnique). He completed his *baccalauréat* degree in three years, graduating at the top of his class in 1877.

After teaching physics at the lycée in Nîmes for a year, Sabatier returned to Paris to conduct doctoral studies at the Collège de France, serving concurrently as an assistant to Pierre Berthelot. He wrote his dissertation on metallic sulfides to earn his doctorate in 1880. He spent the next year conducting research and teaching physics at the Faculté des Sciences in Bordeaux, then transferred to the Faculté des Sciences in Toulouse as an assistant professor of physics in 1882. The next year he added chemistry instruction to his duties. In 1884, when he turned 30, the minimum age required for full professorships in France, the Faculté des Sciences appointed him a full professor of chemistry, a position he retained for the remainder of his career.

Although Sabatier is best known for his research in organic chemistry, he focused his research in his early career on physics and inorganic chemistry. In the latter field he achieved many firsts: He was the first to prepare pure dihydrogen disulfide, the first to make two monosulfides (silicon monosulfide and tetraboron monosulfides), and the first to make selenides of boron and silicon. Working with chromates and dichromates, he correlated their color with their acidity. Other chromate studies focused on hydration, or the addition of water, a technique he also applied to metal chlorides as well as copper compounds. Finally, he distinguished himself in absorption spectroscopy to study chemical reactions.

In 1895 Sabatier inched closer to organic chemistry (the study of carbon-containing substances) by mimicking Ludwig Mond's work on the effects of metal catalysts on nickel carbonyl; working with his assistant, Abbé Jean-Baptiste Senderens, Sabatier similarly passed oxides of nitrogen over finely divided metals to elicit similar results. Then in 1897 he shifted into organic chemistry wholeheartedly by following up on the experiments that FERDINAND-FRÉDÉRIC-HENRI MOISSAN and François Moreau had just abandoned, replacing their acetylene with another organic compound, ethylene. Sabatier and Senderens similarly used powdered nickel as a catalyst, but their choice of ethylene generated ethane. The pair subsequently mixed hydrogen into their ethylene, resulting in hydrogenation and the conversion of benzene vapor into cyclohexane. This represented the first of the work for which he received the 1912 Nobel Prize in chemistry.

Sabatier spent the next 32 years of his career occupied with the catalytic hydrogenation of gaseous hydrocarbons. Continuing to use finely divided nickel as a catalyst, he succeeded in synthesizing methane by the hydrogenation of carbon monoxide. Using the same catalysts at higher temperatures resulted in dehydrogenation, Sabatier discovered, allowing him to transform primary alcohols into aldehydes and secondary alcohols into ketones. He then switched to oxide catalysts, such as manganese oxide, silica, and alumina.

Sabatier contributed more than just teaching and research to the science department at Toulouse University: In 1905 he became its dean, expanding the program to include schools of chemistry, agriculture, and electrical engineering. Sabatier also received more honors than just the Nobel Prize; the Royal Society granted him its 1915 Davy Medal, and the Franklin Institute in Philadelphia awarded him its Franklin Medal in 1933. His home country named him a chevalier in the Légion d'Honneur in 1907, promoting him to commander in 1922. France's Académie des Sciences elected him its first member who did not reside in Paris, having established a special category in 1913 to accom-

modate his inclusion in the honorary society without forcing him to move from Toulouse.

Sabatier continued to lecture actively for a decade after his official retirement in 1929; however, in 1939 failing health forced him to quit even that. On August 14, 1941, he died in Toulouse. He had four daughters with his wife.

⊠ **Sanger, Frederick**
(1918–)
English
Biochemist

Frederick Sanger received the Nobel Prize in chemistry not once, but twice—first in 1958 for his determination of the structure of insulin and then in 1980 for his development of a technique for rapidly sequencing deoxyribonucleic acid (DNA) by fragmentation. He was only the fourth Nobel laureate to receive the prize twice. He used his notoriety later in his career to advance the Human Genome Project by founding the Sanger Centre in 1993 to house research on the project.

Sanger was born on August 13, 1918, in Rendcombe, in Gloucestershire, England. His mother was Cicely Sanger, and his father, after whom he was named, was a physician. He received his early education at the Bryanston School, then matriculated at St. John's College of Cambridge University. He earned his bachelor of arts in natural sciences in 1939. The next year he married Margaret Joan Howe, and the couple had three children, two sons and a daughter. Also in 1940 he commenced his pursuit of doctoral studies, conducting research on the metabolism of the amino acid lysine under Albert Neuberger. Cambridge granted him his Ph.D. in 1943.

At the instigation of Charles Chibnall, who took up Cambridge's chair of biochemistry in 1943, Sanger conducted postdoctoral research on insulin with free amino groups in an attempt to identify and quantify the amino acid content of insulin. Chibnall had identified a significant discrepancy between the expected and the actual number of free amino groups, and Sanger developed a new chromatographic method for identifying groups of amino acids attached to the end of insulin's protein chains. He received support for this research from a Beit Memorial Fellowship for Medical Research (a grant he held from 1944 to 1951).

Sanger published his end-chain results in 1945, and a decade later he published the complete sequence of insulin, a major scientific advancement allowing for the synthesis of insulin. This breakthrough promised to improve the health of diabetics, whose bodies could not produce enough insulin to sustain their needs. He received the 1958 Nobel Prize in chemistry "for his work on the structure of proteins, especially that of insulin," according to the Royal Swedish Academy of Sciences.

In 1951 the Medical Research Council (MRC) hired Sanger, and in 1962 it installed him at the helm of its division of protein chemistry in the newly constructed Laboratory of Molecular Biology. Inspired by his lab colleagues, the likes of FRANCIS HARRY COMPTON CRICK and John Smith, Sanger applied his analytical expertise to nucleic acids, starting with ribonucleic acid (RNA) studies. In collaboration with G. G. Brownlee and B. G. Barrell he developed a rapid method for fragmenting RNA by separating it into small pieces, then overlapping the pieces in sequence.

Sanger tried to apply this same method to DNA with little success until he developed the dideoxy method, which "tagged" fragments of DNA with dideoxy triphosphate. This allowed Sanger to identify the composition of DNA fragments at different points in the strands, which he read on an autoradiograph connected to a gel containing the DNA. As it turned out, this became the most efficient means of analyzing the chemical structure of DNA.

In 1977 Sanger identified the nucleotide sequence in the DNA of the bacteriophage phi-X

174, an accomplishment that represented the first determination of an organism's entire nucleotide sequence. In 1980 Sanger won his second Nobel Prize in chemistry (which he shared with WALTER GILBERT and PAUL BERG) "for their contributions concerning the determination of the base sequences in nucleic acids," according to the Royal Swedish Academy of Sciences.

Throughout his career Sanger received numerous honors in recognition of the significance of his work. In 1951 he won the Corday-Morgan Medal and Prize of the Chemical Society. The Royal Society inducted him into its fellowship in 1954, and that same year he became a fellow of King's College, Cambridge. He became a Companion of the British Empire in 1963. In 1981 he became a Companion of Honor, and in 1986 the Order of Merit was conferred upon him.

Sanger continued to advance his field late into his career. In 1993 he founded the Sanger Centre to support work on the Human Genome Project. Funded in part by the Wellcome Trust and the MRC, the Sanger Centre is housed at the Wellcome Trust Genome Campus in Cambridge, England.

⊠ **Saruhashi, Katsuko**
(1920–   )
Japanese
Geologist, Chemist

Katsuko Saruhashi studied carbon dioxide levels in seawater long before people began to suspect that this gas might increase temperatures on Earth. She also tracked the spread of radioactive debris from atomic bomb tests.

Saruhashi was born in Tokyo, Japan, on March 22, 1920. While she was a student at Toho University, from which she graduated in 1943, she met government meteorologist Yasuo Miyake, who became her friend and mentor. After World War II ended, he hired her as a research assistant in his new Geochemical Research Laboratory, part of the Japanese Transport Ministry's Meteorological Research Institute.

Around 1950 Miyake suggested that Saruhashi measure the concentration of carbon dioxide (CO_2) in seawater. "Now everyone is concerned about carbon dioxide, but at the time nobody was," Saruhashi explains. Indeed, she had to design most of her own techniques for measuring the gas. Her project earned her a doctor of science from the University of Tokyo in 1957, the first doctorate in chemistry awarded by the university to a woman.

In the early 1950s the United States, the Soviet Union, and several other nations tested nuclear bombs at remote sites, filling the air with radioactive debris. Concern about such fallout led the Japanese government to ask Miyake's laboratory in 1954 to measure the amount of radioactive material reaching Japan in rain and the amount found in seawater off the country's coast. Miyake put Saruhashi in charge of this project, which, she says, was the first of its kind. She found that fallout from a U.S. bomb test on the Pacific island of Bikini reached Japan in seawater a year and a half after the test. She later measured fallout in other parts of the world as well. Evidence gathered by Saruhashi and others helped protesters persuade the United States and the Soviet Union to stop aboveground testing of nuclear weapons in 1963.

Saruhashi also continued her measurements of carbon dioxide in seawater, finding that water in the Pacific releases about twice as much CO_2 into the atmosphere as it absorbs from the air. (There is about 60 times more CO_2 dissolved in seawater than in air.) This result suggests that the ocean is unlikely to reduce global warming by absorbing excess CO_2.

Saruhashi was made director of the Geochemical Research Laboratory in 1979. She retired from this post a year later. In 1990, after Miyake's death, she became executive director of the Geochemistry Research Association in Tokyo,

which Miyake had founded in 1972. Among the many honors she has received are election to the Science Council of Japan in 1980, becoming its first woman member, the Miyake Prize for geochemistry in 1985, and the Tanaka Prize from the Society of Sea Water Sciences in 1993.

When Saruhashi retired from the directorship of the geochemical laboratory, in 1980, her coworkers gave her a gift of 5 million yen (about $50,000). She used the money to begin a fund to establish the Saruhashi Prize, given each year since 1981 to a Japanese woman who makes important contributions to the natural sciences. Its first recipient was the population geneticist Tomoko Ohta. Saruhashi says the prize "highlight[s] the capabilities of women scientists. ... Each winner has been not only a successful researcher but ... a wonderful human being as well."

⊠ Seaborg, Glenn Theodore
(1912–1999)
American
Physical Chemist

Glenn T. Seaborg discovered 10 atomic elements, all of them transuranium elements with atomic numbers higher than 92. He also proposed the actinide concept, which reconfigured the periodic table to create a new column to accommodate these new elements. Seaborg shared the 1951 Nobel Prize in chemistry with his collaborator, EDWIN M. MCMILLAN, for their contributions to nuclear science. At the time of his death in 1999, he held Guinness's record for the longest entry in *Who's Who in America*, testament to his broad significance in society.

Glenn Theodore Seaborg was born on April 19, 1912, in Ishpeming, Michigan. His parents, Selma O. Erickson and machinist Herman Theodore Seaborg, had emigrated from Sweden in 1904; Seaborg spoke their mother tongue before he learned English. In search of better educational opportunities, the family moved to the culturally diverse Watts neighborhood of Los Angeles, California, when Seaborg was 10. He graduated from high school as valedictorian of the 1929 class and enrolled in the University of California at Los Angeles (UCLA) to study literature, though he soon changed to a science major. He supported his studies working as a stevedore and a farm laborer and graduated with a degree in chemistry in 1934.

Seaborg traveled north to the Berkeley campus of the University of California to pursue graduate studies, studying chemistry under GILBERT NEWTON LEWIS and physics under Ernest Orlando Lawrence. Working nights to access Lawrence's 27-inch cyclotron, Seaborg conducted studies on the interaction of "fast" neutrons with lead for his dissertation. He concurrently collaborated with Jack Livingood, using the cyclotron to chemically separate radioactive isotopes from their parent elements. The team discovered iodine 131, iron 59, cobalt 60, and technetium 99m. After earning his doctorate in 1937 Seaborg remained in Lawrence's laboratory as an assistant for the next two years, after which Berkeley appointed him an instructor.

In 1941 Seaborg collaborated with Joseph W. Kennedy, Arthur C. Wohl, and McMillan (who had only just discovered neptunium, the first transuranium element) to bombard this 93-element with deuterons (nuclei of the hydrogen isotope deuterium), using Lawrence's newly constructed 60-inch cyclotron. The team thereby discovered the element with the 94th atomic number, plutonium 238 (named after the ninth planet, as neptunium had been named for the eighth), as well as the isotope plutonium 239. The next year Seaborg collaborated with John W. Gofman and Raymond W. Stoughton to isolate the isotope uranium 233.

Seaborg's discovery of these radioactive elements and isotopes placed him at the forefront of the race against the Nazis to construct an atomic bomb. He joined the Manhattan Project on April 19, 1942, leading a group at the University of Chicago's Metallurgical Laboratory that included B. B. Cunningham and L. B. Werner, charged with the task of extracting the isotope plutonium 239 from uranium to fuel nuclear fission. Realizing the unthinkable destructiveness of the atomic bomb, Seaborg joined six other scientists in signing the Franck Report, which in vain recommended inviting the Japanese to view a demonstration of a nuclear explosion. In the midst of such significant events Seaborg married Lawrence's secretary, Helen L. Griggs, on June 6, 1942. The couple eventually had six children—Peter Glenn (who died in 1997), Lynne, David, Steve, Eric, and Dianne.

In 1944 Seaborg posited the actinide concept, proposing that the elements above the atomic weight of 89 belonged in their own column on the periodic table, a revolutionary reconfiguration. After returning from the war effort to Berkeley as a professor and associate director of the Radiation Laboratory (the precursor of the Lawrence Berkeley National Laboratory), Seaborg set about to fill in the periodic table with the transuranium elements whose properties he could now predict according to their position on the new periodic table. Over the next decade and a half he and his colleagues discovered elements 95 through 102—americium, curium, berkelium, californium, einsteinium, fermium, mendelevium, and nobelium, in ascending order. In the midst of this research the Royal Swedish Academy of Sciences awarded Seaborg and McMillan the 1951 Nobel Prize in chemistry.

Seaborg's career broadened its scope after he discovered the majority of the transuranium elements by the late 1950s. In 1958 the University of California at Berkeley appointed him its second vice chancellor, and over the next three years he orchestrated the addition of the College of Environmental Design and the Space Sciences Laboratory. After stepping down from this position in 1961 he tried to exert his influence to bar nuclear testing in the atmosphere, water, and space by promoting the Limited Nuclear Test Ban Treaty. Although this initiative failed to achieve its ultimate goal, it identified Seaborg as a prominent spokesperson for the moral responsibility of the human race to use nuclear power wisely, prompting President John F. Kennedy to appoint him chair of the Atomic Energy Commission. Near the end of his decade-long tenure in this position, he worked on the Non-Proliferation Treaty of 1970.

In 1971 Seaborg returned to Berkeley as University Professor of Chemistry. He continued to contribute to his field and beyond: He helped found the International Organization for Chemical Sciences in Development, which he presided over in 1981; that same year he published *Kennedy, Khrushchev and the Test Ban*; and in 1983 he sat on President Ronald Reagan's National Commission on Excellence in Education, coauthoring the critical report *A Nation at Risk*. In 1991 President George H. W. Bush awarded Seaborg with the nation's highest honor, the National Medal of Science.

In August 1997 element 106, initially named unnilhexium upon its discovery by Seaborg's team in 1974, was officially renamed seaborgium, the first element named after a living person. Seaborg considered this his highest honor. In 1998 the American Chemical Society (of which he had served as president earlier in his career) named Seaborg one of the "Top Seventy-Five Distinguished Contributors to the Chemical Enterprise." While at the society's meeting in Boston to receive this honor on August 24, 1998, Seaborg suffered a stroke. Half a year later, he died while convalescing at his home near Berkeley, in Lafayette, on February 25, 1999.

⊠ Semenov, Nikolai
(1896–1986)
Russian
Physical Chemist, Physicist

Nikolai Semenov became the first Soviet Nobel laureate when he won the prize in chemistry in 1956, which he shared with SIR CYRIL HINSHELWOOD. They each performed research on chemical chain reactions independent of each other. Semenov focused his research on the branched-chain reactions, whereby a reaction elicits further reactions exponentially, thus generating an explosive result very quickly, such as those occurring in combustion.

Nikolai Nikolaevich Semenov was born on April 3, 15, or 16, 1896, in Saratov, Russia. His mother was Elena Dmitrieva, and his father was Nikolai Alex Semenov. At the age of 16, Semenov entered St. Petersburg University to study physics and mathematics. That same year, 1913, Max Bodenstein introduced the notion of a chain reaction into the chemical reaction. Semenov published his first scientific paper, on the collision of molecules and electrons, at the age of 20. His university study coincided with World War I, and his graduation from the renamed Petrograd University coincided with the outbreak of the Russian Revolution in 1917.

The Siberian University of Tomsk hired Semenov for a position in physics after his graduation. In 1920 he returned to Petrograd (renamed Leningrad in 1924, and the name St. Petersburg restored after the collapse of Soviet communism in 1991), where he headed the electron phenomena laboratory at the Physico-Technical Institute, and he remained there for more than a decade. During that time, he married Natalia Nikolaevna Burtseva, a voice teacher, on September 15, 1924. Together, the couple had two children, a son named Yurii Nikolaevich and a daughter named Ludmilla Nikolaevna.

Concurrent with his work at the Physico-Technical Institute, he lectured at the Leningrad Polytechnical Institute, which appointed him to a professorship in 1928. There he organized the Departments of Mathematics and Physics. In 1929 he became a corresponding member of the Academy of Sciences of the U.S.S.R., and in 1932 the academy inducted him into full membership. The previous year the academy had named him director of its Institute of Chemical Physics, a position he retained for the next dozen years.

Semenov's early research into chemical kinetics in the 1920s advanced Bodenstein's notion of chemical chain reaction. Semenov explained the mechanism of particularly explosive chain reactions as a product of branching: Each branch subsequently branches into a new set of reactions and so on, accelerating the chain reaction extremely quickly. This explanation particularly suited combustion and helped scientists better understand this process. His theory of degenerate branching also advanced the understanding of induction periods of oxidation processes.

Semenov summarized this work in the 1934 text *Chemical Kinetics and Chain Reactions,* which was translated into English the very next year, a testament to its significance in the scientific community. When the Soviet Academy of Sciences relocated the Institute of Chemical Physics to Moscow in 1943, Semenov moved with it. Once he was there, Moscow State University appointed him head of the chemical kinetics department in 1944. A decade later, in 1954, he published his second influential text, *Some Problems of Chemical Kinetics and Reactivity,* which he revised four years later. This text went through English, German, and Chinese translation.

In 1956 Semenov received the Nobel Prize in chemistry for his research on chemical reactions. Thereafter, he devoted much of his career to Soviet science and politics (he had become a member of the Communist Party in 1947). In 1958 he served as deputy in the Supreme Soviet, a duty he performed again in 1962 and in 1966. In 1961 he became an alternate to the Central Committee of the Communist Party.

The Soviet government recognized his contributions by awarding him the Order of Lenin seven times, as well as the Stalin prize and the Order of the Red Banner of Labor. However, he was not a mere puppet of the Soviet regime; Semenov actively resisted the communist inclination to divert scientific theorizing to conform to Marxist-Leninist interpretations, instead promoting the model of experimentation as the only conclusive determinant of scientific fact. In fact, when American scientists accused Soviet scientists of holding out on reporting their advances, Semenov defended himself and his colleagues, and as it turned out, the Library of Congress had been receiving the Soviet scientific journals in question all along. Semenov died on September 25, 1986.

⊠ Simon, Dorothy Martin
(1919–)
American
Physical Chemist

Dorothy Martin Simon distinguished herself in industry (rising to the rank of corporate vice president), academia (winning a prestigious Rockefeller fellowship), and government (serving in the National Advisory Committee on Aeronautics administration). Scientifically she advanced the theory of flame propagation and quenching in the theoretical realm. On the practical level she helped to develop the ablative coatings shielding intercontinental ballistic missiles (ICBMs) from the thermal assault of reentry into the atmosphere. The Society of Women Engineers conferred on her its prestigious Achievement Award.

Dorothy Martin was born on September 18, 1919, in Harwood, Missouri. Her mother was Laudell Flynn; her father, Robert William Martin, directed the chemistry department at Southwest Missouri State College. She attended the Greenwood Laboratory School in Springfield, Missouri, graduating at the head of her class. She then matriculated at the college where her father taught, graduating with honors and again at the top of her class in 1940.

Martin next pursued graduate study at the University of Illinois, where she served as an assistant in the chemistry department from 1941 to 1945. For her doctoral dissertation she conducted groundbreaking research on radioactive fallout, specifically investigating deposits of radon and thoron gases. She received her Ph.D. in chemistry in 1945. The next year, on December 6, 1946, she married Sidney L. Simon, a scientist who distinguished himself in industry as a vice president at Sperry Rand.

The year after Simon received her doctorate, she worked as a research chemist at DuPont, investigating the chemical reactions used in the production of Orlon, Du Pont's synthetic fiber. In 1947 she worked as a chemist at Clinton Laboratory, and from 1948 to 1949 she worked at an associate chemist at Argonne National Laboratory in Illinois. During this period she isolated a new isotope of calcium.

The National Advisory Committee on Aeronautics (precursor to the National Aeronautics and Space Administration, or NASA) employed Simon at its Lewis Laboratory as an aeronautical research scientist from 1949 to 1953, when she won a $10,000 Rockefeller Public Service Award and fellowship in recognition of her research on flame properties. She used her fellowship to conduct research in France, Holland, and England, at Cambridge University. When she returned to the United States in 1954, she was promoted to assistant chief of the chemical branch at Lewis Laboratory.

Magnolia Petroleum Company based in Dallas, Texas, hired Simon as a group leader in combustion in 1955. A year later the Avco Corporation hired her as a principal scientist and technical assistant to the president of the research and advanced development division. She remained there for the next quarter century, consistently rising through the ranks. In 1962 she became director of corporate research; that same year Pennsylvania State University tapped Simon to deliver its Marie Curie lecture.

In 1964 Avco promoted Simon to the vice presidency of the defense and industrial products group. Four years later, in 1968, she became corporate vice president and director of research, a position she held for more than a dozen years, until her 1981 retirement. At that time few women had risen so high up the corporate ladder, so Simon paved the way for her sister successors.

Throughout this latter part of her career Simon served the field of science—in academia as well as in the public sector—as a member of numerous committees: From 1972 on she was a member of the committee of sponsored research at the Massachusetts Institute of Technology, from 1978 on she served on the NASA Space Systems and Technology Advisory Committee, and from 1978 to 1981 she was a member of the President's Committee on the National Medal of Science.

In 1966 the Society of Women Engineers (SWE) granted Simon its Achievement Award, one of the most prestigious awards for women in the sciences. In 1971 Worcester Polytechnic Institute conferred on her an honorary doctorate as "perhaps the most important woman executive in American industry today." In 1983 the University of Illinois honored her with its Alumni Achievement Award.

Singer, Maxine
(1931–)
American
Biochemist, Geneticist

Maxine Singer conducted research on ribonucleic acid (RNA) and deoxyribonucleic acid (DNA). Her work with the controversial methods of recombinant DNA (rDNA) led her into the realm of policy making, as she directed her colleagues through the difficult process of determining guidelines to govern this morally complex line of research. She also displayed her acumen for leadership by heading several laboratories in her long career with the National Institutes of Health (NIH); she extended these leadership skills to her position as president of the Carnegie Institution of Washington.

Singer was born Maxine Frank on February 15, 1931, in New York City. Her mother was Henrietta Perlowitz, a hospital admissions officer, children's camp counselor, and model; her father was Hyman Frank, an attorney. She attended Swarthmore College, in Pennsylvania, earning her bachelor of arts degree in 1952. Upon her graduation she married Daniel Singer, and together the couple had four children—Amy Elizabeth, Ellen Ruth, David Byrd, and Stephanie Frank.

Singer pursued doctoral studies at Yale University, in New Haven, Connecticut, in biochemistry, a field that was still in its infancy. Relatively few scientists were working on nucleic acids, Singer's eventual specialty, so she participated in the defining of her field. She earned her Ph.D. in 1957 and served as a U.S. health service postdoctoral fellow working on nucleic acids under Leon Heppel at the National Institute for Arthritis, Metabolism, and Digestive Diseases (NIAMD) of the National Institutes of Health (NIH). After her fellowship ended in 1958, the NIH retained her as a research biochemist, and she has remained affiliated with the institution the rest of her career.

Singer's work at the NIH focused on ribonucleic acid (RNA), in the late 1950s she synthesized RNA polymers for her colleague, Marshall Nirenberg, who was engaged in efforts to crack the genetic code. In her subsequent nucleic acid studies she searched for virus-causing tumors.

Singer spent a yearlong sabbatical, from 1971 to 1972, studying animal viruses in the Department of Genetics at the Weizmann Institute of Sciences in Israel. There she commenced her research on simian virus 40, a line of investigation that led her to study primate genomes, which in turn led to her seminal work on transposable genetics, or recombinant DNA (rDNA).

Maxine Singer's seminal work on recombinant deoxyribonucleic acid (rDNA) led to her organizing the Asilomar Conference, which resulted in the first set of guidelines to govern the morally complex rDNA research. *(Courtesy Carnegie Institution)*

This new area of research carried scientists into a moral quagmire, as it allowed for the manipulation of genetics, a realm formerly reserved for deities or nature, depending on one's spiritual orientation. PAUL BERG, who discovered "gene jumping" in 1972, voluntarily suspended his studies out of ethical considerations.

Singer's sensitive moral compass helped navigate her and her colleagues through this minefield: In 1973 she and Yale University's Dieter Söll cochaired the Gordon Conference devoted to rDNA. The gathered scientists voted to draft a letter addressing the ethics of rDNA research, which the cochairs composed and sent to the National Academy of Sciences; *Science* magazine also published the letter.

In 1974 Singer moved to the NIH's National Cancer Institute, where she headed the section of nucleic acid enzymology of the Division of Cancer Biology and Diagnosis (DCBD). The next year she helped organize the Asilomar Conference, resulting in the first set of guidelines governing rDNA research. By the following year, 1976, she had compiled a list of four principles guiding rDNA research, which stated the following suggestions: the banning of experiments deemed too risky, the safeguarding of less risky rDNA experimentation, the degree of safeguarding corresponding with the degree of risk, and the annual reviewing of these guidelines.

In 1979 the National Academy of Sciences inducted Singer into its fellowship. That same year the NIH promoted her to head the DCBD's laboratory of biochemistry, a position she retained for the next decade. In 1988 she received the highest honor bestowed on civil servants, the Distinguished Presidential Rank Award. That year the NIH named her an emerita researcher (the first person granted such status) so that she could accept the presidency of the Carnegie Institution of Washington, D.C., without severing her affiliation with the NIH. Her subsequent research centered on LINE-1 sequences, a particular repeated DNA sequence.

In the 1990s Singer coauthored three books with Paul Berg: *Genes and Genomes: A Changing Perspective*, a graduate-level textbook published in 1990; *Dealing with Genes: The Language of Heredity*, a general-audience book published in 1992; and *Exploring Genetic Mechanisms*, published in 1997. In 1992 she received the National Medal of Science in recognition of her "outstanding scientific accomplishments and her deep concern for the societal responsibility of the scientist." She continued to display this concern and responsibility by establishing the First Light program, which promotes science education for inner-city youths.

⊠ Soddy, Frederick
(1877–1956)
English
Chemist

Frederick Soddy revolutionized science with his explanations of seemingly inexplicable phenomena: He proposed the theory of nuclear disintegration, elucidating radioactive decay as an elemental transmutation, and also the notion of isotopes, accounting for the existence of identical chemical elements that differed in physical properties. For this latter explanation especially, Soddy won the 1921 Nobel Prize in chemistry.

Soddy was born on September 2, 1877, in Eastbourne, Sussex, England. He was the youngest of seven children born to Hannah Green, who died 18 months after his birth, and Benjamin Soddy, a wealthy London corn merchant. Soddy studied science at Eastbourne College under R. E. Hughes, with whom he cowrote his first scientific paper (on the reaction between dry ammonia and dry carbon dioxide) in 1894, when he was a mere 17 years old. On Hughes's recommendation, he completed a postgraduate year at the University College of Wales at Aberystwyth, where he won an Open Science Postmaster Scholarship to Merton College of Oxford University. He matriculated there in 1895, graduating with first-class honors from the School of Natural Science in 1898.

Soddy remained at Oxford for two years of unremarkable postdoctoral research, after which he traveled to Canada to lobby in support of his candidacy for a post at the University of Toronto, which he failed to receive. On his return trip, however, he stopped over in Montreal, where McGill University offered him a position as a junior demonstrator in LORD ERNEST RUTHERFORD's department. The two men collaboratively developed a theory to explain radioactive decay: They proposed an atomic explanation for radioactive disintegration, positing that the emission of an alpha or beta particle transforms a radioactive element into new substance. This theory of nuclear disintegration, though revolutionary, achieved almost immediate acceptance in the scientific community.

In 1902 Soddy returned to London, working at University College with SIR WILLIAM RAMSAY (who had served as external examiner on his doctoral committee). In 1903 they confirmed spectrographically that radium always produces helium as the by-product of its radioactive decay, a phenomenon predicted by the joint work of Soddy and Rutherford. The next year Soddy embarked on a tour of western Australia as an extension lecturer in physical chemistry and radioactivity for the University of London; upon his return later that year he took up a lectureship in physical chemistry at the University of Glasgow. He produced some of his most significant results there over the next decade. In 1908 he married Winifred Moller Beilby; no children resulted from their union.

In 1911 Soddy proposed the alpha-ray rule that elements decreased by two in atomic number upon emission of an alpha particle. When combined with A. S. Russell's beta-ray rule that elements increased by one in atomic number upon emission of a beta particle, Soddy's finding established what was known as the displacement law two years later. Also in 1913, in the February 28 issue of *Chemical News*, Soddy coined the term *isotope* (which means "same place") to describe elements that are chemically identical but differ in atomic weights. This proposal followed up on his 1910 paper in which he first proposed that radioactive decay can produce elements that are indistinguishable except for their atomic weights. This explanation clarified the confusion resulting from the existence of chemically identical elements such as ionium and radium or thorium and radiothorium that nevertheless exhibited different properties.

In 1914 Soddy ascended to the chair in chemistry at the University of Aberdeen, in Scotland, where he continued to investigate radioactive elements. He demonstrated that the atomic weight of lead could vary significantly from the value that appears on the periodic chart, as lead derives from so many different radioactive reactions. World War I interrupted this line of research, as he contributed to the effort by devising means of converting coal gas into ethylene. After the war Soddy returned to Oxford to inhabit the Dr. Lees Chair in Chemistry.

In 1921 the Royal Swedish Academy of Sciences recognized the significance of Soddy's explanation of isotopes by awarding him the Nobel Prize in chemistry. Soddy tried to translate his laureate status into cachet for advancing his political and social beliefs, but his causes—women's suffrage, Irish autonomy, and the exer-

Frederick Soddy, who won the 1921 Nobel Prize in chemistry for his discovery of isotopes *(E. F. Smith Collection, Rare Book & Manuscript Library, University of Pennsylvania)*

cise of restraint with nuclear energy—proved too advanced to find popular support. Soddy realized that human scientific achievement required commensurate advances in human moral understanding, but ironically, he had trouble convincing his colleagues of the moral responsibility inherent in the discovery of radioactivity, a force more powerful than human reckoning.

Soddy's career came to an abrupt halt when his wife died unexpectedly of a coronary thrombosis in 1936; almost immediately he took up early retirement. His scientific output at that point was practically nil, as he had not researched any original ideas throughout his time at Oxford, but he had been writing prolifically on social causes, which he continued to do in retirement. His cautionary talk of the dangers of nuclear power, which at first had sounded alarmist, quickly took on prophetic airs as World War II bore out his warnings against the destructive applications of radioactivity. He continued to lobby scientists to take more responsibility for how their discoveries were applied. Soddy died on September 22, 1956, in Brighton.

Susan Solomon played a key part in the discovery of the phenomenon of ozone depletion, which prompted a ban on chlorofluorocarbon production by industrialized nations. *(American Geophysical Union, courtesy AIP Emilio Segré Visual Archives)*

⊠   Solomon, Susan
(1956–   )
American
Atmospheric Chemist

Susan Solomon played an instrumental role in the discovery of the phenomenon of ozone depletion from the stratospheric layer of Earth's atmosphere. Her explanation of the process by which chlorofluorocarbons (CFCs) react with ozone, thereby creating a "hole" in the atmosphere that allows dangerous ultraviolet radiation to reach Earth's surface, prompted international cooperation to draft and ratify the Montreal Protocol of 1987, which called for the phaseout of CFC production by industrialized nations by the turn of the century (amended in 1992 to foreshorten this time frame). Interest-

ingly, Solomon did not share the 1995 Nobel Prize in chemistry with PAUL J. CRUTZEN, MARIO MOLINA, and FRANK SHERWOOD ROWLAND, awarded for their work on the devastating effects of CFCs on the ozone layer.

Solomon was born on January 19, 1956, in Chicago, Illinois. Her mother, Alice Rutman, was a fourth-grade school teacher, and her father, Leonard Solomon, sold insurance. She attended the Illinois Institute of Technology in Chicago, where she focused her senior-year research project on atmospheric reactions between ethylene and hydroxyl radical on the planet Jupiter. She graduated with a bachelor of science degree in chemistry in 1977.

That summer Solomon received a student fellowship to study stratospheric ozone at the National Center for Atmospheric Research (NCAR) in Boulder, Colorado, with the center's director of the Atmospheric Chemistry Division, Paul Crutzen, senior scientist Ray Roble, and postdoctoral researcher Jack Fishman. Then in the fall of 1977 she pursued graduate study in chemistry at the University of California at Berkeley under Harold Johnston, who directed her studies of supersonic transport (SST) damage to the atmosphere.

After Solomon received her master of science in 1979, she returned to the NCAR to conduct her dissertation research under Crutzen as a graduate assistant. She also collaborated with Rolando Garcia to develop a coupled, two-dimensional chemical-dynamical model of the stratosphere and mesosphere, specifically tracking the movement of methane and ozone. She earned her Ph.D. from Berkeley in 1981 and worked in Boulder as a research chemist in the Aeronomy Laboratory of the National Oceanic and Atmospheric Administration (NOAA), where she further developed the computer model that numerically simulated the chemical processes affecting stratospheric ozone. In 1984 she published the book *Aeronomy of the Middle Atmosphere: Chemistry and Physics of the Stratosphere and Mesosphere* with coauthor Guy Brasseur. In 1985 the Department of Atmospheric and Ocean Sciences at the University of Colorado at Boulder hired her as an adjoint professor, a position that she has retained ever since.

In 1985 scientists with the British Antarctic Survey announced their discovery of a "hole" in the ozone layer over Antarctica. Solomon stood poised to test her model against actual measurements. She served as the head project scientist for two National Ozone Expeditions (NOZE-1 and NOZE-2) to McMurdo Station on Antarctica from August to November of 1986 and 1987. The data she collected confirmed her theory that heterogeneous chemistry, specifically

the reaction of hydrochloric acid with chlorine nitrate, accounted for the depletion of the ozone.

In between these expeditions Solomon published the congruence between her hypothesis and her findings in an article (coauthored by Garcia, Rowland, and Donald J. Wuebbles), entitled "On Depletion of Antarctic Ozone," that appeared in the June 19, 1986, edition of the prestigious British scientific journal *Nature*. She continued to develop her explanation of the process of ozone depletion, pointing out that chlorine compounds from CFCs reacted with ozone on the surfaces of stratospheric clouds that formed specifically in the polar atmosphere above Antarctica, where temperatures reach their coldest globally. On September 20, 1988, Solomon married Barry Lane Sidwell, who already had a son.

In 1988 the NOAA appointed Solomon program leader of the middle atmosphere group of its Aeronomy Laboratory. Then in 1991 the NOAA promoted her to the rank of senior scientist in the lab, a position she retained thereafter. Despite the fact that she was not included in the awarding of the 1995 Nobel Prize for the work on the role of CFCs in ozone depletion, she did receive much recognition in honor of her contributions to this significant scientific development. She won the 1985 J. B. MacElwane Award of the American Geophysical Union. In 1989 the U.S. Department of Commerce (which oversees the NOAA) bestowed a gold medal on her for her ozone-depletion work. In 1992 *R&D Magazine* named her its Scientist of the Year, and that same year the National Academy of Sciences inducted her into its fellowship.

⊠ **Stanley, Wendell Meredith**
(1904–1971)
American
Biochemist

Wendell Meredith Stanley isolated and crystallized the tobacco mosaic virus (TMV), confirming that it is a nucleoprotein; for this work he shared the 1946 Nobel Prize in chemistry with JOHN HOWARD NORTHROP and JAMES BATCHELLER SUMNER. He also developed the first flu vaccine during World War II. Later he continued his TMV studies, identifying its complete amino acid sequence.

Stanley was born on August 16, 1904, in Ridgeville, Indiana. His parents, Claire Plessinger and James G. Stanley, published two local newspapers, the *Ridgeville News* and *the Union City Eagle*. When his father died in 1920, Stanley's family moved to Richmond, Indiana, where Stanley graduated from high school in 1922. He then enrolled at Earlham College, located in Richmond on land donated by one of his ancestors on the provision that all Stanleys be given special consideration in admission. During a trip to the University of Illinois he met the organic chemist Roger Adams, who influenced his decision to devote his career to chemistry instead of to coaching football. (Stanley had captained the varsity football team his senior year.) He graduated in 1926 with a bachelor of science degree in chemistry and mathematics.

Stanley next pursued graduate studies at the University of Illinois at Champaign-Urbana, serving as a graduate assistant under Adams, who chaired the chemistry department. He conducted studies of the stereochemistry of biphenyl, a carbon and hydrogen molecule, resulting in 1927 in his first published paper. That same year he earned his master's degree in organic chemistry with minors in physical chemistry and bacteriology. He then conducted investigations in search of chemical treatments for leprosy. During this time he met another assistant working under Adams, Marian Staples Jay, with whom he cowrote one of his early scientific papers. The couple married on June 25, 1929, and together they had four children—one son, who was named after Stanley, and three daughters, Marjorie Jean, Dorothy Claire, and Janet Elizabeth.

Also in 1929 Stanley earned his doctorate and remained at the university as a research associate, studying the stereoisomerism of diphenyl compounds, and later as an instructor. In the summer of 1930 he received a postdoctoral fellowship through the National Research Council, which funded a trip to Germany and the University of Munich to study under Heinrich Wieland. Upon his return in 1931 he took up a research assistantship at the Rockefeller Institute for Medical Research (RIMR) in New York City, where he worked under the cell physiologist W. J. V. Osterhout.

In 1932 Simon Flexner and Louis Otto Kunkel invited Stanley to move to Rockefeller Institute's Department of Plant and Animal Pathology in Princeton, New Jersey. There he endeavored to isolate and crystallize TMV. After introducing the virus onto 6-inch-tall Turkish tobacco plants, he chopped and froze samples that he later ground into a liquid and exposed to more than 100 different chemicals, to test its reaction. By 1934 he found that TMV had affinities to protein, prompting him to purify the substance into its crystalline form. In 1935 Rockefeller promoted him to an associateship. He later confirmed that the virus was indeed a nucleoprotein (as other researchers had suggested). In 1937 he became an associate member of the RIMR, and in 1940 he became a full member. For his work with TMV Stanley won the 1946 Nobel Prize in conjunction with John Howard Northrop and James Sumner.

Stanley spent World War II as a member of the U.S. Army Commission on Influenza, developing a centrifuge-purified influenza vaccine using formaldehyde to deactivate the virus while stimulating the body's production of antibodies. After the war he received the Presidential Certificate of Merit in 1948 for his development of the first flu vaccine. That year the University of California at Berkeley hired Stanley to chair its new biochemistry department, while also directing the new virology laboratory, which he had helped to found. A decade later, in 1958, he founded a new virology department.

At Berkeley, besides seeing to his administrative duties, Stanley continued his TMV studies, leading a group in determining the complete amino acid sequence of TMV in 1960. In 1964 his two departments expanded into the Department of Molecular Biology. During this period he also published his two classic books, *The Viruses: Biochemical, Biological, and Biophysical Properties* (1959), which he edited with Frank Macfarlane Burnet, and *Viruses and the Nature of Life* (1961), which he wrote with Evan G. Valens. Throughout his career, he authored more than 150 publications.

Besides the Nobel Prize, Stanley won numerous other awards: the University of Chicago's Rosenburger Medal, Harvard's Isaac Alder Prize, and the city of Philadelphia's John Scott Medal, all in 1938; the American Chemical Society's 1946 Nichols Medal, as well as its 1947 Gibbs Medal; the American Cancer Society's 1963 Medal for Distinguished Service in Cancer Control; and the American Medical Association's 1966 Scientific Achievement Award. He also received the Second Class Order of the Rising Sun from the Japanese government in 1966. Stanley died of a heart attack on June 14, 1971, while he was in Spain chairing a symposium in honor of Dr. Francesc Duran Reynals.

⊠ Staudinger, Hermann
(1881–1965)
German
Organic Chemist

Hermann Staudinger developed the theory of macromolecules, long chains of identical molecules strung together by simple chemical bonds that are ubiquitous in nature and form the foundation of plastics chemistry. The revolutionary quality of this theory elicited the vehement opposition of the scientific community, but Staudinger retained faith in his theory, experimentally proving its validity. He also posited Staudinger's law, a method of quantifying polymers based on the relationship between their viscosity and their molecular weight. He received a belated Nobel Prize in chemistry in 1953, after he had retired.

Staudinger was born on March 23, 1881, in Worms, Germany. His mother was Auguste Wenck, and his father was Dr. Franz Staudinger, a secondary-school philosophy professor. Staudinger attended gymnasium at Worms, graduating in 1899. He then studied at several different universities, as was the practice at the time: He commenced at the University of Halle, studying botany under Professor Klebs, but soon shifted to chemistry studies under Professors Kolb and Stadel at the University of Darmstadt, then under Professor Piloty at the University of Munich. When he returned to Halle in 1901, he completed his chemical studies under Professor D. Vorländer, writing his dissertation on the malonic esters of unsaturated compounds to earn his doctorate in 1903.

Staudinger commenced his professional career as an instructor at the University of Strasbourg, where he conducted research with Johannes Thiele that yielded the discovery of ketenes, or highly reactive and often poisonous gases. In 1907 he wrote a special dissertation on ketenes to earn a promotion to the rank of assistant professor. The next year the Technische Hochschule in Karlsruhe hired him as an associate professor. There he collaborated with C. L. Lautenschläger to prepare polyoxymethylenes, and he also conducted independent research, developing a new method for synthesizing isoprene, a component of synthetic rubber. These investigations represented his first steps toward his theory of macromolecules.

In 1912 he succeeded RICHARD WILLSTÄTTER in the general chemistry chair at the Eidgenössische Technische Hochschule (Swiss Federal Institute of Technology) in Zurich,

Switzerland. There he split his time between a heavy teaching load and research, continuing his ketene investigations while also experimenting with oxalyl chloride and pyrethrin insecticides. During World War I he worked to counteract wartime shortages of certain foodstuffs (such as pepper and coffee) by attempting to produce synthetic substitutes, but none of his methods proved commercially viable.

After the Great War Staudinger again dedicated himself to the study of macromolecules. He first proposed the notion of polymers in 1920, and in a 1924 article in *Berichte der Deutschen Chemischen Gesellschaft*, he coined the term *macromolecule* to describe what he viewed as long carbon chains of identical molecules strung together by simple chemical bonds. The scientific community vehemently opposed Staudinger, believing that he mistook conglomerations of smaller molecules for independent larger molecules.

In 1926 Staudinger moved to the University of Freiburg im Breisgau as a professor of chemistry and plunged himself into the study of macromolecules wholeheartedly. The next year he married Magda Woit, a Latvian plant physiologist who collaborated with her husband in research and writings on macromolecular theory. Staudinger staunchly defended his theory by providing meticulous experimental proof of its validity; for example, he demonstrated that the hydrogenation of rubber creates a saturated hydrocarbon polymer with the same molecular weight as its parent compound. Eventually his theory of macromolecules gained acceptance.

In order to quantify his theory he developed Staudinger's law, a relatively simple equation based on the relationship between a polymer's viscosity and its molecular weight. In testament to the acceptance of his theory the University of Freiburg established the Research Institute for Macromolecular Chemistry, installing Staudinger as its director in 1940. Upon his retirement to emeritus status in 1951 the university renamed it

the State Research Institute for Macromolecular Chemistry.

Two years into his retirement, when Staudinger was 72 years old, the Royal Swedish Academy of Sciences finally recognized the importance of his discovery by awarding him the 1953 Nobel Prize in chemistry. Staudinger had won other honors in his life, namely, the 1931 Leblanc medal from the French Chemical Society and the 1933 Cannizzaro Prize. He died on September 8, 1965, in Freiburg.

⊠ **Stein, William H.**
(1911–1980)
American
Biochemist

William H. Stein collaborated with STANFORD MOORE on pioneering amino acid analyses. The pair developed several apparatuses—namely, the drop-counting automatic fraction collector and the automatic amino acid analyzer—that improved the efficiency and efficacy of their research. These instruments saw continued use in diverse applications in laboratories throughout the scientific world. Stein and Moore later determined the biochemical function of ribonuclease, the first time this information was elucidated for an enzyme. For these achievements they shared the 1972 Nobel Prize in chemistry.

William Howard Stein was born on June 25, 1911, in New York City. He was the middle child of three born to Beatrice Borg and Fred M. Stein, community activists who inculcated in their son the sense of responsibility to help his neighbors, especially those who were needy. Stein attended the Lincoln School of Teachers College, Columbia University, which had a progressive curriculum. He then went off to preparatory school at Phillips Exeter Academy in Exeter, New Hampshire.

Stein matriculated at Harvard University, compiling an unimpressive record in his chemistry studies by his 1933 graduation, and remained there for a year of graduate study in organic chemistry, though he still failed to distinguish himself. Wanting to shift his specialization he transferred to Hans Clark's Department of Biochemistry at Columbia University's College of Physicians and Surgeons. For his dissertation he analyzed the amino acid composition of elastin, a vascular protein believed to play a role in coronary artery disease. He earned his doctorate in 1938. Two years earlier, on June 22, 1936, he had married Phoebe L. Hockstader; the couple had three sons—William H. Jr., David F., and Robert J.

Stein pursued postdoctoral studies at the Rockefeller Institute for Medical Research (RIMR, now Rockefeller University) in Max Bergmann's laboratory, continuing his amino acid analysis. He commenced his collaboration with Stanford Moore in 1939, focusing their analytical research on the amino acids glycine and leucine. World War II broke up the pair of scientists; Moore contributed to the war effort in Washington, D.C., while Stein stayed at Rockefeller, working for the Office of Scientific Research and Development on the physiological effects of mustard gases.

After World War II Moore returned to Rockefeller, where Herbert Gasser, who had succeeded to the directorship after Bergmann's death in 1944, had implemented a newly established protein chemistry program. Stein and Moore resumed their amino acid studies, utilizing innovative applications of existing techniques, while also developing new methods for analyzing these proteins. They separated amino acids from potato starch columns, using first the partition chromatography techniques developed by ARCHER JOHN PORTER MARTIN and RICHARD SYNGE before employing the column chromatography method pioneered by FREDERICK SANGER.

Stein and Moore graduated from potato starch studies to studying columns of ion exchange resins. In order to proceed, however, they needed improved equipment, but none existed, so they had to invent their own instruments to perform the functions they required. They developed their drop-counting automatic fraction collector, an instrument still in use in laboratories throughout the scientific world; they also developed an innovating automatic amino acid analyzer that significantly reduced the time necessary for protein analysis. With this equipment they moved on to even more complex protein analyses, focusing their research on the structure of ribonuclease. They discovered that ribonuclease had a three-dimensional chainlike structure that folds in on itself to catalyze reactions.

During his career Stein promoted the dissemination of scientific information by working at the *Journal of Biological Chemistry*. He was elected to its editorial committee in the 1950s, serving as the committee's chair from 1958 through 1961. In 1962 he joined the journal's editorial board, becoming its associate editor in 1964. In 1968 he succeeded John T. Edsall as the editor of the journal, retaining this position until 1971, when he stepped down due to illness. He also served as the chairman of the United States National Committee for Biochemistry from 1968 to 1969.

In addition to sharing the Nobel Prize in chemistry, Stein and Moore shared multiple other honors: the American Chemical Society's 1964 ACS Award in Chromatography and its 1972 Theodore William Richards Medal, as well as the 1972 Kaj Linderstrøm-Lang Award. Stein died in Manhattan on February 2, 1980.

⊠ **Sumner, James Batcheller**
(1887–1955)
American
Biochemist

James Batcheller Sumner shared the 1946 Nobel Prize in chemistry with JOHN HOWARD NORTHROP and WENDELL MEREDITH STANLEY for their roles in isolating pure crystalline enzymes. Sumner was the first to isolate an enzyme, urease, in 1926, but his discovery was met with incredulity until Northrop's isolation of the enzyme pepsin confirmed Sumner's earlier isolation.

Sumner was born on November 19, 1887, in Canton, Massachusetts. His parents, Elizabeth Rand Kelly and Charles Sumner, were wealthy landowners whose ancestors had emigrated from Bichester, England, in 1636. Sumner attended the Eliot Grammar School and then the Roxbury Latin School. A grouse-hunting expedition when he was 17 ended in injury when a fellow hunter accidentally shot him in the left arm, which required amputation below the elbow. His recovery was exacerbated by the fact that he was left-handed, forcing him to learn to use his right hand. He entered Harvard College in Cambridge, Massachusetts, in 1906 to study electrical engineering, but he soon switched majors to chemistry, earning his degree in 1910.

After working for a brief few months at his uncle's factory, the Sumner Knitted Padding Company, he eagerly accepted a one-term appointment teaching chemistry at the Mt. Alison College in Sackville, New Brunswick, Canada. He spent the next term back in Massachusetts as an assistant in chemistry at the Worcester Polytechnic Institute. In 1912 he entered Harvard Medical School to study biochemistry under Otto Folin. He earned his master's degree the next year, and in 1914 he earned his doctorate. That same year he published a portion of his dissertation, under the title "The Importance of the Liver in Urea Formation from Amino Acids," in the *Journal of Biological Chemistry*.

While on a tour of Europe, the outbreak of World War I temporarily prevented him from leaving Switzerland. There he received a cable informing him that the Cornell University Medical College had hired him as an assistant professor in biochemistry; he arrived in Ithaca, New York, in time to take up this position, which also involved teaching in the College of Arts and Sciences. In 1915 he married Bertha Louise Ricketts, and they had six children (one of whom died young) before divorcing. He remarried twice, to Agnes Paulina Lundkvist in 1931, and to Mary Morrison Beyer in 1943.

Sumner's heavy teaching load left scant time for research, so he decided to focus on a single goal that might distinguish him: the isolation of an enzyme, considered an impossibility at the time. He focused his enzyme research on urease, the catalyst causing the breakdown of urea into ammonia and carbon dioxide. He extracted this enzyme from the jack bean (*Canavalia ensiformis*), spending his first nine years fractionating protein from the bean. In 1921, in the midst of this part of the project, he received an American-Belgian fellowship, which he planned to use to study in Brussels with Jean Effront, an enzyme specialist. However, Effront scoffed at Sumner's effort to isolate urease, considering this task impossible, so Sumner abandoned his plans at collaboration.

Five years later, in 1926, Sumner succeeded in isolating urease in its pure, crystalline form. His announcement, published in the *Journal of Biological Chemistry*, met resistance and ridicule, most notably from the German Nobel laureate in chemistry RICHARD MARTIN WILLSTÄTTER, who may have envied Sumner's having accomplished what he had tried but failed to achieve— the isolation of an enzyme. Not until 1930, when John Northrop isolated the enzyme pepsin by a method similar to Sumner's, was the isolation of urease finally accepted. The year before, Cornell had promoted Sumner to a full professorship in biochemistry.

Sumner finally received recognition for his work: In 1937 he won a Guggenheim Fellowship, which he used to study for five months with Nobel laureate THEODOR SVEDBERG in Uppsala, Sweden. Also in 1937 he won the Scheele Medal in Stockholm, and upon his return to Cornell he collaborated with his student, Alexander L. Dounce, to crystallize the enzyme catalase, proving in the process that it was a protein. He was also the first to crystallize the protein hemagglutinin concanavalin A.

Sumner climaxed his career with the 1946 Nobel Prize in chemistry, which he shared with Northrop and Wendell Stanley. Throughout his professional life he contributed some 125 scientific papers to journals, in addition to authoring several books. Unfortunately, he died within a decade of becoming a Nobel laureate, having developed cancer. He passed away on August 12, 1955.

⊠ **Svedberg, Theodor**
(1884–1971)
Swedish
Chemist

Theodor Svedberg contributed one of the most basic and ubiquitous research tools to science— the ultracentrifuge, which spins samples around an axis at high velocities to separate the chemical constituents. For this innovation Svedberg received the 1926 Nobel Prize in chemistry.

Svedberg was born on August 30, 1884, in Fleräng, Sweden, near Gävle. He was the only child born to Augusta Alsermark and Elias Svedberg, a civil engineer for an ironworks. He attended the Karolinksa School in Örebro; an atypical secondary-school student, he became interested in the advanced work of WALTHER HERMANN NERNST, Richard Zsigmondy, and Georg Bredig. In January 1904 he graduated to the University of Uppsala, where he earned his bachelor of science degree in 1905, after a mere year and a half of undergraduate study.

Svedberg commenced his doctoral work immediately, concentrating his dissertation studies on colloidal solutions. He sought to improve upon the work of Bredig, who had produced fairly pure metallic-colloidal solutions by passing an electric arc between metal electrodes submerged in liquid; instead Svedberg alternated currents through an induction cable with only its spark gap submerged, thereby producing purer and more finely dispersed metal colloids. After he earned his doctorate in 1907, he stayed on at the university as a lecturer in physical chemistry. Within five years the University of Uppsala created Sweden's first chair of physical chemistry for him. In the meanwhile he had married the first of four times, wedding Andrea Andreen in 1909. He married Jan Frodi Dahlquist in 1916, Ingrid Blomquist Tauson in 1938, and Margit Hallen Norback in 1948. His marriages yielded six daughters and six sons altogether.

Svedberg studied Brownian motion (the random collision of tiny particles with molecules discovered by the botanist Robert Brown) in colloidal solutions, noting the effects of temperature, viscosity, and the properties of the originating solvent. He employed an ultramicroscope that used refracted light to illuminate samples invisible under direct light in order to confirm Albert Einstein's theories of Brownian motion. In addition, these ultramicroscope studies confirmed that colloidal solutions conformed to classical physical and chemical laws.

However, the constant collision of colloidal particles prevented Svedberg from determining their sizes and distribution. In his first attempt to solve this problem, conducted in 1923, he and his collaborator, Herman Rinde, suspended colloidal systems on a balance that was specially designed to reduce air-current disturbances in order to measure sediment accumulation. This technique, while effective, proved insufficient to achieve the desired resolution, so Svedberg contemplated other solutions to his problem.

Later in 1923 the University of Wisconsin offered Svedberg an eight-month guest professorship. While he was there, he took advantage to solve his problem of measuring particle size and distribution in colloidal solutions by constructing an ultracentrifuge in collaboration with J. Burton Nichols. Existing centrifuges spun solutions around an axis at high speeds to separate lighter from heavier components, such as milk from cream or red blood cells from plasma, but these machines did not suffice for the separation of tiny colloidal particles. Svedberg's ultracentrifuge sped up the existing process to 30,000 revolutions per minute, multiplying the gravitational force experienced on Earth more than a thousandfold. Svedberg and Nichols determined the particle sizes, distributions, and radii of gold, clay, barium sulfate, and arsenious sulfide.

In 1926 Svedberg received the Nobel Prize in chemistry for his innovation of the ultracentrifuge, and over the next decade he improved it to attain speeds of 45,000 revolutions per minute and gravitational forces of 750,000 times the gravitational pull of Earth. He also expanded its applications, studying the proteins hemoglobin, pepsin, insulin, catalase, and albumin.

In 1930 the University of Uppsala established the Institute of Physical Chemistry and installed Svedberg, the main proponent of the new institution, as its director the following year. Later in the 1930s Svedberg convinced the industrialist Gustaf Werner of the necessity for generating radioactive isotopes; Werner responded by financing the construction of a large cyclotron at the newly founded Gustaf Werner Institute of Nuclear Chemistry. World War II interrupted Svedberg's main line of study, though, redirecting him into the practical realm of rubber synthesis (polychloroprene was the synthetic then in vogue).

Svedberg retired to emeritus status in 1949, when he reached the mandatory age; however, he far from retreated into seclusion, as the Swedish government precluded him from its retirement rule to allow for his appointment as lifetime director of the Gustav Werner Institute, which he had helped to found. He remained there until 1967, when he took it upon himself to resign. Svedberg died on February 25, 1971, in Örebro, Sweden. In his honor the centrifugation coefficient was named after him—the svedberg unit.

⊠ **Synge, Richard**
(1914–1994)
English
Biochemist, Physical Chemist

Richard Synge shared the 1952 Nobel Prize in chemistry with his collaborator, ARCHER JOHN PORTER MARTIN, for their development of the partition chromatography method of chemical separation and analysis, which they had announced to the scientific community in 1944. Their discovery saw immediate application, as other scientists employed the method to analyze plant photosynthesis and deoxyribonucleic acid (DNA) sequencing.

Richard Laurence Millington Synge was born on October 28, 1914, in Liverpool, England. His mother was Katherine Charlotte Swan; his father, Laurence Millington Synge, was a stockbroker. He had two younger sisters. He attended preparatory school from 1928 to 1933 at Winchester College. He then attended Trinity College of Cambridge University on a classics scholarship, though he dropped his classics major to study biochemistry after attending a lecture by biochemist Frederick Gowland Hopkins. He graduated from Trinity in 1936.

Synge remained at Cambridge in its Biochemical Laboratory to pursue graduate studies. While conducting research on the topic that would preoccupy him the rest of his career—the separation of a compound into its constituents, focusing his dissertation on acetyl amino acids—he met his longtime collaborator A. J. P. Martin. But before they could advance far in their joint studies of separation processes, the Wool Industries Research Laboratories in Leeds hired away Martin from Cambridge.

Fortuitously Synge won an International Wool Secretariat scholarship, which he used to join Martin in Leeds. Synge earned his doctorate in 1941, and two years later he married Ann Stephen, a physician and the niece of the author Virginia Woolf. Together the couple had seven children.

Even after the Lister Institute of Preventative Medicine in London hired Synge as a biochemist in 1943, Synge and Martin continued their collaboration. While still at Leeds, they had devised a separation process by combining adsorption chromatography (a process for separating constituent elements according to their attraction to a powder) with countercurrent solvent extraction (which separates elements by their attraction to one of two opposing solutions). The pair's method essentially replaced the adsorption medium of using powder with cellulose paper, which separates constituents by capillary action, or ascending chromatography.

Synge and Martin enhanced this paper chromatography method by dipping the paper into opposing liquids, such as water and alcohol, to further separate molecules by the direction of their attraction. They then sprayed the dried paper with ninhydrin solution, which upon heating, "developed" the molecules into spots that revealed their identity according to their position on the paper. They tweaked the method even further by twisting the paper 90° and repeating the process before development, thus achieving a two-dimensional reading. Synge and Martin further improved the paper chromatography method of separation by constructing a 40-unit extraction machine to streamline analysis. In 1944, using cellulose filtration, they made their final improvement on the method, which they renamed *partition chromatography*.

In 1946 Synge visited ARNE WILHELM KAURIN TISELIUS at the Institute of Biochemistry of the University of Uppsala, Sweden, to study other separation methods, such as electrophoresis and adsorption. He applied what he had learned back at home on several collaborative studies: electrokinetic ultrafiltration with D. L. Mould; rye grass proteins, with Mary Youngson; and isolation of the sporidesmin toxin, with E. P. White. In 1948 he moved to Rowett Research Institute in Aberdeen, Scotland, where he chaired the Department of Protein Chemistry.

In 1952 the Royal Swedish Academy of Sciences awarded Synge and Martin the Nobel Prize in chemistry for their partition chromatography method of separating molecules, which was used in the Nobel Prize–winning work of MELVIN CALVIN, WALTER GILBERT, and FREDERICK SANGER. In 1967 Synge left Scotland for the Food Research Institute of Norwich, which hired him as a biochemist. He retained this position until his 1976 retirement. He did not fully retire, however, as he had been appointed to an honorary professorship in the School of Biological Sciences of the University of East Anglia in 1968, and he retained this position until 1984.

Besides the 1952 Nobel Prize in chemistry, Synge received the 1959 John Price Wetherill Medal from the Franklin Institute. He also belonged to many different scientific societies worldwide, such as the Royal Society and the Royal Irish Academy. He died on August 18, 1994, at the age of 79.

T

Telkes, Maria
(1900–1995)
Hungarian/American
Physical Chemist

Maria Telkes, also known as the Sun Queen, devoted her scientific career to the invention of ways to harness the Sun's power toward practical ends. Early in her career, for example, she won a $45,000 Ford Foundation grant to develop a solar oven; she responded with a design elegant in its adaptability, as it not only cooked foods in ways suited to all cultural methods of preparation but could also dry crops faster than sun-drying them. She applied similar ingenuity to the design of solar houses, solar-powered laboratories, and departments of solar energy research (both academic and industrial). She even devised a means of using the sun to evaporate salt out of seawater to make it safe for drinking.

Telkes was born on December 12, 1900, in Budapest, Hungary, then part of the Austro-Hungarian Empire. Her parents were Aladar and Maria Laban de Telkes. She attended the University of Budapest, studying physical chemistry to earn her bachelor of arts degree in 1920. She remained at the university for graduate study and earned her doctorate in 1924. The next year she traveled to the United States to visit her uncle and stayed there until just before her death.

The year of Telkes's arrival to the United States, 1925, she secured a research position at the Cleveland Clinic Foundation, in Ohio, studying cell pathology and carcinogenesis, as well as other transformative phenomena, for the next dozen years. She first began studying the Sun, indirectly, in 1937, when Westinghouse Electric hired her as a research engineer to study heat energy, specifically searching for means of harvesting it into electric energy. Her inaugural year there she became a citizen of the United States.

Two years later, in 1939, Telkes turned her attention directly to the Sun when the Massachusetts Institute of Technology hired her for its Solar Energy Conversion Project, which endeavored to harness solar energy into electrical energy. Telkes arrived at an innovative answer to the project's central question of how best to transform solar energy to electrical energy: Store the solar energy chemically by crystallizing it into sodium sulfate solution. She utilized this method in developing the energy system for one of the first modern homes powered by solar electricity, the Dover House (named after its location in Dover, Massachusetts), designed by architect Eleanor Raymond for the Boston sculptor Amelia Peabody, in 1948.

As was the case for many of her colleagues, Telkes worked for the United States government in the 1940s; however, she directed her efforts not at war-related destruction but rather the survival of the species: She invented a water desalinator that used solar heat to vaporize seawater, then cool the condensation back into water, minus the salt. The first major recognition of the significance of Telkes's career came in 1952, when the Society of Women Engineers (SWE) granted her its inaugural Achievement Award.

Telkes ended her tenure at MIT in 1953, when New York University's College of Engineering hired her to design its solar energy laboratory. The Curtiss-Wright Company hired her five years later to design the solar heating and energy-storage system powering its Princeton, New Jersey, lab. She became head of the same solar energy research lab she had helped build and there developed such commercial applications as solar-powered dryers and heaters, as well as grander schemes such as sending thermoelectric generators into space to collect energy.

Telkes continued to concern herself with space. While working at the Cryo-Therm company from 1961 to 1963, she developed materials strong enough to withstand the pressures of space and the subsequent splashdown into the sea for the Apollo and Polaris project's temperature-sensitive instruments. Afterward she took on the directorship of yet another solar energy laboratory, this time with the Melpar company. She ended her career back in academia in 1969, working at the Institute of Energy Conversion of the University of Delaware.

In 1977 Telkes won the Charles Greely Abbot Award from the American section of the International Solar Energy Society. The next year the University of Delaware granted her emerita status; however, she continued to contribute to solar energy advancement, helping design the solar energy system for the Carlisle House (in Carlisle, Massachusetts) in 1980. She returned to Budapest before her death on December 2, 1995.

⊠ **Tiselius, Arne Wilhelm Kaurin**
(1902–1971)
Swedish
Physical Biochemist

Arne Tiselius won the 1948 Nobel Prize in chemistry for his development of a method for separating compounds by electrophoresis, or the study of molecular motion in chemical compounds elicited by an electric field. The prize also recognized his identification of globulins in the protein of blood serum.

Arne Wilhelm Kaurin Tiselius was born on August 10, 1902, in Stockholm, Sweden. His father, Hans Abraham J. Tiselius, was a mathematician who worked for an insurance company. When his father died in 1906, Tiselius's mother, Rosa Kaurin, moved the family to her husband's hometown of Göteborg. There Tiselius attended secondary school at the local gymnasium, graduating in 1921. He enrolled at the University of Uppsala to study physical chemistry under THEODOR SVEDBERG, who was on the cusp of winning the Nobel Prize in chemistry. In 1924 Tiselius earned his master's degree in chemistry, physics, and mathematics from the University of Uppsala, where his father had earned his degree in mathematics.

Tiselius stayed on at Uppsala to conduct doctoral research in Svedberg's laboratory, enabling his mentor to focus on developing his ultracentrifuge (for which he won the 1926 Nobel) by taking over Svedberg's electrophoresis studies, which investigated charged colloidal particles by observing their migration under the influence of an electric field. Tiselius focused his attention on proteins, as electrophoresis further separated elements that remained homogenous even after ultracentrifuge, allowing for a more complete account of the constitution of these complex compounds.

Tiselius published his first paper (coauthored by Svedberg), "A New Method for Determination of the Mobility of Proteins," in the

September 1926 edition of the *Journal of the American Chemical Society*. He submitted his dissertation four years later to earn his doctorate. That same year he married Ingrid Margareta Dalén on November 26, 1930, and together the couple eventually had two children, a son named Per and a daughter named Eva.

Upon conferring his doctorate, the University of Uppsala retained Tiselius by appointing him to a docent position in the chemistry department. He continued his protein studies, employing yet another method to identify the constituent elements more discriminatingly than possible with ultracentrifugation: chromatography. This technique shines light of a specific frequency through the sample, which reflects, refracts, or diffuses the light in patterns that identify its constituents—its so-called chromatic signature. He applied this method to the translucent mineral zeolite, exploiting its unique property of retaining its crystalline structure when exchanging its crystalline waters with those of other substances, and even upon vacuum dehydration and rehydration.

Tiselius spent 1934 through 1935 as a visiting researcher under Hugh S. Taylor at the Frick Chemical Laboratory of Princeton University in New Jersey. He continued his zeolite studies, devising a method to measure the diffusion of water molecules through zeolite crystals. In discussion with Karl Landsteiner and Leonor Michaelis, he realized that he could apply this method to his earlier protein studies, so he constructed an electrophoretic separation apparatus. Upon his return to Uppsala he improved on its design, then investigated protein serum in horse blood plasma, yielding readings of four electrophoretically distinct protein bands. He dubbed three of these *alpha, beta,* and *gamma globulins;* the fourth and fastest-moving band contained antibodies and albumin.

Continuing to improve upon this technique, Tiselius discovered a way to avoid "tailing," or the corruption of one constituent by another, by adding to the elution a substance with a higher adsorption affinity than those in the solution, which aided in maintaining separation. Developed in 1943, he called this technique "displacement analysis." Throughout the 1940s he continued to develop other electrophoretic techniques, including paper electrophoresis and zone electrophoresis. The Royal Swedish Academy of Sciences recognized this work by awarding him its 1948 Nobel Prize in chemistry.

By this time Tiselius had expanded his career to contribute more generally to the field of science. In 1944 the Swedish government had assigned him as an adviser on scientific affairs, and in 1946 the Swedish Natural Science Research Council had appointed him to a four-year stint as its chairman, during which time he helped create the Science Advisory Council to the Swedish government. The year before he received the Nobel, in 1947, the Nobel Foundation appointed him as its vice president as well as to a seat on the Nobel Committee for Chemistry. He accepted another vice presidency that year for the International Union of Pure and Applied Chemistry's section for biological chemistry; in 1951 he ascended to the presidency of the union. In 1960, while presiding over the Nobel Foundation, Tiselius founded the Nobel Symposium, congregating Nobel laureates to discuss how to apply the advances made in all five of the prize fields to the betterment of humankind. Tiselius retired in 1968.

Besides the Nobel Prize, Tiselius won the Royal Swedish Scientific Society's 1926 Bergstedt Prize, the Franklin Institute's 1956 Franklin Medal, and the University of Zurich's 1961 Paul Karrer Medal in Chemistry. On October 28, 1971, Tiselius suffered a heart attack that took his life the following morning, on October 29.

⊠ **Todd, Baron Alexander**
(1907–1997)
Scottish
Chemist

Baron Alexander Todd won the 1957 Nobel Prize in chemistry for his nucleotide research, synthesizing nucleosides and nucleotides, as well as adenosine di- and triphosphate (ADP and ATP). Todd also identified tetrahydrocannabinol (THC) as the psychoactive ingredient of hashish and marijuana.

Alexander Robertus Todd was born on October 2, 1907, in Glasgow, Scotland. His parents, Jane Lowrie and Alexander Todd, worked hard to elevate their three children above their working-class status. To encourage Todd's interest in science, for example, they bought their eight-year-old a chemistry set and later secured his entry into the Allan Glen School, a secondary school specializing in science, which Todd attended from 1918 to 1924. Todd went on to study chemistry at the University of Glasgow, earning an academic scholarship after his stellar first year (for which his proud father had refused the "charity" of financial aid). Todd wrote his thesis under T. E. Patterson on the reaction between phosphorus pentachloride with ethyl tartrate and its diacetyl derivative to earn his bachelor of science degree in chemistry with first-class honors in 1928.

A Carnegie research scholarship supported another year at Glasgow, but Patterson's research topic—optical rotary dispersion—failed to inspire Todd, so he departed for Germany to pursue doctoral studies in organic chemistry under Walther Borsche at the University of Frankfurt. He focused his research on the bilious chemical apocholic acid, earning his Ph.D. in 1931. Todd won an 1851 Exhibition Senior Studentship, which supported doctoral research under SIR ROBERT ROBINSON at Oxford University. Although officially a postdoctoral student, Todd earned a second doctorate in 1934 by conducting research on the synthesis of anthocyanins, or water-soluble plant pigmentation.

Todd then conducted postdoctoral research as an assistant in medical chemistry at the University of Edinburgh under a Medical Research Council grant. He worked to elucidate the structure of vitamin B_1, an anti-beriberi necessary for the metabolism of carbohydrates. German and American teams beat Todd's team to the synthesis of thiamine, as the vitamin was called; in the long run, though, the simple elegance of the British solution proved more practical for industrial applications. Todd also focused his research on the chemistry of vitamin E, an antioxidant that prevents the departure of electrons. During his stint in Edinburgh, he met a fellow postgraduate researcher from the pharmacology department, Alison Dale, the daughter of the Nobel laureate Henry Hallett Dale. The couple married in January 1937; together they had three children—Alexander (born in 1939), Helen (born in 1941), and Hilary (born in 1945).

At Edinburgh Todd also searched for the psychoactive ingredient of the plant *Cannabis sativa* (marijuana). He thought he had found his answer in cannabinol, until he identified this product isolated from the plant's resin as pharmacologically inactive. Todd returned to these studies at the University of Manchester, where he was named to the Sir Samuel Hall chair of chemistry and the director of the chemical laboratories in 1938, after having spent a year at London's Lister Institute of Preventive Medicine and another year as a reader in biochemistry at the University of London. While attempting to synthesize cannabinol at Manchester, he discovered an intermediary substance, tetrahydrocannabinol (THC), which produced the same effects in experimental rabbits as did hashish, suggesting that he had isolated the active ingredient that makes marijuana and hashish produce a euphoric high in users.

Verfication of the psychoactivity of THC fell to others, though, as Todd spent World War II as a member (and later as chair) of the Chemical Committee. He conducted research to develop a practical production method for the sneeze gas diphenlyamine chloroarsine, designed a nitrogen mustard factory to produce the blistering agent, and advanced penicillin research. In 1943 Todd accepted an appointment to the chair in organic chemistry at Christ College of Cambridge University (after he had refused its offer of a chair in biochemistry earlier in the year).

Todd focused his work in the 1940s and 1950s on the synthesis of nucleosides (which combine a sugar with a heterocyclic nitrogen compound), nucleotides (which add a phosphate group to the sugar in nucleosides), and such nucleic acids as ribonucleic acid (RNA) and deoxyribonucleic acid (DNA). His study on the latter proved instrumental in the elucidation of the structure of DNA by James Watson and FRANCIS HARRY COMPTON CRICK. In other studies Todd managed to synthesize adenosine in 1949, and subsequently he synthesized adenosine diphosphate, nucleotides that govern energy production and storage in plants and muscles. For these studies he received the 1957 Nobel Prize in chemistry.

While Todd continued to conduct research—in particular determining the structure of vitamin B_{12}, the antipernicious anemia agent, in collaboration with DOROTHY CROWFOOT HODGKIN in 1955—he devoted his later career to administration and concerns in the field of science more generally. The British government named him chair of its advisory council on scientific policy in 1952, a post he retained until 1964. From 1963 to 1978 he served as master of Christ College, and from 1975 to 1980 he presided over the Royal Society, whose fellowship he had joined in 1942. He retired officially in 1971.

Besides the Nobel Prize, Todd received numerous other honors. Queen Elizabeth II knighted him in 1954, and in 1962 he was named Baron Todd of Trumpington. In 1977 he was named a member of the Order of Merit. He published his autobiography, A Time to Remember, in 1983. Todd died on January 10, 1997, in Cambridge.

U

Urey, Harold
(1893–1981)
American
Physical Chemist

Harold Urey won the 1934 Nobel Prize in chemistry for his isolation of deuterium, hydrogen's sole isotope. Urey went on to develop new means of preparing deuterium and later replaced hydrogen with deuterium in several compounds (the first was deuteromethane), a method later developed for isotope labeling.

Harold Clayton Urey was born on April 29, 1893, in Walkerton, Indiana. His mother was Cora Rebecca Reinoehl; his father, the Reverend Samuel Clayton Urey, was a schoolteacher and minister who died when Urey was six years old. Unable to afford college tuition, Urey taught in country schools in Indiana from 1911 to 1912 and in Montana until 1914, when he had finally earned enough to afford in-state tuition at the University of Montana. He graduated with a bachelor of science degree in zoology in 1917, then worked as a research chemist for two years at the Barrett Chemical Company in Philadelphia, manufacturing explosives for World War I through its end.

After the war Urey returned to the University of Montana as an instructor in chemistry for two years. In 1921 he entered the doctoral program in physical chemistry at the University of California at Berkeley, where he studied under GILBERT NEWTON LEWIS. He conducted spectroscopic research for his dissertation on the heat capacity and entropy of gases to earn his Ph.D. in 1923. He then conducted postdoctoral research under Niels Bohr at the Institute for Theoretical Physics at the University of Copenhagen, in Denmark, as the American-Scandinavian Foundation Fellow from 1923 through 1924.

Upon his return to the United States in 1925, Urey served as an associate in chemistry at Johns Hopkins University. On June 12, 1926, he married Freida Daum, a bacteriologist, in Lawrence, Kansas, and the couple eventually had four children—Gertrude Elizabeth, Freida Rebecca, Mary Alice, and John Clayton. In 1929 Columbia University in New York City appointed Urey an associate professor of chemistry. The next year he coauthored with A. E. Ruark his first book, entitled *Atoms, Molecules and Quanta*, which became a classic text in the field of physical chemistry.

In 1931 Urey performed his most famous experiment, in collaboration with Ferdinand Brickwedde and George M. Murphy: The trio isolated hydrogen's sole isotope by evaporating 4 liters of liquid hydrogen down to 1 milliliter of so-called heavy hydrogen, or what they called "deuterium." They confirmed their discovery by

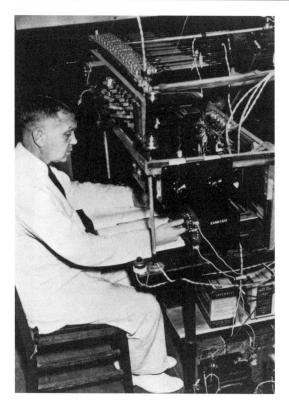

Harold Urey, who won the 1934 Nobel Prize in chemistry for his discovery of the isotope deuterium *(Argonne National Laboratory, courtesy AIP Emilio Segré Visual Archives)*

spectroscopic analysis, which revealed a clear line at the mass of 2, or twice that of hydrogen. In 1933 Columbia named Urey its Ernest Kempton Adams Fellow; that year he also became editor of the *Journal of Chemical Physics*, a post he held for the next seven years. In 1934 Columbia promoted him to a full professorship in the chemistry department.

The Royal Swedish Academy of Sciences recognized Urey's discovery of deuterium by granting him the 1934 Nobel Prize in chemistry; that same year he also won the American Chemical Society's Willard Gibbs Medal. He went on to develop with E. W. Washburn a new method for preparing deuterium by the electrolytic decomposition of water. In 1940 he dropped his line of research to contribute to the war effort as director of war research at Columbia's atomic bomb project. That year he won the Davy Medal of the Royal Society of London, and later in his wartime research he won the 1943 Franklin Medal of the Franklin Institute.

After World War II Urey moved to the Enrico Fermi Institute of Nuclear Studies at the University of Chicago as a Distinguished Service Professor of Chemistry, and then in 1952 he was named the Martin A. Ryerson Professor. That year he published *The Planets: Their Origin and Development,* in which he discussed how the interstellar disbursement of isotopes bears in on the origins of life on Earth. In 1956 he served for one year as the George Eastman Visiting Professor at the University of Oxford in England. In 1958 he took up the emeritus position of professor at large at the University of California at La Jolla, where he stayed for the remainder of his career. In his late research he correlated shifts in global temperature by analyzing the concentration of isotopes in fossils. He died of a heart attack in La Jolla on January 5, 1981.

V

Virtanen, Artturi Ilmari
(1895–1973)
Finnish
Biochemist

Artturi Ilmari Virtanen discovered a process for preserving the nitrogenous content of fodder used as cattle feed, which thereby retained its nutritional value. The AIV method, as it was called (after his initials), also maintained the protein, carotene, and vitamin C content in stored fodder. The milk, in turn, produced by dairy cattle fed with this fodder had a high protein and vitamin A content. In addition, he discovered the necessary presence of the red pigment leghemoglobin (related to the human blood pigment hemoglobin) in leguminous plant roots in order for them to synthesize nitrogenous compounds. For this line of investigation Virtanen won the 1945 Nobel Prize in chemistry.

Virtanen was born on January 15, 1895, in Helsinki, Finland. His mother was Serafina Isotalo, and his father was Kaarlo Virtanen. He received his secondary school education at the classical lyceum in Vyborg, Russia. He matriculated at the University of Helsinki, where he studied chemistry, biology, and physics. He earned his master of science degree in 1916, then served as first assistant in the Central Laboratory of Industries at Helsinki. After a year he

returned to the university for doctoral studies, earning his doctorate in 1919.

Virtanen served as a chemist in the Laboratory of Valio of the Finnish Cooperative Dairies' Association from 1919 to 1920. Later that year he traveled to Zurich, Switzerland, to study physical chemistry under G. Wiegner. Also in 1920 he married Lilja Moisio; together the couple had two sons, Kaarlo and Olavi. In 1921 he returned to direct the coop lab and later studied bacteriology in Stockholm under C. Barthel. From 1923 through 1924 he studied enzymology under H. von Euler in Stockholm. In 1924 he became a *dozent,* or lecturer, at the University of Helsinki.

During 1924 Virtanen demonstrated the necessary presence of cozymase in lactic and propionic acid fermentations; he also experimented on the phosphorylation of sugar. In these studies he noticed the similarity between the primary stages of different fermentation processes and decided to investigate the phenomenon in more depth. These studies also led him to hypothesize that most plant cell proteins are in fact enzymes.

In 1925 this line of investigation veered into the examination of the nitrogenous substances of plants, specifically their production in the root nodules of leguminous plants, and began to show promise in solving the commonly known problem of fodder's low nutritional value.

As the nitrogenous content of fodder decreased over storage time, the fodder's nutritional value also became reduced, but after some experimentation Virtanen discovered that this nitrogenous decrease slowed in the presence of acidity. In this vein he continued to experiment with different levels of acidity, arriving at a mixture of hydrocholoric and sulfuric acids that created the desired degree of acidity to preserve fodder's nitrogenous content, as well as its protein, carotene, and vitamin C content. The AIV method was introduced on Finnish farms in 1929. Cattle fed with this AIV fodder produced protein- and vitamin A–rich milk. This method of fodder preservation was subsequently adapted in many other countries.

In 1929 Virtanen announced the results of another line of investigation that he had been pursuing concurrently: He elucidated the sugar fermentation of dioxyaceton to glycerol and glyceric acid in the presence of phosphates as elicited by coliform bacteria, the first sugar fer-mentation to be chemically elucidated in full. He had conducted these studies in collaboration with H. Karström, who focused his doctoral dissertation on the topic.

In 1931 the Finland Institute of Technology in Helsinki hired Virtanen as a professor of biochemistry, a position he retained the remainder of the decade, until he moved to a similar position as a professor of biochemistry at the University of Helsinki in 1939. The year before he had collected his fodder findings into the text *Cattle Fodder and Human Nutrition with Special Reference to Biological Nitrogen Fixation.*

In 1945 Virtanen won the Nobel Prize in chemistry for his work on the preservation of the nitrogenous content of cattle fodder. He died on November 11, 1973, in Helsinki. His name lives on in more than just the initials of the fodder preservation method that he developed. The University of Kuopio named its A. I. Virtanen Institute after him as well.

W

⊠ Wallach, Otto
(1847–1931)
German
Organic Chemist

Otto Wallach won the 1910 Nobel Prize in chemistry for his pioneering studies of essential oils, or terpenes, a class of complex compounds that were little understood before his investigations. This class is now called isoprenoids and forms the foundation of the perfume industry.

Wallach was born on March 27, 1847, in Königsberg, Germany. His mother was Otillie Thoma; his father, Gerhard Wallach, was a civil servant who rose to the rank of auditor general of Potsdam. Wallach attended gymnasium in Potsdam and in 1867 entered the University of Göttingen to study chemistry under FRIEDRICH WÖHLER, Rudolf Fittig, and Hans Hübner. He spent one semester at the University of Berlin studying under A. W. Hoffmann and G. Magnus, then returned to Göttingen to complete coursework for his doctoral degree in a mere five semesters. He wrote his dissertation under Hübner on the position of isomers in the toluene series to earn his doctorate in 1869.

For the next year Wallach worked in Berlin as an assistant to H. Wichelhaus, collaborating on the nitration of beta-naphthol. In 1870 he traveled to the University of Bonn to work with

August Kekulé, until he was conscripted into military service to serve in the Franco-Prussian War. He returned from the war to Berlin, where the newly established firm Aktien-Gesellschaft für Anilin-Fabrikation (Agfa) hired him; however, he was physically unable to stand the polluted environment of the factory and quit his industry job to return to academia.

In 1872 Wallach returned to the University of Bonn, where he served first as an assistant in the organic chemistry laboratory before his promotion to the rank of *privatdozent,* or lecturer. In 1876 the university appointed him an extraordinary professor. In 1879 the chair of the Department of Pharmacology departed, and the university named Wallach his replacement. Despite the fact that he had little direct experience in that particular field, Wallach threw himself into pharmacological studies, bringing himself up to speed as quickly as possible.

In 1880 Wallach discovered the transformation of azoxybenzene into 4-hydroxy azobenzene in sulphuric acid, or what became known as the Wallach rearrangement. He also discovered the iminochlorides by the action of phosphorous pentachloride on the acid amides. It was at this point in Wallach's career that Kekulé, his former mentor, happened to mention the pharmacology department's supply of essential oils, which was sitting unused in a cupboard. Although Kekulé doubted the possibility of analyzing the oils, Wallach set out to do just that, armed with his persistence and experimental exactitude.

Wallach meticulously analyzed many essential oils, distilling them repeatedly and eliciting reactions with the reagents hydrogen chloride and hydrogen bromide to separate and isolate their components. He published his first paper in 1884, questioning the degree of diversity in this class of complex compounds; a year later he pronounced with confidence that many of the essential oils from different plants were in fact identically structured.

In 1889 the University of Göttingen appointed Wallach to the chair that Victor Meyer had vacated and his mentor Wöhler had inhabited, as well as the concurrent position of director of the Chemical Institute of Göttingen. He held these positions for the next quarter century, continuing his terpene studies while also conducting diverse other researches: the conversion chloral into dichloroacetic acid, the amide chlorides, imide chlorides, aminides, and gloxalines; and azo dyes and diazo compounds. In 1895 his terpene studies culminated with his identification of the structure of the compound alpha-terpineol.

In 1909 Wallach collected his accumulated findings from his essential oil research into a 600-page tome, *Terpene und Campher,* which he dedicated to his students. The next year the Royal Swedish Academy of Sciences recognized the significance of his career-long study of terpenes by awarding him the 1910 Nobel Prize in chemistry. Besides this honor Wallach also received the Imperial Order of the Eagle in 1911, the Davy Medal of the Royal Society of London in 1912, and the Royal Order of the Crown in 1915.

In 1915 Wallach retired when six of his assistants were killed in World War I, but he retained his affiliation with the University of Göttingen, where he continued to conduct research throughout his retirement. A lifelong bachelor, Wallach died on February 26, 1931.

⊠ Werner, Alfred
(1866–1919)
French/Swiss
Chemist

Alfred Werner is considered the father of coordination chemistry, the theory he introduced in 1893 proposing the existence of a new set of chemical bonds (ligands) radiating from a central atom to form a geometric framework of compounds. Over the next two decades he studied and prepared new compounds, culminating in his 1911 discovery of optically active isomer complexes. In 1913 he won the Nobel Prize in chemistry.

Werner was born on December 12, 1866, in Mulhouse, Alsace, France (Germany annexed the territory in 1871). His parents were Jeanne Tesche and J. A. Werner, an iron foundry foreman. He performed his first independent chemistry experiments at the age of 18. While serving in the German military from 1885 to 1886, he attended lectures by Engler at the Karlsruhe Technische Hochschule. The next year he enrolled at the Zurich Polytechnic in Switzerland, studying under Arthur Hantzsch. After he earned his diploma in technical chemistry in 1889, he remained at the polytechnic as an assistant in Professor Lunge's laboratory while conducting collaborative research in organic chemistry with Hantzsch.

In 1890 Werner and Hantzsch coauthored a paper demonstrating the isomerism of the oximes as analogous to the carbon double bonds proposed by Jacaobus Van't Hoff and Joseph Le Bel 16 years earlier. Werner then pursued his doctoral studies in inorganic chemistry (the field of his most significant contributions) through the University of Zurich. He traveled to Paris in 1890 to conduct doctoral research under Pierre-Eugène-Marcelin Berthelot at the Collège de France. While there in 1891 he published a paper on the theory of affinity and valence, redefining August Kekulé's notion of constant valence with the idea that affinity exerts an attractive force from the center of the atom to its surface uniformly.

In 1892 Werner returned to Zurich to submit his doctoral thesis, which continued to lay the groundwork for his coordination theory. In the dissertation he proposed the notion of tervalent nitrogen molecules, a three-cornered configuration of valence bonds emanating from a nitrogen atom that forms the fourth corner. After earning his doctorate he lectured at the polytechnic.

A year later, in the *habilitation* thesis he wrote to qualify for university-level teaching, Werner stated the coordination theory he had been working toward. He proposed a set of secondary bonds (besides ionic and covalent bonds) consisting of ligands emanating from a central atom in any of a small number of integers (commonly four, six, and eight) to form the geometric superstructure of so-called coordination compounds. The electrovalency of these compounds depended on the charge of their ligands: Water and ammonia, neutral themselves, maintained the electrovalency in coordination compounds, while chloride and cyanide transformed the charge of the compounds. His coordination theory thus extended into stereochemistry, as well as suggesting an extension into isomerism.

The University of Zurich immediately appointed him an associate professor (succeeding Victor Merz), and he lectured in organic chemistry. The next year, in 1894, he married Emma Giesker, a German from Zurich, and together the couple had two children, Alfred and Charlotte. In 1895 the university promoted him to a full professorship.

Werner devoted the rest of his career to the preparation of new compounds, focusing on those with coordination numbers of four and six (as well as others). He published more than 150 single-authored papers on his findings, not to mention numerous coauthored papers. In 1904 he published two monographs, *New Ideas in Inorganic Chemistry* and *Textbook of Stereochemistry*.

By 1911 he studied a series of more than 40 optically active complexes to discover a six-coordinate isomer at its base. Two years later Werner received the 1913 Nobel Prize in chemistry (the first awarded in the branch of inorganic chemistry) for his work in coordination chemistry. Besides the Nobel Prize, he also received the Leblanc Medal of the Société Chimique of France.

By the time Werner became a Nobel laureate, he had developed arteriosclerosis, a condition that forced him to abandon his chemistry lecturing in 1915. By 1919 he could no longer hold his professorship. At the end of the year, on November 15, 1919, he died in Zurich at the age of 52.

⊠ **Wieland, Heinrich Otto**
(1877–1957)
German
Organic Chemist

Heinrich Wieland won the 1927 Nobel Prize in chemistry for his investigation of the chemistry of natural substances. His other research ranged widely within this field: He isolated crystalline cyclopeptides of phalloidine and amanitine, allowing him to synthesize these hallucinatory poisons found in the death cap, a "magic" *Amanita* mushroom. He also studied the structure of the opiate morphine and the poison strychnine; his related curare studies and lobelia alkaloid synthesis are considered scientific masterpieces. His experimentation also encompassed biochemistry, as he studied (at one of his student's suggestion) the composition and synthesis of pterin compounds, discovering that this pigment contained in butterfly wings is related to folic acid, a key to human nutrition.

Heinrich Otto Wieland was born on June 4, 1877, in Pforzheim, Germany. His parents, who hailed from the Württemberg region, were Elise Blom and Dr. Theodor Wieland, a pharmaceutical chemist. Starting in 1896, Wieland toured several universities, as was educational custom at the time, studying at the Universities of Munich, Berlin, and Stuttgart before finishing up his education with doctoral studies back at Munich's Baeyer Laboratory under Johannes Thiele to earn his degree in 1901.

In 1904 Wieland received the venia legendi, a certification that allowed him to teach, and became a *privatdozent*, or lecturer, at the University of Munich. In 1908 he married Munich-native Josephine Bartmann, and together the couple had four children—Wolfgang, who became a doctor of pharmaceutical chemistry; Theodor, who became a professor of chemistry at the University of Frankfurt; Otto, who became a professor of biochemistry at the University of Munich; and Eva, who married University of Munich biochemistry professor (and 1964 Nobel laureate in medicine or physiology) Feodor Lynen.

In one of the first of his 400 publications, Wieland announced in 1909 his method of preparing fulminic acid from ethyl alcohol (ethanol), nitric acid, and mercury. In 1911 he oxidized diphenylamine, resulting in tetraaryl-hydrazine, which he then heated to break it down into toluene, a scientific first. In 1913 Wieland moved internally, to the University Chemical Laboratory to become a senior lecturer. The year before he had commenced his bile studies, eventually discovering its structural link to cholesterol (as well as to poisons found on the skins of certain toads, interestingly).

In 1917 Wieland left the university but remained in Munich at the Technische Hochschule; however he almost immediately took a year-long leave to honor FRITZ HABER's request for assistance in chemical warfare research at the Kaiser Wilhelm Institute of Chemistry in Berlin-Dahlem. Upon returning to the Technische Hochschule after World War I, he began his work in biological oxidation, or the release of energy through the conversion of substances such as glucose to carbon dioxide and

water. He proved oxidation to be catalytic dehydrogenation by using palladium to catalyze this reaction anaerobically. Otto Warburg's proof of oxidation as an aerobic process directly contradicted Wieland's results and conclusion, prompting the two scientists to vie for primacy of their respective theories; as it turns out, both scientists were correct, as biological oxidation takes place one way in the presence of oxygen and another way in its absence.

In 1921 Wieland accepted a professorship at the University of Freiburg, only to return to the University of Munich within four years, when RICHARD MARTIN WILLSTÄTTER handpicked him as director of the Baeyer Laboratory and professor of chemistry. Two years later Wieland won the 1927 Nobel Prize in chemistry for his assertion of the basic skeletal structure of the steroid molecule, but his findings came into question, prompting him to collaborate with H. King and O. Rosenheim, who performed X-ray analyses that led to the revision of the steroid structure in 1932.

Wieland spent 20 years as editor of the scientific journal *Justus Liebigs Annalen der Chemie*. He retired to emeritus status at the University of Munich in 1950. Besides the Nobel Prize Wieland also won the Order of Merit and the Otto Hahn Prize. He died on August 5, 1957, in Starnberg. After his death, the Heinrich Wieland Prize was established, awarding 50,000 deutsche marks for the best original scientific work in chemistry, biochemistry, or physiology of nutrition (specifically fats and lipids).

Wilkinson, Sir Geoffrey
(1921–1996)
English
Inorganic Chemist

Geoffrey Wilkinson shared the 1973 Nobel Prize in chemistry with the German chemist ERNST OTTO FISCHER, who simultaneously established the "sandwich" structure of a newly discovered iron compound. Wilkinson made his discovery, which is considered one of the most important contributions to chemistry, in collaboration with the organic chemist ROBERT BURNS WOODWARD.

Wilkinson was born on July 14, 1921, in the village of Springside, near Todmorden in West Yorkshire, England. He was named after his father and grandfather, both of whom were master house painters and decorators. His maternal uncle, an organist and choirmaster who had married into a family that owned a chemical company, introduced Wilkinson to chemistry by bringing the youngster to visit the company's laboratory. In 1932 Wilkinson won a county scholarship to the Todmorden Secondary School (which also produced the Nobel laureate Sir John Cockcroft).

In 1939 Wilkinson won a Royal Scholarship to the University of London's Imperial College of Science and Technology. He graduated at the top of the class of 1941 and remained there to contribute to the war effort by conducting research under H. V. A. Briscoe. In 1942 F. A. Paneth recruited him as a scientific officer in a joint British-American-Canadian atomic energy commission. On January 11, 1943, he set sail on the RMS *Andes* from Greenock, Scotland, bound for Canada; he participated in the secret uranium project in Montreal and later in Chalk River until 1946. He then became the first non-American to gain clearance from the U.S. Atomic Energy Commission to work on nuclear research, joining the Radiation Laboratory at the University of California at Berkeley under GLENN THEODORE SEABORG. During his four-year tenure there he conducted research on nuclear taxonomy while also discovering more isotopes of elements than had any chemist before him.

The Massachusetts Institute of Technology (MIT) hired Wilkinson in 1950 as a research associate, studying the transition metal complexes (such as olefin complexes) as he had in his undergraduate days. After only a year at MIT he was offered an assistant professorship at Harvard University, where he conducted some nuclear studies on excitation functions for protons on cobalt. In 1952 he married Lise Sølver, the only daughter of Professor Svend Aage Schou, the rector of Denmark's Pharmaceutical High School. The couple eventually had two daughters.

In early 1952 a paper by T. J. Kealy and P. L. Pauson on dicyclopentadienyliron in the British journal *Nature* encouraged him to investigate the structure of this new organic compound of iron. On January 30, 1952, at about four o'clock in the afternoon, the idea for a possible structure struck Wilkinson: A central metal could be "sandwiched" by two parallel five-membered rings of cyclopentadiene to create an aromatic compound with a sextet of electrons. Amazingly his Harvard colleague Robert Burns Woodward happened upon the same notion at about the same time, so the two collaborated on a paper entitled "The Structure of Iron Biscyclopentadienyl," published in March 1952. For easier reference they named this compound *ferrocene*, eliciting a flurry of research on metallocenes.

In 1954 Wilkinson synthesized a rhodium metal complex, now known as Wilkinson's catalyst, which has proven instrumental in industrial applications. Before returning to England in 1955 he spent a nine-month sabbatical as a John Simon Guggenheim Fellow in the Copenhagen, Denmark, laboratory of Jannik Bjerrum. In June 1955 Imperial College created Britain's first chair of inorganic chemistry for Wilkinson, who thereby became one of the college's youngest professors ever, at the age of 34. He remained there for the rest of his career as the Sir Edward Frankland Chair of Inorganic Chemistry, conducting research in the Johnson Matthey Laboratories—endowed and built expressly for him.

In 1962 Wilkinson published the first edition of *Advanced Organic Chemistry*, coauthored by one of his former students from the United

States, F. A. Cotton. "Cotton and Wilkinson," as the classic text was commonly referred to, went through six editions. The Royal Society inducted him into its fellowship in 1965. In 1973 Wilkinson shared the Nobel Prize in chemistry with E. O. Fischer; both scientists had almost simultaneously discovered the sandwich structure of metallic compounds some two decades earlier. In 1976 he was knighted. In 1982 Wilkinson edited the nine-volume encyclopedia *Comprehensive Organometallic Chemistry;* in 1995 he edited the 14-volume supplement. He died a year later, on September 26, 1996.

⊠   **Willstätter, Richard Martin**
(1872–1942)
German
Chemist

Richard Martin Willstätter elucidated the composition of chlorophyll, the plant pigment that contributes to the photosynthetic process, revealing it to comprise several separate components. His experiments required him to revive the chromatographic technique first used by the Russian botanist Mikhail Tswett, though Willstätter is often credited with the method as he first applied it to the separation of chemicals. For these researches he won the 1915 Nobel Prize in chemistry.

Willstätter was born on August 13, 1872, in Karlsruhe, Germany, son of a textile merchant. He attended secondary school at the Technische Hochschule in Nuremberg, graduating at age 18 to the University of Munich to study chemistry under ADOLF VON BAEYER. He pursued doctoral studies under A. Einhorn, writing his dissertation on the structure of cocaine (which came to the attention of industrialist Carl Duisberg, who supported much of his work thereafter) to earn his doctorate in 1894.

Willstätter remained in Munich, assisting in Baeyer's laboratory. In 1896 the university appointed him to a lectureship. He devoted his independent research to tropine and atropine, determining their structures to synthesize them by 1898. In 1902 he succeeded J. Thiele as an "extraordinary professor" at the university.

In the summer of 1905 he immigrated to Switzerland to take up a full professorship at the Technische Hochschule in Zurich, where he spent the next seven years productively and happily. In 1905, while studying tropine alkaloids, he prepared cyclo-octatetraene, determining its properties analogous to benzene. During this period, however, both his wife, Sophie Lester, and his son, Ludwig, died. They were survived by his daughter, Ida Margarete.

At the jubilee celebrations for the University of Berlin in 1912, Kaiser Wilhelm established the Society for the Promotion of Scientific Knowledge, which offered Willstätter the directorship of its newly founded Kaiser Wilhelm Institute of Chemistry in Berlin-Dahlem. He installed himself in a well-appointed research laboratory and held an honorary professorship at the University of Berlin. In the few brief years before the eruption of World War I he performed his landmark studies of chlorophyll, a line of investigation he had begun in Zurich.

Willstätter, along with his team of colleagues, applied the technique of chromatography, developed by Mikhail Tswett in 1906, to study chlorophyll; however, Willstätter adapted this technique by passing a chlorophyll solution through a chalk column to separate the compound into its constituent parts. He discovered that chlorophyll is not a homogenous but rather a quadpartite substance, consisting of two green components (blue-green chlorophyll *a* and yellow-green chlorophyll *b* in a ratio of 3:1) and two yellow components (carotene and xanthophyll in a ratio of 2:1). In their most surprising finding, they realized that chlorophyll contains magnesium, pyrrole residues, and phytol (a long-chain alcohol); later, a connection between chlorophyll and the blood pigment hemoglobin was demonstrated.

World War I interrupted these studies, which required expensive materials made scarce by the wartime economy and blockades; at the behest of FRITZ HABER, Willstätter turned his attention to the design of an effective gas mask. In 1916 he returned to the University of Munich to succeed his former mentor, Baeyer, as a full professor of chemistry. During his decade-long tenure there he extended his plant pigmentation studies to examine the process of photosynthesis, as well as the activity of enzymes (though he failed to prove his hypothesis that enzymes are not proteins), adsorbents, metal hydroxides, hydrogels, the degradation of cellulose, and silicic acids. Many of these amounted to seminal studies in the modern discipline of biochemistry.

In 1924 Willstätter abruptly announced his premature retirement in protest of the anti-Semitic practices and policies of the German government (he was of Jewish descent himself). He continued to advise his successor, Heinrich Wieland, as well as some of his former students, in telephone consultations. During his retirement he turned down generous offers from institutions outside the German regime.

In 1938 pursuit by the Gestapo forced Willstätter to flee to Switzerland with the assistance of his former student A. Stoll. He settled in Muroalto, near Locarno, where he composed his autobiography, *From My Life*, edited by Stoll and published posthumously in 1949 (translated into English in 1965). He died of a heart attack in Locarno on August 3, 1942.

⊠ Windaus, Adolf Otto Reinhold
(1876–1959)
German
Organic Chemist

Adolf Windaus won the 1928 Nobel Prize in chemistry for his study of sterols, elucidating their relationship to vitamins (especially vitamin D, or calciferol). His early studies of choles-

terol elucidated much about this substance, which was little understood at the time; Windaus established the relationship between sterols and cholesterol and later demonstrated that sterols represent the building blocks of natural substances.

Adolf Otto Reinhold Windaus was born on December 25, 1876, in Berlin, Germany. His mother was Margarete Elster, and his father, Adolf Windaus, hailed from a line of drapery manufacturers. Windaus graduated from the Französisches Gymnasium in Berlin in 1895, then pursued medical studies at the Universities of Freiburg and Berlin. Two years later, in 1897, he passed his *physicum* examinations, qualifying him to continue studying medicine.

Inspired by the chemistry lectures delivered by EMIL FISCHER in Berlin, Windaus returned to Freiburg to study chemistry while continuing with his concurrent medical studies. He also minored in zoology. He wrote his dissertation on the poisonous effects of the *Digitalis* plant on the heart (which turned out to be related to the steroid cholesterol) under the supervision of H. Kiliani. He earned his doctorate during the winter of 1899–1900, then returned to Berlin for postdoctoral studies under Fischer and OTTO DIELS, with whom he collaborated throughout his life.

In 1901 Windaus again returned to Freiburg to continue working under Kiliani, picking up on the line of investigation he had pursued in his doctoral thesis, which led toward the elucidation of the structure of cholesterol. His findings suggested that sterols, the parent substance of cholesterol, also recur as progenitors of all other natural substances. He proposed this theory in his *habilitation* paper, entitled "On Cholesterol," which earned him a lectureship in 1903.

In 1906 Innsbruck University in Austria hired Windaus as an assistant professor of applied medical chemistry, promoting him to a full professorship in 1913. Two years later he succeeded OTTO WALLACH as professor of chemistry at the

University of Göttingen, back in Germany, where he also served as the director of the laboratory for general chemistry (previously known as the Wöhler Institute). That year, 1915, he married Elisabeth Resau; the couple had three children—Günter (born in 1916), Gustav (born in 1918), and Margarete (born in 1921).

In 1919 Windaus succeeded in transforming cholesterol into cholanic acid, which HEINRICH OTTO WIELAND had derived from bile acid, thus demonstrating bile's relationship to sterols. In 1925 he applied himself to the question of the antirachitic vitamin D (calciferol) at the instigation of the physiologist Alfred Hess, who invited Windaus to New York to collaborate. Hess had irradiated rat food to prevent rats' development of rickets, leading to the hypothesis that the ultraviolet light transformed the cholesterol into vitamin D. Windaus discovered it was not cholesterol but more precisely a minor impurity of dehydrocholesterol that was being transformed Windaus won the 1928 Nobel Prize in chemistry for this work. Then in 1936 he managed to isolate pure vitamin D from cod-liver oil, explicating the process of photochemical synthesis from sterol.

Other significant studies included collaborative research with Franz Knoop, in which the pair attempted to convert sugar into amino acids by introducing ammonia, thus establishing a relationship between sugar and protein. Imidazole derivatives resulted from this reaction, leading Windaus to demonstrate that the protoprotein histidine is an imidazole alanine. Pursuing a similar line of investigation, Windaus discovered the hormone histamine, or imidazole-ethylalanine.

In other vitamin studies, supported by the I. G. Farbenindustry, he proved that vitamin B_1 contains not an imidazole ring but rather a thiazole and a pyrimidine ring, explaining the vitamin as an antineuritic. He also investigated the stereochemistry of hydrogenated ring systems and later proposed the use of chemotherapy for cancer treatment. Windaus retired from the University of Göttingen in 1944. His students

included the 1939 Nobel laureate in chemistry ADOLF BUTENANDT.

Besides the Nobel Prize, Windaus received many other honors, including the 1938 Pasteur Medal, the 1941 Goethe Medal, the 1951 Grand Order of Merit, and the 1956 Grand Order of Merit with Star. He also received honorary doctorates from most of the institutions he worked at in his career, the Universities of Freiburg, Munich, and Göttingen. Windaus died in Göttingen on June 9, 1959.

⊠ **Wittig, Georg**
(1897–1987)
German
Chemist

Georg Wittig shared the 1979 Nobel Prize in chemistry with Herbert Brown for their discoveries of organic synthesis reactions; the Wittig reaction converted carbonyl compounds to alkenes by reacting with a phosphorous reagent. He used this reaction to synthesize vitamin D_3; muscalure, the housefly sex pheromone; and squalene, a synthetic precursor to cholesterol. Other significant research included his discovery of the halogen-metal exchange reaction, which he arrived at concurrently with H. Gilman. He also contributed to his field by directing the chemistry institutes at several institutions, including the Universities of Tübingen and Heidelberg.

Georg Friedrich Karl Wittig was born on June 16, 1897, in Berlin, Germany. He attended secondary school at the Wilhelms Gymnasium and then studied at the Universities of Kassel and Marburg-Lahn. He earned his doctorate from the faculty of chemistry at the latter institution in 1926 and simultaneously passed his *habilitation* to qualify for university teaching as a lecturer. He retained this position for the next half-dozen years. In 1930 he published a textbook on stereochemistry.

In 1932 the Technische Hochschule in Braunschweig appointed Wittig as head of its chemistry department. He remained there for half a decade, then moved on to the University of Freiburg in 1937 as an associate professor. In 1942 he used control experiments on Diels-Alder adducts to prove the short-lived status of dehydrobenzol, a different method than that employed by D. Roberts the same year to arrive at the same conclusion. In 1944 the University of Tübingen appointed him a professor and director of its Institute of Chemistry. He turned down an offer to succeed HERMANN STAUDINGER in a similar position at the University of Freiburg.

In 1949 Wittig discovered the reaction that would eventually be named after him (though he didn't report it until 1954) when he combined a carbonyl compound (aldehyde or ketone) with an organic phosphorous compound (alkylidene-triphenylphosphorane) to create a so-called ylide (pronounced Ill id). The ylide is neutrally charged, as the positive charge of the phosphorous counterbalances the carbanion, which itself acts as a good nucleophile transforming the aldehyde or ketone carbonyl group into a four-membered carbon ring arranged as an alkene and trimethylphosphonium oxide.

What distinguishes the Wittig reaction is this generation of a carbon double bond; the alklidene group of the reagent reacts with the carbonyl's oxygen atom to create a hydrocarbon double bond. The Wittig reaction also distinguishes itself by occurring in neutral, mild conditions. What's more, the architecture of the double bond is unambiguous, making it stable and identifiable, as opposed to the dehydration of alcohol, for example, which yields a varied product.

The Wittig reaction has various practical applications, for example, the synthesis of the housefly sex pheromone muscalure. Other uses include the precholesterol synthetic squalene, as well as vitamin D_3. The Wittig-Horner reaction, which uses a phosphite ester instead of a phosphine to create a more reactive ylide, is a variant reaction that is easier and cheaper to produce than its parent reaction. In 1956, soon after Wittig had announced his discovery, he accepted an offer for a dual post similar to the one he already held. He became professor and director of the department of organic chemistry at the University of Heidelberg, where he remained for the rest of his career.

In 1959 Wittig was named the first Sprague Lecturer (a position established by James Sprague under the support of the Merck Pharmaceutical Company). Earlier honors had already been bestowed upon him—the 1953 Adolf von Baeyer Memorial Medal of the German Chemical Society and a 1957 Silver Medal from the University of Helsinki as well as an honorary doctorate from the Sorbonne that same year—and future honors still awaited him—including the honorary doctorates in 1962 from the Universities of Tübingen and Hamburg, as well as the 1967 Otto Hahn Award for chemistry and physics, the 1973 Roger Adams Award from the American Chemical Society, and the 1975 Karl Ziegler Prize.

In 1979, a dozen years after he had retired to emeritus status in 1967, the Royal Swedish Academy of Sciences capped off his career by granting him the Nobel Prize in chemistry jointly with Herbert Brown for their discovery of reagents (Brown's was boron based) in important organic reactions. Wittig died on August 26, 1987.

⊠ Wöhler, Friedrich
(1800–1882)
German
Chemist

Friedrich Wöhler effectively killed the vitalist theory of organic chemistry when he prepared urea, an organic compound, from ammonium cyanate, an inorganic substance. He also helped to bring order to the chaos of organic chemistry with his discovery of benzoyls, which gave rise to the discovery of other similar compound radicals that served as the building blocks of many compounds.

Wöhler was born on July 31, 1800, in Eschersheim, near Frankfurt am Main, in the Hesse region of Germany. His father was a veterinary surgeon under the crown prince of Hesse-Kassel. Wöhler commenced his medical studies at Marburg University in 1820 but within a year transferred to the University of Heidelberg. There he studied under Leopold Gmelin, who encouraged him to study chemistry; he heeded his mentor's advice, though not before gaining his medical degree in 1823.

Wöhler commenced his chemical studies in the laboratory of JÖNS JAKOB BERZELIUS (with whom he maintained contact thereafter) in Stockholm, Sweden, where he spent the years of 1823 and 1824. In 1825 he landed a position teaching at a technical school in Berlin; it was during his tenure there that he made his landmark discoveries. In 1827 he developed a method for isolating metallic aluminum by heating potassium and anhydrous aluminum chloride to generate a chemical reaction. He utilized this method to prepare various aluminum salts, and the next year he used it to isolate beryllium.

In 1828 Wöhler tolled the death knell of the vitalist theory of organic chemistry when he synthesized an organic compound, urea, out of inorganic ingredients. The vitalist theory, which was the original basis for the distinction between organic (vital) and inorganic (nonvital) chemistry, held that natural (vital) organisms could not be synthesized from unnatural (nonvital)-substances. Wöhler overturned this theory by simply heating the inorganic substance ammonium thiocyanate, transforming it into urea (carbamide). "I must tell you that I can make urea without the use of kidneys, either man or dog," he reportedly said of his discovery.

In 1831 Wöhler moved to a teaching position at a technical school in Kassel. The year before he had commenced his collaboration with Justus von Liebig, a relationship that lasted many years despite the fact the von Liebig often failed to credit his collaborator for their mutual discoveries. In 1830 they proved cyanates and fulminates to be polymers. Two years after that they studied the oil of bitter almonds, discovering it to contain not hydrocyanic acid, as was previously believed, but benzaldehyde. They identified the atom cluster at the foundation of benzaldehyde, as well as many other compounds. They named this cluster, which formed the basis of many organic compounds, *benzoyl*. This discovery helped bring order to the chaotic field of organic chemistry. "Organic chemistry almost drives me mad," Wöhler stated in 1835, continuing that "to me it appears like a primeval tropical forest full of the most remarkable things, a dreadful endless jungle into which one does not dare enter for there seems to be no way out."

In 1836 Wöhler took up a full professorship in chemistry in the medical faculty of the University of Göttingen, succeeding Friedrich Strohmeyer, who had established a solid reputation for his laboratory. Under Wöhler's direction the laboratory became renowned throughout Europe as the preeminent teaching lab. Wöhler retained his position as director of the lab and professor at the university the remainder of his career. In 1837 he collaborated with von Liebig in investigations on uric acid derivatives. He also discovered quinone and hydroquinone, and then combined equimolar amounts of the two to create the molecular complex quinhydrone.

In inorganic chemistry Wöhler isolated boron and silicon, as well as preparing silicon nitride and hydride. After discovering calcium carbide he demonstrated that it reacted with water to generate acetylene (ethyne). His discovery of vanadium and niobium came after others had beat him to the punch; however, he was able to prepare pure titanium and demonstrated its similarity to carbon and silicon.

After 1845 Wöhler conducted scant original research, instead focusing on his teaching. It is estimated that he taught around 8,000 students throughout his career. He survived his wife, who died prematurely. Wöhler died in Göttingen on September 23, 1882.

⊠ Woodward, Robert Burns
(1917–1979)
American
Organic Chemist

Robert Burns Woodward won the 1965 Nobel Prize in chemistry in recognition of the totality of his contributions to the field, too numerous to focus on just one. He made major breakthroughs almost every year of his career, elucidating the structure and synthesizing steroids, alkaloids, antibiotics, tranquilizers, and vitamins, among other substances.

Woodward was born on April 10, 1917, in Boston, Massachusetts, the only child born to Margaret Burns, a native of Glasgow, Scotland. His father, Arthur Woodward, died a year after Woodward's birth, in October 1918. Woodward attended the public schools in suburban Quincy. In 1933, at the age of 16, he entered the Massachusetts Institute of Technology (MIT), which expelled him the next year for academic underperformance.

The year after that, 1935, Woodward reenrolled and swiftly earned his bachelor of science degree in 1936, and his doctorate the very next year, thus completing his entire postsecondary education in a mere three years. He married Irja Pullman in 1938, and they had two daughters—Siiri Anne (born in 1939) and Jean Kirsten (born in 1944). He later married Eudoxia Muller in 1946, and they had a daughter, Crystal Elizabeth, in 1947 and a son, Eric Richard Arthur, in 1953.

After spending a brief postdoctoral stint at the University of Illinois, Woodward returned to Boston, landing a year-long postdoctoral fellowship in 1937 at Harvard University. He remained there for the rest of his career. By 1941 he had become an instructor in chemistry (after serving as a member of the society of fellows over the previous three years). That year he also proposed what became known as the Woodward rules for the ultraviolet spectra of steroids.

In 1944 Woodward firmly established his reputation in the field by synthesizing quinine (in collaboration with William Doering). Harvard promoted him to the rank of assistant professor that year. During World War II he participated in the study of the structure of penicillin, finally elucidating it in 1947. In the meanwhile Harvard had again promoted him, to an associate professorship, in 1946. In 1949 he determined the structure of strychnine.

In 1950 Woodward ascended to a full professorship. The following year he conducted graceful syntheses of the steroids cholesterol and cortisone. In 1952 he collaborated with his Harvard colleague SIR GEOFFREY WILKINSON to propose the "sandwich" structure of ferrocene (interestingly, each arrived at the same conclusion independently on the same campus and then coauthored the paper announcing their theory). In 1953 Harvard named Woodward to an endowed chair as the Morris Loeb Professor of Chemistry. The next year he synthesized yet another steroid, lanosterol, and the alkaloid strychnine, as well as lysergic acid, the basis of the hallucinogenic drug LSD. He continued his pioneering path in 1956, becoming the first to synthesize a tranquilizing drug, reserpine.

In 1960 Harvard named him to a different endowed chair as the Donner Professor of Science, a post he retained thereafter. That year he prepared the plant pigment chlorophyll. In 1961 he followed up on his 1941 Woodward rules with the octant rule, which he formulated with his colleagues C. Djerassi, A. Moscowitz, and W. Moffitt. In 1962 he and his team generated the antibiotic tetracycline, and in 1963 the Woodward Research Institute in Basel, Switzerland, appointed him its director. From 1964 to 1965 he collaborated with Roald Hoffmann to formulate the rules governing the conservation of orbital symmetry. In 1965 he returned to antibiotic research by creating cephalosporin C.

That same year Woodward won the 1965 Nobel Prize in chemistry for his cumulative achievements in the field. He did not announce any major advances over the next several years as he devoted himself to the synthesis of vitamin B_{12}, or cyanocobalamin, in collaboration with the Swiss chemist Albert Eschenmoser. They announced success in 1971, and by 1976 they completed a total synthesis. In the meantime he spent the year of 1973–74 serving as the Todd Professor of Chemistry and Fellow of Christ's College of Cambridge University in England.

Besides the Nobel, Woodward received numerous other honors, including the 1959 Davy Medal of the Royal Society of London, the 1964 National Medal of Science, the 1968 Lavoisier Medal of the Société Chimique de France, and the 1970 Order of the Rising Sun, Second Class, from the emperor of Japan. Woodward was renowned for his love of coffee, cigarettes, and whiskey. He died at the age of 62 on July 8, 1979, in Cambridge, Massachusetts. Had he been alive, Woodward almost certainly would have shared the 1981 Nobel Prize in chemistry with Hoffmann (as well as K. Fukui) for their work on the conservation of orbital symmetry, also known as the Woodward-Hoffman rules.

⊠ **Wrinch, Dorothy Maud**
(1894–1976)
Argentine/English/American
Biochemist, Mathematician

Dorothy Maud Wrinch was considered the "woman Einstein" for her interdisciplinary brilliance. She was the first woman to receive a doctor of science degree (her second doctorate) from Oxford University. She is best remembered

as the first scientist to propose a structure for protein, what she called the "cyclol theory," which she generated based on mathematical formulas. The theory created controversy, and it was later discovered to apply only to certain alkaloids.

Wrinch was born on September 13, 1894, in Rosario, Argentina. Her parents, Hugh Edward Hart and Ada Minnie Souter Wrinch, eventually returned to their homeland of England, where Wrinch attended Surbiton High School, in a London suburb. In 1913 she matriculated at Girton College of Cambridge University. In 1916, she earned the title of "wrangler" (the highest rank attainable) in mathematics on her final examination to earn her bachelor of arts degree. She remained at Girton to earn her first of five graduate degrees, a master of arts in mathematics, in 1918.

Wrinch spent the next two years lecturing in pure mathematics at University College London. While there she earned two more advanced degrees, a master of science in 1920 and a doctor of science in 1921. The latter year she returned to Girton as a research scholar. Two years later, in 1923, she moved to Balliol College of Oxford University and also married John William Nicholson. She gave birth to their only child, a daughter named Pamela, in 1927.

In 1924 Wrinch earned her fourth graduate degree, a master of arts degree from Oxford. Five years later, in 1929, she became the first woman to earn a doctor of science degree from Oxford. In 1930 she published *The Retreat from Parenthood* under the pen name Jean Ayling. Then in 1931 she secured a three-year research fellowship at Oxford. Over those three years, though, she traveled twice to study on the Continent, first at the University of Vienna from 1931 through 1932 and later at the University of Paris from 1933 to 1934.

Throughout her tenure at Oxford, from 1923 to 1939, Wrinch served as a lecturer in mathematics, as director of studies for women, and as a member of the physical sciences department. She also tutored mathematics at the five women's colleges in the region during this period. Her early publications, dating from 1918 through 1932, numbered 36, split almost evenly between pure and applied mathematics (20) and scientific methodology and philosophy (16).

In 1935 the Rockefeller Foundation granted Wrinch a six-year research fellowship to apply mathematical models to the study of biochemical structures. Within this first year she proposed her cyclol theory of protein structure based on mathematical constructs. Specifically she drew a parallel between mathematical symmetry and chemical covalent bonding between two adjacent amino acids. This interdisciplinary theory drew fire from the scientific community for its radical revisionism; LINUS CARL PAULING particularly attacked Wrinch's cyclol theory in favor of his own alpha helix theory. Ironically, both theories proved incorrect, though the former applies only to certain alkaloids.

In addition to this professional discord, strife filled Wrinch's personal life, as her marriage unraveled due to her husband's alcoholism. Their marriage was dissolved in 1938. While serving as a research fellow at Somerville College of Oxford University from 1939 to 1941, she visited the United States, where she concurrently served as a lecturer in chemistry at Johns Hopkins University in Baltimore, Maryland. She and her daughter remained in the United States thereafter. In 1941 she married Otto C. Glaser (who died in 1950). Over the next year she served as a visiting professor in natural sciences at Amherst College and Mount Holyoke College, both in Massachusetts, while also lecturing in physics at Smith College in the same state. She retained the last position for the next dozen years, until Smith appointed her to a visiting professorship in 1954. She retained this position the remainder of her career.

Considered the "woman Einstein" for her interdisciplinary brilliance, Dorothy Maud Wrinch was in 1921 the first woman to receive a doctor of science degree from Oxford University, one of four graduate degrees she earned. *(College Archives, Smith College)*

In this latter period of her career Wrinch published three books that exemplified the breadth of her expertise: *Fourier Transforms and Structure Factor*, published in 1946; *Chemical Aspects of the Structure of Small Peptides: An Introduction*, published in 1960; and *Chemical Aspects of Polypeptide Chain Structure and the Cyclol Theory*, published in 1965. A prolific writer, she published 192 professional papers throughout her career on topics as diverse as X-ray crystallography, mineralogy, and the structure of globular protein molecules. Wrinch retired to Woods Hole, Massachusetts, in 1971, and she died in 1976.

Z

Ziegler, Karl
(1898–1973)
German
Organic Chemist

Karl Ziegler won the 1963 Nobel Prize in chemistry with GIULIO NATTA for their work on high polymers, specifically discovering a catalyst that polymerized very long and very standardized polymer chains that proved very useful for industrial applications. He also discovered that heavy metals, such as nickel, catalyzed the polymerization of ethylene at atmospheric pressure (previous processes had required high pressures).

Ziegler was born on November 26, 1898, in Helsa, near Kassel, Germany. His father was a Lutheran minister. Ziegler studied chemistry under Karl von Auwers at the University of Marburg-Lahn, earning his doctorate in 1920. In 1923 he submitted his *habilitation* thesis on triaryl methyl radicals to qualify as a lecturer at the University of Frankurt am Main. The year before he married Maria Kurtz; together the couple had two children—Marianne Witte, who became a medical doctor, and Erhart Ziegler, who became a physicist and patent attorney.

After only a brief stay at Frankfurt, Ziegler moved to the University of Heidelberg to serve as a lecturer until 1927, when the university promoted him to a professorship. He extended his *habilitation* thesis research into the study of free radicals of trivalent carbon, studying their reaction to the substitution in hexaphenyl ethane on their dissociation equilibrium. He also prepared new radicals through reactions between alkali metals and organo-metallic compounds on the one hand and ethers and organic halides on the other, demonstrating that the metal alkyls polymerized isoprene and butadiene. He continued this line of research for more than a quarter of a century.

Starting in 1928 Ziegler conducted concurrent research on the catalysis of organo-alkali and aluminum compounds, continuing this line of investigation throughout his career. Proceeding from the study of lithium alkyls to the similar organo-aluminum compounds, he managed to generate reactions between aluminum alkyls and ethylene to create hydrocarbons containing from four to 40 molecules of carbon, a significant number despite the fact that he sought to create even higher polymers. This line of research established a new branch of organo-metallic chemistry.

In 1933 Ziegler discovered a method for producing large carbon rings later utilized by the perfume industry to synthesize musks. (Ziegler always kept potential industrial applications at the forefront of his mind in his experimentation.) In 1936 the University of Halle-Saale appointed Ziegler a professor and director of its Chemical Institute. Also in that year he lectured at the University of Chicago as a visiting professor.

In 1943 Ziegler became director of the Kaiser Wilhelm Institute for Coal Research (later renamed the Max Planck Institute) in Mülheim an der Ruhr. Ziegler's most significant discoveries came a decade later, in 1953. In collaboration with his student E. Holzkamp, he heated ethylene with aluminum trialkyl, unexpectedly resulting in the creation of dimer butylenes. While searching for an explanation, they realized that the autoclave used in the experiment had previously been hydrogenating with the catalyst colloidal nickel, which left deposits on the apparatus; they thus discovered nickel to be the catalyst of this surprise reaction. Furthermore they conducted their experiment at normal atmospheric pressure, instead of the high pressures previously required to polymerize ethylene.

Also in 1953 Ziegler and Natta collaborated on the research that earned them the Nobel Prize in chemistry a decade later, their work on high polymers. They discovered that a catalyst similar to those Ziegler had used previously, triethyl aluminum combined with titanium tetrachloride, polymerized isoprene into long, standardized chains that simulated the structure of rubber. The exactitude and regularity of the polymers created by these stereospecific catalysts made them extremely valuable in industrial applications for the synthesis of plastics, fibers, and films.

Besides the Nobel Prize Ziegler also won the Liebig Medal from the Verein Deutscher Chemiker and the Lavoisier Medal of the Société Chimique de France, among numerous others. In 1971 the Royal Society of London inducted him as a foreign member. Throughout his career he supervised 150 doctoral dissertations. Near the end of his career he endowed the Ziegler Fund at the Max Planck Institute at Mülheim with 40 million deutsche marks in royalties from the Ziegler polyethylene process. He retired from his position as director of the insti-

tute in 1969 and died in Mülheim on August 11 or 12, 1973.

⊠ **Zsigmondy, Richard Adolf**
(1865–1929)
Austrian/German
Chemist

Richard Adolf Zsigmondy invented the ultramicroscope, an instrument that increased the magnification capabilities of a standard microscope by illuminating the sample plate vertically instead of horizontally. This invention facilitated his research on the colloidal chemistry of glass, allowing him to view individual particles in a colloidal solution. For this work, as well as for his invention of the ultramicroscope, Zsigmondy received the 1925 Nobel Prize in chemistry.

Zsigmondy was born on April 1, 1865, in Vienna, Austria, just before formation of the Austro-Hungarian Empire. His mother was Irma von Szakmary; his father, Adolf Zsigmondy, was a dentist who had invented several surgical instruments and had published various scientific works. Zsigmondy's interest in chemistry was sparked at an early age when he read chemistry textbooks by Adolf Stoeckhardt (*Schule der Chemie*) and JÖNS JAKOB BERZELIUS. He also conducted chemical experiments in a small laboratory that he set up in his home.

Zsigmondy attended the University of Vienna for one year, studying quantitative analysis under E. Ludwig of the medical faculty, before transferring to Vienna's Technische Hochschule. In 1887 he moved to the University of Munich, where he studied organic chemistry under W. von Miller. He earned his doctorate in organic chemistry in 1889 and remained at Munich as an assistant in von Miller's laboratory. He subsequently went to Berlin to work in a similar position, as an assistant in the physicist August Kundt's laboratory, studying inorganic inclusions in glass.

In 1893 Zsigmondy qualified as a lecturer in chemical technology at the Technische Hochschule in Graz, where he also taught and conducted research on luster colors for glass and china and, more generally, on the chemistry of colloids. In 1897 the glass manufacturing company Schott und Genossen in Jena, Germany, hired Zsigmondy to continue his research on the colloid chemistry of glass. There he invented Jena milk glass, which became renowned. He also conducted more general research on colloids, recognizing that colloidal inclusions lend color or opacity to glass.

Along this line of research Zsigmondy experimented on the red fluids resulting from the reduction of gold sols discovered by MICHAEL FARADAY, recognizing them to be colloidal analogs of ruby glass. He devised a replicable means of preparing these red fluids, which suspended particles separated by electrical charge. In 1898 he demonstrated that gold sols are a mixture of tiny gold and stannic acid particles. He also demonstrated that the color of these sols could change due to coagulation, which changed the particle size. He remained with the glassworks until 1900, when he established his own private laboratory to conduct independent studies (though this did not preclude him from doing work for Schott und Genossen).

In 1903 Zsigmondy collaborated with H. F. W. Siedentopf to develop the ultramicroscope. Instead of illuminating samples by a light source shining parallel to the microscope's axis, as was customary with conventional microscopes, the pair designed a perpendicular illumination system, which used the Tyndall effect of scattering light to resolve individual particles as small as 3 nanometers in diameter. He could also estimate their size by counting the number of particles in a specific volume. Despite the fact that Zsigmondy was no longer directly affiliated with the glassworks, Schott und Genossen director Ernst Abba gave the chemist carte blanche to use the company's laboratories and materials in designing and constructing the prototypical ultramicroscope.

That same year, 1903, Zsigmondy married Laura Luise Müller, who was the daughter of the pathological anatomy lecturer Wilhelm Müller. Together the couple had two daughters, Kathe and Annemarie (who married Zsigmondy's young colleague ERICH HÜCKEL in 1925). In 1907 the University of Göttingen appointed Zsigmondy a professor and director of its institute of inorganic chemistry based on the reputation he had created for himself by inventing the ultramicroscope. At the university institute he studied silica and soap gels, as well as ultrafiltration.

In 1925 Zsigmondy won the Nobel Prize in chemistry for his invention of the ultramicroscope and his advances in colloidal chemistry. In his career he published two significant books, *Lehrbuch der Kolloidcmemie* and *Über das Kolloide Gold,* as well as a collection of papers on his theories of adsorption, *Kolloidforschung in Einzeldarstellungen,* which his son-in-law Hückel contributed to. He retired in mid-1929 and died a few months later, on September 24, in Göttingen.

Entries by Country of Birth

ARGENTINA
Wrinch, Dorothy Maud 238

AUSTRALIA
Cornforth, Sir John
    Warcup 43
Nyholm, Sir Ronald
    Sydney 157

AUSTRIA
Kuhn, Richard 118
Perutz, Max Ferdinand 170
Zsigmondy, Richard
    Adolf 242

AUSTRIA-HUNGARY
Cori, Gerty Theresa
    Radnitz
    (Czechoslovakian) 41
Heyrovský, Jaroslav
    (Czechoslovakian) 92
Le Beau, Désirée (Polish) 124
Pregl, Fritz (Slovenian) 174
Prelog, Vladimir
    (Bosnian) 175
Ružička, Leopold
    (Croatian) 191

BELGIUM
Baekeland, Leo Hendrik 11

CHINA
Lee, Yuan Tseh
    (Taiwanese) 126

DENMARK
Brønsted, J. N. 25

EGYPT
Hodgkin, Dorothy
    Crowfoot 97
Maria the Jewess 135

FINLAND
Virtanen, Artturi Ilmari 225

FRANCE
Ampère, André-Marie 2
Becquerel, Antoine-Henri 15
Friedel, Charles 72
Gay-Lussac, Joseph-Louis 75
Grignard, François-
    Auguste-Victor 81
Joliot-Curie, Frédéric 105
Joliot-Curie, Irène 107
Lavoisier, Antoine-
    Laurent 123
Lehn, Jean-Marie 127
Leloir, Luis Federico 128
Moissan, Ferdinand-
    Frédéric-Henri 142

Pasteur, Louis 165
Proust, Joseph-Louis 179
Ramart-Lucas, Pauline 181
Sabatier, Paul 195
Werner, Alfred 228

GERMANY
Alder, Jurt 1
Baeyer, Adolf von 13
Bergius, Friedrich 19
Bosch, Carl 23
Buchner, Eduard 27
Bunsen, Robert Wilhelm 28
Butenandt, Adolf 30
Diels, Otto 57
Eigen, Manfred 59
Fischer, Emil 65
Fischer, Ernst Otto 66
Fischer, Hans 68
Haber, Fritz 83
Hahn, Otto 85
Herzberg, Gerhard 91
Hückel, Erich 100
Staudinger, Hermann 209
Wallach, Otto 227
Wieland, Heinrich Otto 229
Willstätter, Richard
    Martin 232
Windaus, Adolf Otto
    Reinhold 233

Wittig, Georg   234
Wöhler, Friedrich   235
Ziegler, Karl   241

PRUSSIA
Nernst, Walther Hermann 150

GREAT BRITAIN

England
Aston, Francis William   7
Barton, Derek H.R.   14
Cavendish, Henry   39
Clark, Josiah Latimer   40
Crick, Francis Harry
    Compton   45
Dalton, John   51
Daniell, John Frederic   52
Davy, Sir Humphry   54
Faraday, Michael   63
Franklin, Rosalind   70
Harden, Arthur   87
Haworth, Sir Walter   89
Hinshelwood, Sir Cyril   95
Hodgkin, Dorothy
    Crowfoot   97
Ingold, Sir Christopher   103
Kendrew, Sir John
    Cowdery   114
Martin, Archer John
    Porter   136
Newlands, John
    Alexander Reina   151
Norrish, Ronald G. W.   154
Porter, Sir George   172
Priestley, Joseph   176
Robinson, Sir Robert   186
Sanger, Frederick   196
Soddy, Frederick   204
Synge, Richard   214
Wilkinson, Sir Geoffrey   230

Scotland
Graham, Thomas   79
Ramsay, Sir William   182
Todd, Baron Alexander   219

HUNGARY
Polanyi, Michael   171
Telkes, Maria   217

Austria-Hungary
Cori, Gerty Theresa Radnitz
    (Czechoslovakian)   41
Heyrovský, Jaroslar
    (Czechoslovakian)   92
Le Beau, Désirée (Polish)   124
Pregl, Fritz (Slovenian)   174
Prelog, Vladimir
    (Bosnian)   175
Ružičkar, Leopold
    (Croatian)   191

ITALY
Avogadro, Lorenzo Romano
    Amedeo Carlo   9
Natta, Giulio   149

JAPAN
Fukui, Kenichi   73
Saruhashi, Katsuko   197

LATVIA
Ostwald, Wilhelm   163

LITHUANIA
Klug, Sir Aaron   117

MEXICO
Molina, Mario   143

NETHERLANDS
Crutzen, Paul J.   47
Debye, Peter   55

NEW ZEALAND
Rutherford, Ernest, Lord   189

NORWAY
Hassel, Odd   88
Onsager, Lars   160

POLAND
Curie, Marie   48

RUSSIA
Karrer, Paul   113
Mendeleyev, Dmitri
    Ivanovich   140
Prigogine, Ilya   177
Semenov, Nikolai   200

SPAIN
Ochoa, Severo   159

SWEDEN
Arrhenius, Svante August   5
Berzelius, Jöns Jakob   20
Nobel, Alfred   153
Svedberg, Theodor   213
Tiselius, Arne Wilhelm
    Kaurin   218

UNITED STATES
Anfinsen, Christian
    Boehmer   4
Berg, Paul   17
Bishop, Hazel Gladys   21
Brady, St. Elmo   24
Calvin, Melvin   33
Carothers, Wallace Hume   34
Carr, Emma Perry   36
Carver, George
    Washington   37
Elion, Gertrude Belle   60
Flory, Paul   69
Gilbert, Walter   76
Good, Mary Lowe   78
Hill, Henry Aaron   94
Hoobler, Icie Gertrude
    Macy   98
Jones, Mary Ellen   108
Just, Ernest   110
King, Reatha Clark   115
Langmuir, Irving   121
Lewis, Gilbert Newton   130

Libby, Willard Frank	132	Richards, Ellen Henrietta
Massie, Samuel Proctor	137	Swallow	183
McMillan, Edwin M.	139	Richards, Theodore
Moore, Stanford	145	William	185
Mulliken, Robert S.	146	Rowland, Frank
Northrop, John Howard	155	Sherwood	188
Osborn, Mary Jane	162	Seaborg, Glenn
Pauling, Linus Carl	166	Theodore	198
Pennington, Mary Engle	168	Simon, Dorothy Martin	201
		Singer, Maxine	202
		Solomon, Susan	206
		Stanley, Wendell
		Meredith	207
		Stein, William H.	210
		Sumner, James
		Batcheller	211
		Urey, Harold	223
		Woodward, Robert Burns	237

Entries by Country of Major Scientific Activity

ARGENTINA
Leloir, Luis Federico — 128

AUSTRALIA
Nyholm, Sir Ronald Sydney — 157

AUSTRIA
Pregl, Fritz — 174

BELGIUM
Prigogine, Ilya — 177

CANADA
Herzberg, Gerhard — 91
Rutherford, Ernest, Lord — 189

CZECHOSLOVAKIA
Heyrovský, Jaroslav — 92

DENMARK
Brønsted, J. N. — 25

EGYPT
Maria the Jewess — 135

FINLAND
Virtanen, Artturi Ilmari — 225

FRANCE
Ampère, André-Marie — 2
Becquerel, Antoine-Henri — 15
Curie, Marie — 48
Friedel, Charles — 72
Gay-Lussac, Joseph-Louis — 75
Grignard, François-Auguste-Victor — 81
Joliot-Curie, Frédéric — 105
Joliot-Curie, Irène — 107
Lavoisier, Antoine-Laurent — 123
Lehn, Jean-Marie — 127
Moissan, Ferdinand-Frédéric-Henri — 142
Pasteur, Louis — 165
Proust, Joseph-Louis — 179
Ramart-Lucas, Pauline — 181
Sabatier, Paul — 195

GERMANY
Alder, Kurt — 1
Baeyer, Adolf von — 13
Bergius, Friedrich — 19
Bosch, Carl — 23
Buchner, Eduard — 27
Bunsen, Robert Wilhelm — 28
Butenandt, Adolf — 30
Crutzen, Paul J. — 47
Debye, Peter — 55
Diels, Otto — 57
Eigen, Manfred — 59
Fischer, Emil — 65
Fischer, Ernst Otto — 66
Fischer, Hans — 68
Haber, Fritz — 83
Hahn, Otto — 85
Hückel, Erich — 100
Kuhn, Richard — 118
Nernst, Walther Hermann — 150
Ostwald, Wilhelm — 163
Polanyi, Michael — 171
Staudinger, Hermann — 209
Wallach, Otto — 227
Wieland, Heinrich Otto — 229
Willstätter, Richard Martin — 232
Windaus, Adolf Otto Reinhold — 233
Wittig, Georg — 234
Wöhler, Friedrich — 235
Ziegler, Karl — 241
Zsigmondy, Richard Adolf — 242

GREAT BRITAIN

England
Aston, Francis William — 7

Barton, Derek H. R. 14
Cavendish, Henry 39
Clark, Josiah Latimer 40
Cornforth, Sir John
    Warcup 43
Crick, Francis Harry
    Compton 45
Dalton, John 51
Daniell, John Frederic 52
Davy, Sir Humphry 54
Faraday, Michael 63
Franklin, Rosalind 70
Graham, Thomas 79
Harden, Arthur 87
Haworth, Sir Walter 89
Hinshelwood, Sir Cyril 95
Hodgkin, Dorothy
    Crowfoot 97
Ingold, Sir Christopher 103
Kendrew, Sir John
    Cowdery 114
Klug, Sir Aaron 117
Martin, Archer John
    Porter 136
Newlands, John
    Alexander Reina 151
Norrish, Ronald G. W. 154
Perutz, Max Ferdinand 170
Polanyi, Michael 171
Porter, Sir George 172
Priestley, Joseph 176
Ramsay, Sir William 182
Robinson, Sir Robert 186
Rutherford, Lord Ernest 189
Sanger, Frederick 196
Soddy, Frederick 204
Synge, Richard 214
Todd, Baron Alexander 219
Wilkinson, Sir Geoffrey 230
Wrinch, Dorothy Maud 238

Scotland
Graham, Thomas 79
Synge, Richard 214

ITALY
Avogadro, Lorenzo Romano
    Amedeo Carlo 9
Natta, Giulio 149

JAPAN
Fukui, Kenichi 73
Saruhashi, Katsuko 197

NORWAY
Hassel, Odd 88

RUSSIA
Mendeleyev, Dmitri
    Ivanovich 140
Semenov, Nikolai 200

SPAIN
Proust, Joseph-Louis 179

SWEDEN
Arrhenius, Svante August 5
Berzelius, Jöns Jakob 20
Nobel, Alfred 153
Svedberg, Theodor 213
Tiselius, Arne Wilhelm
    Kaurin 218

SWITZERLAND
Debye, Peter 55
Karrer, Paul 113
Prelog, Vladimir 175
Ružička, Leopold 191
Staudinger, Hermann 209
Werner, Alfred 228

UNITED STATES
Anfinsen, Christian
    Boehmer 4
Baekeland, Leo Hendrik 11
Berg, Paul 17
Bishop, Hazel Gladys 21
Brady, St. Elmo 24
Calvin, Melvin 33
Carothers, Wallace Hume 34

Carr, Emma Perry 36
Carver, George
    Washington 37
Cori, Gerty Theresa
    Radnitz 41
Crutzen, Paul J. 47
Debye, Peter 55
Elion, Gertrude Belle 60
Flory, Paul 69
Gilbert, Walter 76
Good, Mary Lowe 78
Hill, Henry Aaron 94
Hoobler, Icie Gertrude
    Macy 98
Jones, Mary Ellen 108
Just, Ernest 110
King, Reatha Clark 115
Langmuir, Irving 121
Le Beau, Désirée 124
Lee, Yuan Tseh 126
Lewis, Gilbert Newton 130
Libby, Willard Frank 132
Massie, Samuel Proctor 137
McMillan, Edwin M. 139
Molina, Mario 143
Moore, Stanford 145
Mulliken, Robert S. 146
Northrup, John Howard 155
Ochoa, Severo 159
Onsager, Lars 160
Osborn, Mary J. 162
Pauling, Linus Carl 166
Pennington, Mary Engle 168
Richards, Ellen Henrietta
    Swallow 183
Richards, Theodore
    William 185
Rowland, Frank
    Sherwood 188
Seaborg, Glenn
    Theodore 198
Simon, Dorothy
    Martin 201
Singer, Maxine 202

Solomon, Susan   206
Stanley, Wendell
    Meredith   207
Stein, William H.   210
Sumner, James Batcheller   211
Telkes, Maria   217
Urey, Harold   223
Woodward, Robert Burns   237
Wrinch, Dorothy Maud   238

ENTRIES BY YEAR OF BIRTH

FIRST CENTURY

Maria the Jewess 135

1700–1749

Cavendish, Henry 39
Lavoisier, Antoine-
  Laurent 123
Priestley, Joseph 176

1750–1774

Dalton, John 51
Proust, Joseph-Louis 179

1775–1799

Ampère, André-Marie 2
Avogadro, Lorenzo Romano
  Amedeo Carlo 9
Berzelius, Jöns Jakob 20
Daniell, John Frederic 52
Davy, Sir Humphry 54
Faraday, Michael 63
Gay-Lussac, Joseph-Louis 75

1800–1809

Graham, Thomas 79
Wöhler, Friedrich 235

1810–1819

Bunsen, Robert Wilhelm 28

1820–1829

Clark, Josiah Latimer 40
Pasteur, Louis 165

1830–1839

Baeyer, Adolf von 13
Friedel, Charles 72
Mendeleyev, Dmitri
  Ivanovich 140
Newlands, John Alexander
  Reina 151
Nobel, Alfred 153

1840–1849

Richards, Ellen Henrietta
  Swallow 183
Wallach, Otto 227

1850–1859

Arrhenius, Svante August 5
Becquerel, Antoine-Henri 15
Fischer, Emil 65
Moissan, Ferdinand-
  Frédéric-Henri 142
Ostwald, Wilhelm 163
Ramsay, Sir William 182
Sabatier, Paul 195

1860–1869

Baekeland, Leo Hendrik 11

Buchner, Eduard 27
Carver, George
  Washington 37
Curie, Marie 48
Haber, Fritz 83
Harden, Arthur 87
Nernst, Walther
  Hermann 150
Pregl, Fritz 174
Richards, Theodore
  William 185
Werner, Alfred 228
Zsigmondy, Richard
  Adolf 242

1870–1879

Aston, Francis William 7
Bosch, Carl 23
Brønsted, J. N. 25
Diels, Otto 57
Grignard, François-
  Auguste-Victor 81
Hahn, Otto 85
Lewis, Gilbert Newton 130
Pennington, Mary Engle 168
Rutherford, Ernest, Lord 189
Soddy, Frederick 204
Wieland, Heinrich Otto 229
Willstätter, Richard
  Martin 232

Windaus, Adolf Otto
    Reinhold   233

1880–1889

Bergius, Friedrich   19
Brady, St. Elmo   25
Carr, Emma Perry   36
Debye, Peter   55
Fischer, Hans   68
Haworth, Sir Walter   89
Just, Ernest   110
Karrer, Paul   113
Langmuir, Irving   121
Ramart-Lucas, Pauline   181
Robinson, Sir Robert   186
Ružička, Leopold   191
Staudinger, Hermann   209
Sumner, James Batcheller   211
Svedberg, Theodor   213

1890–1899

Carothers, Wallace Hume   34
Cori, Gerty Theresa
    Radnitz   41
Hassel, Odd   88
Heyrovský, Jaroslav   92
Hinshelwood, Sir Cyril   95
Hoobler, Icie Gertrude
    Macy   98
Hückel, Erich   100
Ingold, Sir Christopher   103
Joliot-Curie, Irène   107
Mulliken, Robert S.   146
Norrish, Ronald G. W.   154
Northrop, John Howard   155
Polanyi, Michael   171
Semenov, Nikolai   200
Urey, Harold   223
Virtanen, Artturi Ilmari   225
Wittig, Georg   234
Wrinch, Dorothy Maud   238
Ziegler, Karl   241

1900–1909

Alder, Kurt   1
Bishop, Hazel Gladys   21
Butenandt, Adolf   30
Herzberg, Gerhard   91
Joliot-Curie, Frédéric   105
Kuhn, Richard   118
Le Beau, Désirée   124
Leloir, Luis Federico   128
Libby, Willard Frank   132
McMillan, Edwin M.   139
Natta, Giulio   149
Ochoa, Severo   159
Onsager, Lars   160
Pauling, Linus Carl   166
Prelog, Vladimir   175
Stanley, Wendell
    Meredith   207
Telkes, Maria   217
Tiselius, Arne Wilhelm
    Kaurin   218
Todd, Baron Alexander   219

1910–1919

Anfinsen, Christian   4
Barton, Derek H. R.   14
Calvin, Melvin   33
Cornforth, Sir John
    Warcup   43
Crick, Francis Harry
    Compton   45
Elion, Gertrude Belle   60
Fischer, Ernst Otto   66
Flory, Paul   69
Fukui, Kenichi   73
Hill, Henry Aaron   84
Hodgkin, Dorothy
    Crowfoot   97
Kendrew, Sir John
    Cowdery   114
Martin, Archer John
    Porter   136

Massie, Samuel Proctor   137
Moore, Stanford   145
Nyholm, Sir Ronald
    Sydney   157
Perutz, Max Ferdinand   170
Prigogine, Ilya   177
Sanger, Frederick   196
Seaborg, Glenn
    Theodore   198
Simon, Dorothy Martin   201
Stein, William H.   210
Synge, Richard   214
Woodward, Robert Burns   237

1920–1929

Berg, Paul   17
Eigen, Manfred   59
Franklin, Rosalind   70
Jones, Mary Ellen   108
Klug, Sir Aaron   117
Osborn, Mary J.   162
Porter, Sir George   172
Rowland, Frank
    Sherwood   188
Saruhashi, Katsuko   197
Wilkinson, Sir Geoffrey   230

1930–1939

Crutzen, Paul J.   47
Gilbert, Walter   76
Good, Mary Lowe   78
King, Reatha Clark   115
Lee, Yuan Tseh   126
Lehn, Jean-Marie   127
Singer, Maxine   202

1940–1949

Molina, Mario   143

1950–1959

Solomon, Susan   206

CHRONOLOGY

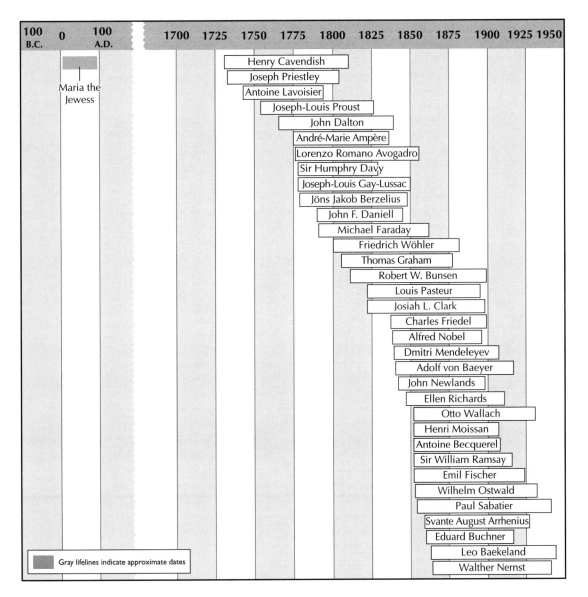

100 B.C.	0	100 A.D.		1700	1725	1750	1775	1800	1825	1850	1875	1900	1925	1950

Maria the Jewess

Henry Cavendish
Joseph Priestley
Antoine Lavoisier
Joseph-Louis Proust
John Dalton
André-Marie Ampère
Lorenzo Romano Avogadro
Sir Humphry Davy
Joseph-Louis Gay-Lussac
Jöns Jakob Berzelius
John F. Daniell
Michael Faraday
Friedrich Wöhler
Thomas Graham
Robert W. Bunsen
Louis Pasteur
Josiah L. Clark
Charles Friedel
Alfred Nobel
Dmitri Mendeleyev
Adolf von Baeyer
John Newlands
Ellen Richards
Otto Wallach
Henri Moissan
Antoine Becquerel
Sir William Ramsay
Emil Fischer
Wilhelm Ostwald
Paul Sabatier
Svante August Arrhenius
Eduard Buchner
Leo Baekeland
Walther Nernst

Gray lifelines indicate approximate dates

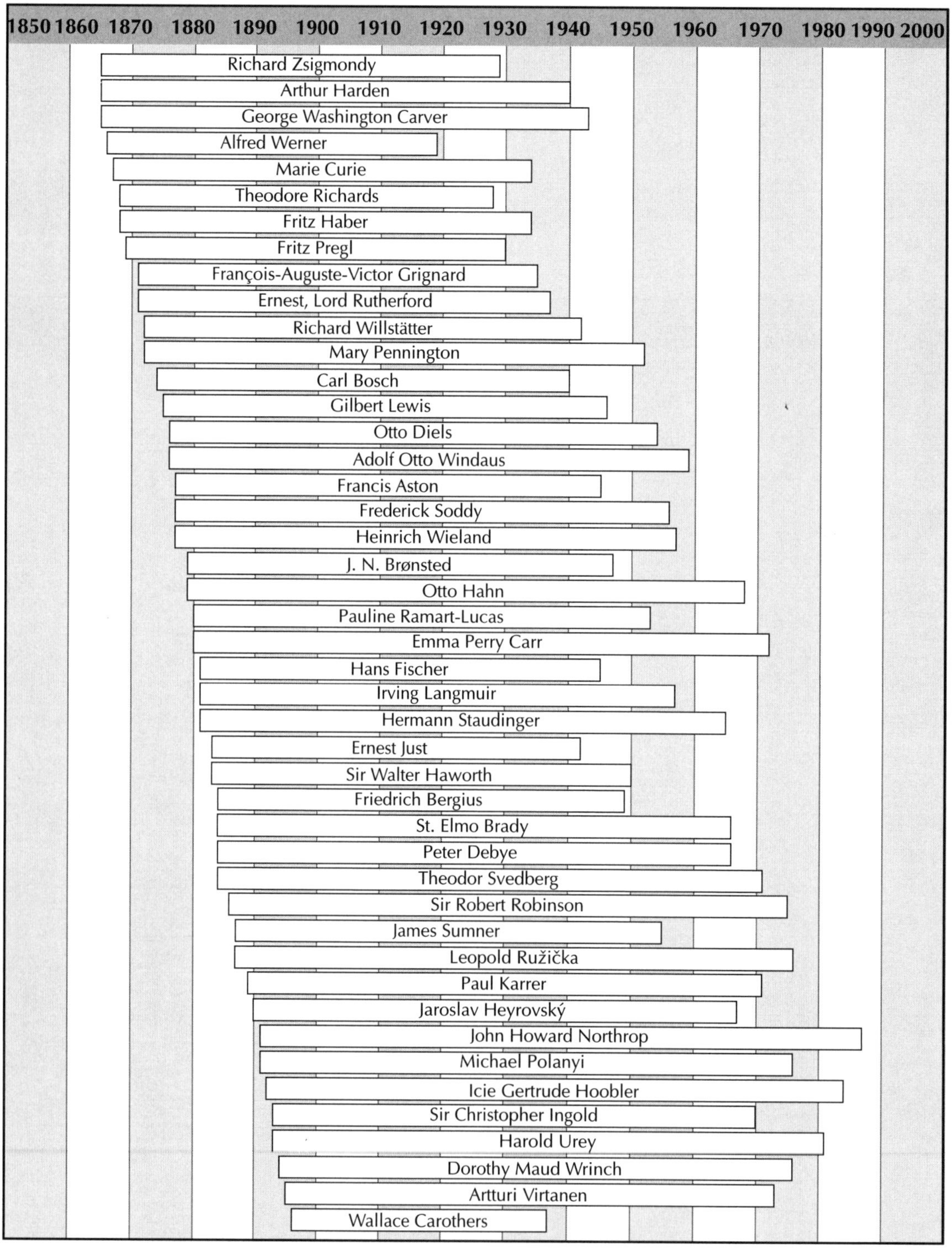

	1850	1860	1870	1880	1890	1900	1910	1920	1930	1940	1950	1960	1970	1980	1990	2000

Richard Zsigmondy

Arthur Harden

George Washington Carver

Alfred Werner

Marie Curie

Theodore Richards

Fritz Haber

Fritz Pregl

François-Auguste-Victor Grignard

Ernest, Lord Rutherford

Richard Willstätter

Mary Pennington

Carl Bosch

Gilbert Lewis

Otto Diels

Adolf Otto Windaus

Francis Aston

Frederick Soddy

Heinrich Wieland

J. N. Brønsted

Otto Hahn

Pauline Ramart-Lucas

Emma Perry Carr

Hans Fischer

Irving Langmuir

Hermann Staudinger

Ernest Just

Sir Walter Haworth

Friedrich Bergius

St. Elmo Brady

Peter Debye

Theodor Svedberg

Sir Robert Robinson

James Sumner

Leopold Ružička

Paul Karrer

Jaroslav Heyrovský

John Howard Northrop

Michael Polanyi

Icie Gertrude Hoobler

Sir Christopher Ingold

Harold Urey

Dorothy Maud Wrinch

Artturi Virtanen

Wallace Carothers

	1850	1860	1870	1880	1890	1900	1910	1920	1930	1940	1950	1960	1970	1980	1990	2000

Gerty Cori

Erich Hückel

Robert S. Mulliken

Nikolai Semenov

Irène Joliot-Curie

Sir Cyril Hinshelwood

Ronald Norrish

Odd Hassel

Georg Wittig

Karl Ziegler

Frédéric Joliot-Curie

Richard Kuhn

Maria Telkes

Linus Pauling

Kurt Alder

Arne Tiselius

Lars Onsager

Giulio Natta

Adolf Butenandt

Wendell Stanley

Gerhard Herzberg

Severo Ochoa

Luis Leloir

Hazel Bishop

Vladimir Prelog

Désirée Le Beau

Edwin McMillan

Baron Alexander Todd

Willard Libby

Paul Flory

Dorothy Hodgkin

Archer Martin

William Stein

Melvin Calvin

Glenn Seaborg

Stanford Moore

Richard Synge

Max Perutz

Henry Aaron Hill

Christian Anfinsen

Francis Crick

Sir Ronald Nyholm

Robert Burns Woodward

Sir John Kendrew

Sir John Cornforth

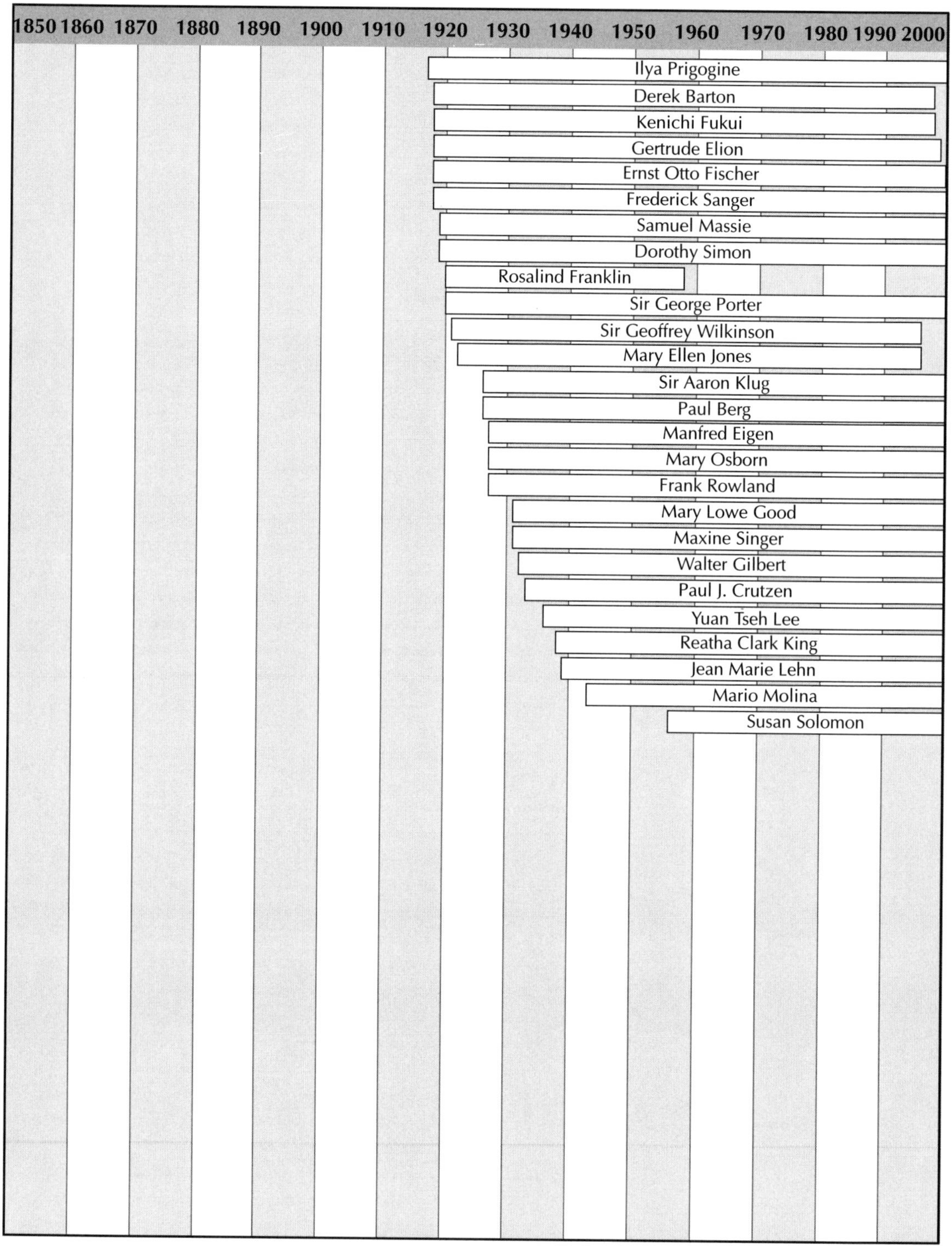

	1850	1860	1870	1880	1890	1900	1910	1920	1930	1940	1950	1960	1970	1980	1990	2000

Ilya Prigogine

Derek Barton

Kenichi Fukui

Gertrude Elion

Ernst Otto Fischer

Frederick Sanger

Samuel Massie

Dorothy Simon

Rosalind Franklin

Sir George Porter

Sir Geoffrey Wilkinson

Mary Ellen Jones

Sir Aaron Klug

Paul Berg

Manfred Eigen

Mary Osborn

Frank Rowland

Mary Lowe Good

Maxine Singer

Walter Gilbert

Paul J. Crutzen

Yuan Tseh Lee

Reatha Clark King

Jean Marie Lehn

Mario Molina

Susan Solomon

Abir-Am, P. G., and Dorinda Outram, eds. *Uneasy Careers and Intimate Lives: Women in Science 1789–1979*. New Brunswick, N.J.: Rutgers University Press, 1987.

American Men and Women of Science, 1995–96: A Biographical Directory of Today's Leaders in Physical, Biological, and Related Sciences. New York: R. R. Bowker, 1994.

Asimov, Isaac. *Asimov's Biographical Encyclopedia of Science and Technology: The Lives and Achievements of 1,510 Great Scientists from Ancient Times to the Present, Chronologically Arranged*. New York: Doubleday, 1982.

Bailey, B. *The Remarkable Lives of 100 Women Healers and Scientists*. Holbrook, Mass.: Bob Adams, 1994.

Bailey, M. J. *American Women in Science*. Santa Barbara, Calif.: ABC-CLIO, 1994.

Barrett, Eric C., and David Fisher, eds. *Scientists Who Believe: Twenty-One Tell Their Own Stories*. Chicago: Moody Press, 1984.

The Biographical Dictionary of Scientists. New York: P. Bedrick Books, 1983–1985.

Bridges, Thomas C., and Hubert H. Tiltman. *Master Minds of Modern Science*. New York: L. MacVeagh, 1931.

Concise Dictionary of Scientific Biography. New York: Scribner's, 1981.

CWP at the University of California–Los Angeles. "Contributions of 20th Century Women to Physics." (Available online.) URL: http://www.physics.ucla.edu/~cwp.

Current Biography. New York: H. W. Wilson, 1940–.

Daintith, John, et al. eds. *Biographical Encyclopedia of Scientists*. Philadelphia: Institute of Physics Publishing, 1994.

Darrow, Floyd L. *Masters of Science and Invention*. New York: Harcourt, 1923.

Dash, Joan. *The Triumph of Discovery: Women Scientists Who Won the Nobel Prize*. Englewood Cliffs, N.J.: Julian Messner, 1991.

Debus, A. G., ed. *World Who's Who in Science*. Chicago: Marquis Who's Who, 1968.

Defries, Amelia D. *Pioneers of Science*. London: Routledge and Sons, 1928.

Dictionary of American Biography. New York: Scribner's, 1928–.

Dictionary of Scientific Biography. New York: Scribner's, 1970–1980.

Elliott, Clark A. *Biographical Dictionary of American Science: The Seventeenth through the Nineteenth Centuries*. Westport, Conn.: Greenwood Press, 1979.

Eric Weisstein's Treasure Troves of Science. (Available online.) URL: http://www.treasure-troves.com

Feldman, Anthony. *Scientists and Inventors*. New York: Facts On File, 1979.

Gaillard, Jacques. *Scientists in the Third World.* Lexington: University Press of Kentucky, 1991.

Grinstein, Louise S., Rose K. Rose, and Miriam H. Rafailovich, eds. *Women in Chemistry and Physics: A Biobibliographic Sourcebook.* Westport, Conn.: Greenwood Press, 1993.

Herzenberg, Caroline L. *Women Scientists from Antiquity to the Present: An Index.* West Cornwall, Conn.: Locust Hill, 1986.

Howard, Arthur Vyvyan. *Chamber's Dictionary of Scientists.* New York: Dutton, 1961.

Hutchings, D., and E. Candlin. *Late Seventeenth Century Scientists.* Oxford, N.Y.: Pergamon Press, 1966.

Ireland, Norma O. *Index to Scientists of the World from Ancient to Modern Times: Biographies and Portraits.* Boston: Faxon, 1962.

James, Laylin K., ed. *Nobel Laureates in Chemistry, 1901–1992.* Washington, D.C.: American Chemical Society/Chemical Heritage Foundation, 1993.

Jones, Bessie Zaban, ed. *The Golden Age of Science: Thirty Portraits of the Giants of 19th Century Science by Their Scientific Contemporaries.* New York: Simon and Schuster, 1966.

Kass-Simon, G., and Patricia Farnes. *Women of Science: Righting the Record.* Bloomington: Indiana University Press, 1990.

Kessler, James H., et al., eds. *Distinguished African American Scientists of the 20th Century.* Phoenix, Ariz.: Oryx Press, 1996.

Krapp, Kristine M. *Notable Black American Scientists.* Detroit: Gale Research, 1998. *Makers of Modern Science: A Twentieth Century Library Trilogy.* New York: Scribner's, 1953.

McGraw-Hill Modern Men of Science: 426 Leading Contemporary Scientists. New York: McGraw-Hill, 1966–1968.

McGraw-Hill Modern Scientists and Engineers. New York: McGraw-Hill, 1980.

McGrayne, Sharon Bertsch. *Nobel Prize Women in Science: Their Lives, Struggles, and Momentous Discoveries.* Secaucus, N.J.: Carol Publishing, 1993.

McMurray, Emily J., and Donna Olendorf, eds. *Notable Twentieth Century Scientists.* Detroit: Gale Research, 1995.

Millar, David, et al., eds. *The Cambridge Dictionary of Scientists.* New York: Cambridge University Press, 1996.

Murray, Robert H. *Science and Scientists in the Nineteenth Century.* New York: Macmillan, 1925.

Nobel E-Museum. (Available online.) URL: http://www.nobel.se.

North, J. *Mid-Nineteenth-Century Scientists.* Oxford, N.Y.: Pergamon Press, 1969.

Olby, Robert C., ed. *Late Eighteenth Century European Scientists.* Oxford, N.Y.: Pergamon Press, 1966.

A Passion to Know: 20 Profiles in Science. New York: Scribner's, 1984.

Pelletier, Paul A. *Prominent Scientists: An Index to Collective Biographies.* New York: Neal-Schuman, 1994.

Porter, Roy, ed. *The Biographical Dictionary of Scientists.* New York: Oxford University Press, 1994.

Schlessinger, Bernard S., and June H. Schlessinger. *Who's Who of Nobel Prize Winners.* Phoenix, Ariz.: Oryx Press, 1991.

School of Mathematics and Statistics, University of St. Andrews, Scotland. "Mactutor History of Mathematics Archive." (Available online.) URL:http://www-history.mcs.st-andrews.ac.uk/history/.

Scott, Michael Maxwell. *Stories of Famous Scientists.* London: Barker, 1967.

Shearer, B. F., and B. S. Shearer, eds. *Notable Women in the Physical Sciences.* Westport, Conn.: Greenwood Press, 1997.

Siedel, Frank, and James M. Siedel. *Pioneers in Science.* Boston: Houghton, 1968.

Uglow, Jennifer S. *International Dictionary of Women's Biography.* New York: Continuum, 1982.

Unterburger, Amy L., ed. *Who's Who in Technology*. Detroit: Gale Research, 1989.

Van Sertima, Ivan, ed. *Blacks in Science: Ancient and Modern*. New Brunswick, N.J.: Transaction Books, 1983.

Van Wagenen, Theodore F. *Beacon Lights of Science: A Survey of Human Achievement from the Earliest Recorded Times*. New York: Thomas Y. Crowell, 1924.

Weisberger, Robert A. *The Challenged Scientists: Disabilities and the Triumph of Excellence*. New York: Praeger, 1991.

Who's Who in Science and Engineering. 1994–1995. New Providence, N.J.: Marquis Who's Who, 1994.

Who's Who in Science in Europe. Essex, England: Longman, 1994.

Who's Who of British Scientists, 1980–81. New York: St. Martin's, 1981.

Williams, Trevor I., ed. *A Biographical Dictionary of Scientists*. 4th ed. Glasgow, Scotland: HarperCollins, 1994.

Wyndham, D. Miles, ed. *American Chemists and Chemical Engineers*. Washington, D.C.: American Chemical Society, 1976–1994.

Youmans, William J., ed. *Pioneers of Science in America: Sketches of Their Lives and Scientific Work*. New York: Arno Press, 1978.

Yount, Lisa. *A to Z of Women in Science and Math*. New York: Facts On File, 1998.

Zuckerman, Harriet. *Scientific Elite: Nobel Laureates in the United States*. New York: Free Press, 1979.

Note: Page numbers in **boldface** indicate main topics. Page numbers in *italic* refer to illustrations.

A

Abbotsbury Laboratory 137
Abelson, Philip 140
Aberdeen, University of 205
absolute zero 150, 151
"The Absorption Spectra of Some Derivatives of Cyclopropane" (Carr) 37
Academic Assistance Council 191
Academy of Science (India) 114
Academy of Sciences (France) 52, 81, 82, 106, 114, 123, 143, 196
Academy of Sciences (USSR) 200–201
acids and bases, defining 26, 130, 131
actinide concept 198, 199
"The Activity Coefficients of Ions in Very Dilute Solutions" (Brønsted) 26
Adams, Roger 208
adenosine diphosphate (ADP) 220
adenosine triphosphate (ATP) 220
Advanced Organic Chemistry (Cotton, Wilkinson) 231–232
aeronautics 201, 202
Aeronomy of the Middle Atmosphere (Brasseur, Solomon) 207
affinity 164
Agnes Faye Morgan Research Award 79
agriculture 23–24, 37–38, 83, 84, 225–226
Agronomic Institute 142
Aktien-Gesellschaft für Anilin-Fabrikation (Agfa) 227

Albert Einstein College of Medicine 162
Albert Lasker Medical Research Award 18, 78
Albert Medal 29
Albright and Wilson Company 14
alchemy 135–136
"Alcoholic Fermentation without Yeast Cells" (Buchner) 27–28
Alder, Kurt **1–2**, 57
Alexander Hamilton Medal 156
alkaloids 186, 187, 237
alkylation 72–73
AlliedSignal Corporation 79
alpha particles 190, 191, 205
aluminum 93
Amelia Earhart Award 79
American Academy of Arts and Sciences 163, 186
American Academy of Sciences 45
American Association for the Advancement of Science 79, 123, 145, 186
American Association for Cancer Research 62
American Association of University Professors 110
American Association of University Women 181, 182, 184–185
American Balsa Company 169
American Cancer Society 4, 209
American Chemical Society 12, 34, 37, 79, 95, 123, 125, 163, 186, 200
    Award in Chromatography 211

    Creative Work in Synthetic Chemistry Award 15
    Esselen Award 145
    Faraday Medal 57, 123
    Garvan Medal 36, 37, 43, 62, 79, 100, 169–170
    Irving Langmuir Prize 123
    Kaj Linderstrøm-Lang Award 211
    Kendall Award 57
    Kirkwood Medal 60
    Langmuir Prize 168
    Linus Pauling Medal 60, 92
    Nichols Medal 12, 57, 209
    Oesper Prize 34
    Peter Debye Award 127, 148
    Priestley Award/Medal 15, 34, 57, 70, 79
    Roger Adams Award 235
    Theodore William Richards Medal 211
    Willard Gibbs Medal 7, 57, 92, 148, 209, 224
American Council on Education 117
American Electrochemical Society 12
American Geophysical Union 207
American Home Economics Association 185
American Institute of Chemistry Gold Medal 79
American Institute of Electrical Engineers 41
American Institute of Nutrition 100
American Institute of Refrigeration 170

American Medical Association 209
American Philosophical Society 92
American Physical Society 123
American Scandinavian Foundation 4
American Society for Biochemistry and
    Molecular Biology 110
American Society of Biological
    Chemists 18, 78, 146
American Society of Chemical
    Engineering 12
American Society of Plant Physiologists
    34
American Society of Refrigerating
    Engineers 170
American University 147
Amherst College 239
amino acids 65, 66, 136, 145–146
ammonia synthesis 23–24, 83, 84
Ampère, André Marie 2, 2–4, 9–10
Anfinsen, Christian Boehmer 4–5,
    146
angiotensin 128
Annalen der Naturalphilosophie 164
antibiotics 44, 97, 98, 237–238
"The Application of Flash Techniques
    to the Study of Fast Reactions"
    (Norrish, Porter) 155
argon 182, 183
Argonne National Laboratory 202
Arkansas, University of 78–79
Arkansas, University of, at Pine Bluff
    137–138
Armour and Company 108–109
aromaticity 100, 101
Arrhenius, Svante August 5, 5–7, 100
arsenic 29
ascorbic acid (vitamin C) 89–90, 168
Association for the Advancement of
    Science (U.K.) 158, 187, 191
Association of German Chemists 2
Association of Medical School
    Departments of Biochemistry (U.S.)
    110
Association of Science Education
    (U.K.) 157
Aston, Francis 7, 7–9
The Astonishing Hypothesis (Crick)
    46–47
atmospheric chemistry 47–48,
    188–189, 206–207

Atomic Energy Commission (France)
    106
Atomic Energy Commission (U.S.)
    132–133, 188–189, 200, 231
Atomic Spectra and Atomic Structure
    (Hertzberg) 92
atomic theory 51–52, 121, 122,
    189–190, 191
atomic weights 20, 51, 52, 152, 183,
    185, 186, 199 *See also* periodic table
    of the elements
Atoms, Molecules, and Quanta (Ruark,
    Urey) 223
Atoms for Peace Award 139, 140
Austro-American Rubber Works 125
Avco Corporation 202
Avogadro, Lorenzo Romano Amedeo
    Carlo 9, 9–10
Award in Chromatography 211
AZT (azidothymidine) 60, 62

B
Bache and Company 22
bacterial growth, chemical reactivity of
    96
Badische Anilin und Soda Fabrik
    (BASF) 23, 24
Baekeland, Leo Hendrik 11–12, 12
Baeyer, Adolf von 13–14, 27, 65
Bakelite 11, 12
Barnard Medal 17
Barrett Chemical Company 223
Barton, Derek 14–15, 88, 89
*Basic Methods for Experiments in Eggs of
    Marine Animals* (Just) 111
"Basic Reactions Occurring during
    Reclaiming of Rubber I" (Le Beau)
    125
Beckman Center for Molecular and
    Genetic Medicine 18
Becquerel, Antoine-Henri 15–17, 16,
    49
Beit Memorial Fellowship for Medical
    Research 197
benzene 63–64, 101
benzoyls 236
Berg, Paul 17–19, 18, 203, 204
Bergius, Friedrich 19–20
Bergmann, Max 145
Bergstedt Prize 219

Berlin, University of 1, 13, 31, 56, 57,
    66, 86, 88, 150, 151, 227, 232, 233
Berlin Institute of Technology 13
Berthelot, Pierre 195
Berthollet, Claude-Louis 75, 179, 180
Berzelius, Jöns Jacob 20, 20–21
beta-carotene 113
beta particles 190, 191, 205
Biochemical Society (U.K.) 88
Biogen N.V. 78
biological oxidation 230
biological resistance (bioresistance)
    119
The Biology of the Cell Surface (Just)
    111
Birkbeck College 15, 71, 118
Birks, John W. 48
Birmingham University (Mason
    College) 7, 91
Bishop, Hazel Gladys 21–23
Black, James 60
Blodgett, Katherine B. 122
Bodenstein, Max 200
Bohr, Niels 191
bonding 131, 146, 147, 167–168,
    228, 229
Bonn, University of 65, 227
Boots Pure Drug Company 136–137
boron 75, 76
Bosch, Carl 19, 20, 23–24
Brady, St. Elmo 24–25
Brandeis University 109
Breslau, University of 19–20, 28
Brickwedde, Ferdinand 223–224
The Bridge 164
Bristol, University of 91–92, 98, 182
Brønsted, Johannes Nicolaus 25–27
Brookhaven National Laboratory 188
Brooklyn Polytechnic 46
Brown, Herbert 234
Brownian motion 213
Brown University 161
Brussels, University of 146
Buchner, Eduard 27–28, 87–88
Budapest University 171, 217
Buenos Aires, University of 128, 130
Bunsen, Robert Wilhelm 28, 28–29
Bunsen Society for Applied Physical
    Chemistry 164
Butenandt, Adolf 30–31, 192

C

Cahn, Robert 175, 176
Caisse Nationale des Sciences 105
California, University of, at Berkeley
    6, 33, 34, 66, 99, 126–127, 127,
    130, 132, 139–140, 144, 156, 162,
    199, 200, 207, 223, 231
California, University of, at Irvine
    144, 188, 189
California, University of, at La Jolla
    48, 224
California, University of, at Los
    Angeles 133, 199
California, University of, at San Diego
    168
California Institute of Technology
    (CalTech) 139, 144, 167
Calvin, Melvin 33–34, 131
Cambridge University 77, 97–98, 98,
    137, 146
    Biochemical Laboratory 214–215
    Cavendish Laboratory 8, 40,
        45–46, 115, 162, 170, 190,
        191
    Christ College 221, 238
    Emmanuel College 154–155, 173
    Girton College 238
    King's College 197
    Newnham College 71
    Peterhouse College 39, 136
    St. John's College 196
    Strangeways Laboratory 45
    Trinity College 8, 114, 118, 190,
        214
Canadian Association of Physicists
    Gold Medal 92
Canizzaro Prize 114, 123, 210
Canterbury College 190
Cape Town, University of 117
Capital City Commercial College 34
carbamoyl phosphate 108, 109
carbohydrates 20, 33–34, 42–43, 65,
    66, 90, 128, 129
carbon dating 132–133
carbon dioxide 39, 197–198
Carnegie, Andrew 41
Carnegie Institution 204
Carnegie scholarships 92, 220
carotenoids 113, 118, 119, 136
Carothers, Wallace Hume 34–36, 35

Carr, Emma Perry 36, 36–37
Carver, George Washington 37–38
catalysis 163–164, 241–242
catalytic hydrogenation 195, 196
Cattle Fodder and Human Nutrition
    with Special Reference to Biological
    Nitrogen Fixation (Virtanen) 226
Cavendish, Henry 39, 39–40
Cavendish experiment 39, 40
Cavendish Laboratory See under
    Cambridge University
Center for the Study of Democratic
    Institutions 168
Central Arkansas, University of 78
Central College for Women 99
Central Laboratory of Industries
    (Helsinki) 225
Centre National de la Recherche
    Scientifique 106
centrifuge, ultra 213–214
Certificate of Merit (U.S.) 156
chain reactions 95–96, 200–201
Chandler Medal 156
Charles, Jacques 75
Charles University 93, 94
Charlottenberg Technical College 83
Chemical Anthropology (Hoobler, Kelly)
    100
Chemical Aspects of Polypeptide Chain
    Structure and the Cyclol Theory
    (Wrinch) 239
Chemical Aspects of the Structure of
    Small Peptides (Wrinch) 239
Chemical Institute of Göttingen 228
The Chemical Kinetics of the Bacterial
    Cell (Hinshelwood) 96
Chemical Kinetics and Chain Reactions
    (Semenov) 201
Chemical Manipulation (Faraday) 64
Chemical Manufacturers Association
    138
Chemical Panel for Atomic Energy 91
chemical physics, founder of 146, 147
chemical reaction rates, measuring
    163–164
"Chemical Reactions Produced by Very
    High Light Intensities" (Norrish,
    Porter) 155
Chemical Society of Japan 74

Chemical Society of London 29, 80,
    96, 155, 158
Chemical Society of the United
    Kingdom 7, 26, 60, 91, 158
chemical warfare 83, 84
Chemistry Consultative Committee
    157
The Chemistry of Cooking and Cleaning
    (Richards) 185
Chemistry in Microtime (Porter) 174
"The Chemistry of Phenothiazine"
    (Massie) 138
Chibnall, Charles 196–197
Chicago, University of 36, 48, 92, 95,
    99, 108, 116, 126, 132, 147, 148,
    178, 188, 199, 209, 224, 242
Children's Fund of Michigan 99
chlorine gas 83, 84
"Chlorine</208>An Introduction"
    (Porter) 174
chlorine, liquefying 63–64
chlorofluorocarbons (CFCs) 47,
    143–144, 188, 189, 206–207
chlorophyll 69, 232–233, 237
cholesterol 58, 233–234, 237
Christopher Ingold Lectureship 104
chromatography 136–137, 146, 214,
    215
CIBA-GEIGY Corporation 117, 176
Cincinnati, University of 70
City University of New York (CUNY)
    116
Clark, Josiah Latimer 40–41
Clark College 116
Cleveland Clinic Foundation 217
Cleve, Per Theodor 6
Clifton College 114
cloud seeding 122
coal hydrogenation 19–20, 23
Collections of Czechoslovak Chemical
    Communications (Heyrovsk?) 94
College of Agriculture (Berlin) 28
College Chemistry (Pauling) 168
Collège de Cherbourg 81
Collège de France 3, 106, 128, 142,
    195, 228
Collège Marazin 123
Collège des Quatre Nations 123
Collège Saint-Louis 165
Collège Sévigné 107

colloid chemistry 80, 124–125, 213, 242–243
Cologne University 2
Colorado, University of 99
Colorado State University 48
Columbia School of Mines 121
Columbia University 22, 78, 116, 129, 132, 155–156, 156, 211, 223, 224
combustion 123, 124, 155, 201
Combustion Institute, Bernard Lewis Gold Medal 155
Companion of the British Empire 115, 171, 197
Companion of the Order of Australia 45
Comprehensive Organometallic Chemistry (Wilkinson, ed.) 232
conformational analysis 14–15, 88, 89
"The Conformation of the Steroid Nucleus" (Barton) 14–15
Connecticut, University of 162–163
Consden, R. 137
Considérations générales sur la nature des acides (Lavoisier) 124
Considerations on the Mathematical Theory of Games (Ampère) 3
constant composition, law of 179, 180
The Constitution of Sugars (Haworth) 90
coordination compounds 157, 228, 229
Copenhagen, University of 26, 223
Copley Medal 29, 45, 54, 55, 81, 97, 142, 171, 187
Corday-Morgan Medal and Prize 158, 197
Cori, Gerty Theresa Radnitz **41–43**, *42*
Cornell University 56–57, 60, 70, 86, 212–213
Cornforth, John Warcup **43–45**
cosmetics 22
Cotton, F. A. 232
A Course of Six Lectures on the Chemical History of a Candle (Faraday) 64
Crafts, James Mason 72
Cram, Donald J. **127–128**
Creative Work in Synthetic Chemistry Award 15

Crick, Francis Harry Compton *45,* **45–47,** 71
Crutzen, Paul J. **47–48**
Cryo-Therm 218
Crysalline Enzymes (Herriott, Kunitz, Northrup) 156
"The Crystal Structure of Molybdenite" (Pauling) 167
Curie, Marie **48–50,** *49*
Curtiss-Wright Company 218
cyclol theory 238, 239
Czech Institute of Technology 175
Czechoslovak Academy of Sciences 94
Czech State Prize 94

D

Dalton, John 10, **51–52,** *52,* 75, 76, 87
Daniel Giraud Elliot Medal 156
Daniell, John Frederic **52–54**
Daniel Osiris Prize 50
dark space 7–8
Darmstadt, University of 209
Dartmouth College 110
Davy, Humphry **54–55,** 63, 76, 176
Davy Medal 7, 29, 34, 45, 50, 69, 88, 91, 97, 142, 143, 153, 155, 173, 183, 187, 196, 224, 228, 238
Dealing with Genes (Berg, Singer) 19, 204
Debray, Henri 142
Debye, Peter *55,* **55–57,** 100, 161–162
Debye-Hückel theory of electrolytic dissociation 55, 100, 161–162
definite proportions, law of 179, 180
Delaware, University of 218
"The Dependence between the Properties of the Atomic Weights of the Elements" (Meneleyev) 141–142
deuterium 131, 223–224
Dewey & Almy Chemical Company 95, 125
Die Chemie de Pyrrols (Fischer, Orth) 69
Diels-Alder reaction 1–2, 58
Diels, Otto 1, 2, **57–58,** 233
Die quantitative organische Mikroanalyse (Pregl) 175

Dijon Lycée 165
dipole moments 55, 56, 161–162
Discover magazine Scientist of the Year 48
dissipative structures, theory of 177–178
Distinguished Presidential Rank Award (U.S.) 204
DNA (deoxyribonucleic acid) 45, 46, 70–72, 76–78, 159, 160, 196, 197, 202, 203–204 *See also* rDNA (recombinant DNA)
Döbereiner, Johann 152
"The Double Bond" (Carothers) 35
The Double Helix (Watson) 72
drop-counting automatic fraction collector 210, 211
Duke University 62
DuPont 34, 35, 70, 124, 202
dyes 13–14, 23, 65, 187
dynamite 153–154

E

Earlham College 208
Eastbourne College 204
ecdysone 31
École Centrale 3
École des Mines 72
École Normale Spéciale 81
École Normale Supérieure 165, 166, 195, 196
École Polytechnique 3, 16, 75
École des Ponts et Chaussées 16, 75
École Supérieure de Pharmacie 142, 143
École Supérieure de Physique et de Chimie Industrielle 105
ecology 183, 185
Edinburgh, University of 80, 177, 220
Eggs (Pennington) 169
Ehrenzeichen für Wissenschaft und Kunst 171
E. & H.T. Anthony Company 11
Eidgenössische Technische Hochschule (Institute of Technology, Zurich) 119, 176, 192, 193, 209–210
Eigen, Manfred **59–60,** 154, 173
Einführung in die Organische Chemie (Diels) 57
Einstein, Albert 56

electric-arc furnace 142
electricity 39, 40, 63, 64
"The Electro-Affinity of Aluminum"
    (Heyrovsk?) 93
electrochemistry 41
electrolytic dissociation 55, 56, 100
electromagnetic induction 63–64
electromagnetic theory 2, 3
electromagnetic wave detector 190
electron diffraction 167–168
electron dot theory 131
electron sharing 130–131, 146, 147
electrophoresis 218, 219
elements 123–124 *See also* isotopes
    new 20, 29, 48, 50, 54, 75, 76,
        85–86, 139, 140, 182, 183,
        198, 199
    periodic table 140, 141–142,
        152, 153, 183, 185, 186, 199
Elements of Agricultural Chemistry
    (Davy) 55
Elements of Chemical Philosophy (Davy)
    55
Eli Lily Prize in Biochemistry 18
Elion, Gertrude Belle **60–62,** *61*
Ellen Swallow Richards Research Prize
    181, 182
*The End of Certainty, Time, Chaos and
    the New Laws of Nature* (Prigogine,
    Stengers) 179
Enrico Fermi Award 87
*Environmental Consequences of Nuclear
    War 1985* (Birks, Crutzen) 48
enzymes 4, 66, 119, 146, 155, 156,
    210, 211, 212
Erlangen, University of 66, 87
Eschenmoser, Albert 238
Essay on Chemical Proportions
    (Berzelius) 21
Esselen Award 145
essential oils 227–228
ethylene 241, 242
European Molecular Biology
    Laboratory 115
euthenics 183–185
*Experimental Researches in Chemistry
    and Physics* (Faraday) 64
Experimental Researches in Electricity
    (Faraday) 64

*Experimental Studies of Decomposition
    and Combustion of Hydrocarbons*
    (Haber) 83–84
"Experiments on Air" (Cavendish)
    39–40
*Experiments and Observations on
    Different Kinds of Air* (Priestly) 177
Exploring Genetic Mechanisms (Berg,
    Singer) 19, 204

F

Faculté des Sciences (Toulouse) 195
Faculty of Sciences (Lyon) 81, 82
Faraday, Michael 55, **63–64,** *64*
Faraday effect 64
Faraday Medal 7, 57, 60, 123, 187
Faraday Society 96
Farman, Joe 189
fast chemical reactions 59, 155, 173
Federation of American Societies of
    Experimental Biology 146, 163
Femtosecond Chemistry (Porter) 174
Ferme Générale 123–124
fermentation 27–28, 87–88, 166,
    225–226
field theory 64
Finland Institute of Technology 226
First Berlin Chemical Institute 68
Fischer, Emil 57, *65,* **65–66**
Fischer, Ernst Otto **66–68,** 230–232
Fischer, Hans **68–69**
Fisica dei corpi ponderabili (Avogadro)
    10
Fisk University 24, 25, 138
flash photolysis 155, 172–173
Florida, University of 68
Florida State University 148
Flory, Paul **69–70**
fluorine 142, 143
fluorine flame calorimetry 115, 116
Le fluor et ses composés (Moissan) 143
flu vaccine 208
fodder preservation 225–226
food chemistry 37–38, 99, 168–170,
    183–185, 225–226
Food Materials and Their Adulterations
    (Richards) 185
Food Research Institute 215
Food Research Laboratory 169
Ford Foundation 217

Le four électrique (Moissan) 143
Fourier Transforms and Structure Factor
    (Wrinch) 239
Frankfurt, University of 220, 241
Franklin, Rosalind 46, **70–72,** 117, 118
Franklin Institute
    Franklin Medal 12, 57, 196, 219,
        224
    John Price Wetherill Medal 215
Fredericiana Technology College 83
free energy 130–131
free radicals 91, 92, 155, 241
Freiburg, University of 144, 210, 230,
    233, 235
Frémy, Edmund 142, 143
Fridtjof Nansen Award 89
Friedel, Charles **72–73**
From My Life (Willstätter) 233
Fukui, Kenichi **73–74**

G

Garcia, Rolando 207
Garvan Medal 36, 37, 43, 62, 79, 100,
    169–170 *See under* American
    Chemical Society
gases
    chlorine 83, 84
    combining volumes, law of 75,
        76
    diffusion, Graham's law 80
    mixing of 52
    molecular makeup of 3, 9–10
    new 176, 177, 182, 183
    paper chromatography 136–137
The Gases of the Atmosphere (Ramsey)
    183
Gates and Crellin Laboratories 168
Gay-Lussac, Joseph-Louis **75–76,** *76*
Geiger, Hans 191
General Chemistry (Pauling) 168
General Electric Research Laboratory
    121–123
General Mills Foundation 116, 117
Genes and Genomes (Berg, Singer)
    18–19, 204
Genoa, University of 149
Gentry, Ron 126
Geochemical Research Laboratory 198
Geochemistry Research Association
    198

Georg-August University 59
Georg Speyer Haus 113
German Chemical Society 119
    Adolf von Baeyer Memorial Medal
       58, 235
    Alfred Stock Memorial Prize 68
    August Wilhelm von Hofmann
       Medal 183
    Liebig Medal 20, 69, 242
German nobility, induction into 14
German Physical Society, Otto Hahn
    Prize 60
germ theory of disease 165, 166
Ghent, University of 11
Gilbert, Walter 76–78, 77
Glasgow, University of 15, 80, 159,
    182, 205, 220
glass 242–243
Glaxo Wellcome (Wellcome Research
    Laboratories) 61–62, 137
glycogen 87
glycogen storage disorders 43
glycolysis 28
Goethe Medal 234
Good, Mary Lowe 78–79
Gordon, A. H. 13
Gordon Conference 203–204
Göttingen, University of 29, 30, 56,
    90, 91, 100, 121, 130, 150, 151,
    227, 228, 234, 236, 243
Göttingen Academy Prize 68
Government Higher Normal School of
    Science (France) 11
"Grafting the Cacti" (Carver) 38
Graham, Thomas 79–81
Grand Cross for Federal Services of
    West Germany 31, 58
Grand Order of Merit (Germany) 234
Graz, University of 125, 174, 175
greenhouse effect 6–7
Grignard, François-Auguste-Victor
    81–82
Grignard reagents 81–82
*Growth, Function, and Regulation in
    Bacterial Cells* (Dean, Hinshelwood)
    96
Guggenheim Fellowships 147, 212,
    231
Guldberg Medal 97
Gunnerus Medal 89

Gustaf Werner Institute of Nuclear
    Chemistry 214
György, Paul 119

H
Haber, Fritz 23, 83–85, 84
Hahn, Martin 27
Hahn, Otto 85, 85–87
half-life 131–132, 190
Halle, University of 209
Halle-Saale, University of 241–242
Halogen-metal exchange reaction 234
Hantzsch, Arthur 228
Harden, Arthur 87–88
Harradence, Rita 44
Harvard University 14, 35, 69, 77, 78,
    127, 128, 130, 147, 164, 185–186,
    210, 211, 212, 231, 237
Harvard University Medical School 4,
    212
hashish 220
Hassel, Odd 88–89
Haworth, Walter 89–91
Hazel Bishop Laboratories 22
heavy hydrogen 223–224
heavy water 131
Heiber, Walter 67
Heidelberg, University of 13, 29, 83,
    119, 234, 235, 236, 241
Heinrich Wieland Prize 230
helium 183
Helmholtz Medal 66
Helsinki, University of 225, 226
hemoglobin 170–171
Herschbach, Dudley 126
Herty Medal 79
Herzberg, Gerhard 91, 91–92
Hess, Alfred 234
Heyrovský, Jaroslav 92–94, 93
Hidden Hunger (Hoobler, Williams)
    99–100
high-vacuum electron tubes 121
Hill, Henry Aaron 94–95
Hillebrand, William F. 183
Hinshelwood, Cyril 95–97
History of Electricity (Priestly) 177
Hitchings, George Herbert 60, 61–62
Hodgkin, Dorothy Crowfoot 97,
    97–98, 221
Hoffman, Roald 73–74, 238

Holzkamp, E. 242
home economics 183–185
Home Sanitation (Richards) 185
Hoobler, Icie Gertrude Macy 98–100
hormones 30–31, 31, 186, 187, 192
host-guest chemistry 127–128
Houssay, Bernardo A. 43, 128, 129
Howard University 24, 25, 110–111,
    138
Hückel, Erich 56, 100–101
Huggins, M. L. 70
Hughes, E. D. 103–104
Hughes, R. E. 204
Hughes Medal 8
Human Genome Project 18, 78, 196,
    197
Humboldt, Alexander von 75
Hunter College 60
hydrocarbon reactions (alkylation)
    72–73
hydrogen, discovery of 39, 122
hydrogenation, coal and wood 19–20
hydrogen and oxygen interactions 96
hydrogen welding torch 122
hygrometer, dew-point 53
Hypercycles (Eigen, Schuster) 60

I
I. G. Farben 1, 24, 234
Illinois, University of 24–25, 35,
    201–202, 237
Illinois, University of, at Champagne-
    Urbana 208
Illinois Institute of Technology 206
immunization 165, 166, 208
Imperial College of Science,
    Technology and Medicine 14, 15,
    90, 103, 174, 231
Imperial Order of the Eagle 228
Imperial Order of the Rising Sun
    (Japan) 179, 209, 238
incandescent lamp, gas-filled 121
Incendiary Projects Committee 155
Industrial Chemical Research Center
    149
"The Influence of Nitrogen Oxides on
    the Atmosphere Ozone Content"
    (Crutzen) 47
influenza vaccine 208

Ingold, Christopher **103–104**, 175, 176

Innsbruck, University of 174, 233

Institute of Chemical Physics (USSR) 201

Institute of Cytophysiology (Copenhagen) 17

Institute of France 40

Institute of Physical Chemistry and Electrochemistry (Karlsruhe) 84

Institute of Physics Duddell Medal and Prize 8

Institute of Preventative Medicine (U.K.) *See* Lister Institute

Institute of Technology (Eidgenössische Technische Hochschule, Zurich) 119, 176, 192, 193, 209–210

Instituto de Investigaciones Bioquímcas, Fundación Campomar 129

insulin 97, 98, 196–197

Integrated Laboratory Sequence (Good) 79

International Congress on Pure and Applied Chemistry 114

International Council of Scientific Unions 115

International Education Board 26

International Organization for Chemical Sciences in Development 200

International Solar Energy Society 218

International Solvay Institutes of Physics and Chemistry 178

International Union of Biochemistry 160

International Union of Pure and Applied Chemistry 219

International Union of Pure and Applied Physics 115

Introduction to the Mathematical Study of the Natural Sciences (Nernst) 151

Introduction to Quantum Mechanics (Pauling) 168

Introduction to the Study of Chemical Philosophy (Daniell) 53

Introduction to Thermodynamics of Irreversible Processes (Prigogine) 178

iodine 76

ions 6

Iowa State College of Agriculture and Mechanic Arts 38

Iowa State University 138

irreversible processes 177–178

Irving Langmuir Prize 123

isoprenoids (terpenes) 90, 227–228

isotopes 7, 8, 86, 131, 147, 204, 205, 223–224, 231

Is Science Necessary? (Perutz) 171

Italian Chemical Society 158

I Wish I'd Made You Angry Earlier (Perutz) 171

J

James, A. T. 137

James, R. W. 117

Japan Medal of Honor 74

Jardin des Plantes 76

Jenner Institute *See* Lister Institute

Jet Propulsion Laboratory 144

John Moores University 171

Johns Hopkins University 5, 161, 223, 239

Johnson C. Smith University 94–95

Joliot-Curie, Frédéric **105–106**, *107*, 107–108

Joliot-Curie, Irène 105–106, *107*, **107–108**

Jones, Mary Ellen **108–110**, *109*

Journal of Biological Chemistry 210, 211

Journal of Chemical Physics 224

Just, Ernest **110–111**

Justus Liebigs Annalen der Chemie (Wieland) 230

K

Kaiser Wilhelm Institute 24, 30–31, 83, 84, 85, 86, 111, 159, 172, 230, 232, 242 *See also* Max Planck Society

Kaj Linderstrøm-Lang Award 211

Karlson, Peter 31

Karl Ziegler Prize 235

Karolinan Children's Hospital 42

Karrer, Paul **113–114**

Kassel, University of 29, 234

Keith Medal 80

Kekulé, Friedrich 13, 227–228

Kendall Award 57

Kendrew, John Cowdery **114–115**, 170–171

Kennedy, Joseph W. 199

Kennedy, Khrushchev and the Test Ban (Seaborg) 200

kerotakis 135–136

Kiel, University of (Christian Albrecht University) 1, 27, 57–58

Kimball Union Academy 110

kinetics 95–96, 154–155, 171, 172, 200–201

Kinetics of Chemical Change (Hinshelwood) 96

King, Reatha Clark **115–117**, *116*

King's College 53, 71, 197

Kirchoff, Gustav 28, 29

Kirkwood Medal 60

Klug, Aaron **117–118**

knighthoods (U.K.) 45, 55, 91, 96, 115, 118, 157, 158, 173, 183, 187, 191, 221, 232

Knoop, Franz 234

Kornberg, Arthur 18, 159

Kuhn, Richard **118–119**

Kunitz, Moses 156

Kuopio, University of 226

Kurtze, Walter 59

Kyoto Imperial University 73, 74

Kyoto Institute of Technology 74

L

Laboratoire de Synthèse Atomique 106

Langmuir, Irving **121–123**, *122*, 130, 131

Langmuir Prize 168

Langston University 138

Lappe, Franz 24

Lasker Medical Research Award 18, 78

Lavoisier, Antoine-Laurent *123*, **123–124**

Lavosier Medal 97, 238, 242

"The Law of Octaves, and the Causes of Numerical Relations among the Atomic Weights" (Newlands) 152

Lawrence, Ernest Orlando 199

Laws of the Game (Eigen, Winkler) 60

Le Beau, Désirée **124–125**

Leblanc Medal 210, 229

LeBreton, Pierre 126

Lee, Yuan Tseh **126–127**

Leeds University 103, 173

Legion of Honor (France) 31, 81, 128, 179, 180, 196
Lehn, Jean-Marie **127–128**
Lehrbuch der Kolloidchemie (Zsigmondy) 243
Leipzig, University of 23, 56, 101, 130, 151
Leloir, Luis Federico **128–130**, *129*
Lemelson-MIT Lifetime Achievement Award 62
Leningrad Polytechnic Institute 200
Lewis, Gilbert Newton **130–131**, *131*
Lewis Laboratory 202
Lewis-Langmuir electron dot theory 131
Lewis Thomas Prize 171
Libby, Willard Frank *132*, **132–133**
Liebig, Justus von 236
Life Itself (Crick, Orgel) 46
Lille University 166
Linderstrøm-Lang, Kaj 4
Linus Pauling Institute of Science and Medicine 168
Linus Pauling Medal 60, 92
lipopolysaccharide 162
lipstick 22
Lister Institute (Jenner Institute, Institute of Preventative Medicine) 87, 88, 137, 215, 220
Liverpool, University of 187
Liversidge Lecturer 158
Liversidge Medal 155
London, University of 88, 220
  Birkbeck College 15, 71, 118
  Imperial College 14, 15, 90, 103, 174, 231
  King's College 71
  University College London 45, 80, 85, 93, 103, 104, 157, 182–183, 238
Longstaff Medal 91, 104, 187
Lorentz Medal 57
Louisiana State University 79
Lowenthal Fellowship 36
Lowry, Thomas 25, 26
luminescence 16–17
Lunar Society 177

M

macromolecules, theory of 209–210
Madrid, University of 159

Maeyer, Leo de 59
magnetic fields 16
Magnolia Petroleum Company 202
Málaga College 159
Malvern College 7
Manchester, University of (U.K.) 33, 87, 90, 172, 187, 191, 220
Manchester Academy (U.K.) 51
Manchester College (U.S.) 69
Manchester Literary and Philosophical Society (U.K.) 51–52
Manhattan Project 33, 132, 138, 140, 199
Man of the Year (Australia) 45
Marburg, University of 30, 68, 85, 101, 236
Marburg-Lahn, University of 234, 241
Marcel Benoist Award 176
Maria the Jewess **135–136**
marijuana 220
Marine Biology Laboratory, Woods Hole 110
Marsburg, University of 29
Marsden, Ernest 191
Martin, Archer John Porter **136–137**, 214, 215
Marvel Benoist 114
Mary E. Woolley Fellowship 36
Massachusetts General Hospital 109
Massachusetts Institute of Technology (MIT) 68, 95, 124, 125, 130, 140, 144, 147, 183, 184, 202, 217, 231, 237
Massie, Samuel Proctor *137*, **137–139**
mass spectrometers 8
Maxam, Allan 78
Max Planck Institute 48, 59, 86, 118, 119, 163 *See also* Kaiser Wilhelm Institute
Max Planck Society 31, 119
McDonald, Doug 126
McGill University 85–86, 190, 191, 204
McMillan, Edwin M. *139*, **139–140**, 199
"The Mechanisms of Carbon Monoxide Reactions of Nickel II Salts..." (E. O. Fischer) 67
Medaille d'Or 179

Medical Pneumatic Institution 54, 55
Medical Research Council (U.K.) 44, 45, 115, 117, 118, 137, 170, 171, 197
Meitner, Lise 86
Meldola Medal and Prize 104
Mellon Institute of Industrial Research 70
"Memoir on Combustion in General" (Lavoisier) 124
Mendel, Lafayette B. 99
Mendeleyev, Dmitri Ivanovich **140–142**, *141*, 152, 153
Merrill-Palmer School 99
metabolism 42–43, 87–88, 128, 129
Metal (pi)-Complexes (Fischer, Werner) 67
Meteorological Essays (Daniell) 53
Meteorological Observations and Essays (Dalton) 51
"Methods for Investigation of Ionic Reactions in Aqueous Solutions..." (Eigen) 60
Methods of Chemical Nomenclature (Lavoisier) 124
methotrexate 162
Metropolitan State University 116, 117
Miami, University of 162
Michigan Technological University 33
microbiology 165–166
microchemistry 174–175
microscope, ultra 242, 243
Midwest Rubber Reclaiming Company 125
Milan Polytechnic Institute 149
Minneapolis Award 117
Minnesota, University of 33
Mittasch, Alwin 23
Miyake, Yasuo 197–198
Modern Thermodynamics (Prigogine, Kondepudi) 179
Moissan, Henri **142–143**
The Molecular Basis of Evolution (Anfinsen) 4
molecular beam apparatus 126–127
molecular biology 45–47, 76–78
molecular chemistry 19–20
"Molecular Scientists and Molecular Science" (Mulliken) 148

Molecular Spectra and Molecular Structure (Hertzberg) 92
molecular spectroscopy 91–92
molecular three-dimensionality 14–15, 88–89
Molina, Mario **143–145**, 189
Montana, University of 223
Montecatini Company 150
Moore, Stanford 4, **145–146**, 210, 211
morphine 2, 15, 187, 229
Moscow State University 201
Mount Holyoke College 36–37, 239
Muller-Hill, Benno 77
Mulliken, Robert S. 73, **146–148**, *147*
Munich, University of 14, 31, 56, 65, 67, 68, 85, 118–119, 209, 230, 232, 233, 242
Muñoz, J. M. 128
Munson, Paul 109
Murphy, George M. 223–224
Musée d'Histoire Naturelle 16, 142
My 132 Semesters of Chemistry Studies (Prelog) 176
myoglobin 114, 115, 170, 171

N

Nacional Autónoma de México 144
Nancy Chemical Institute 82
NASA Medal for Exceptional Scientific Advancement 145
National Academy (Italy) 114
National Academy of Sciences (U.S.) 17, 18, 35, 60, 62, 70, 74, 110, 114, 138, 146, 163, 173, 189, 204, 207
National Advisory Committee on Aeronautics (U.S.) 201, 202
National Aeronautics and Space Administration (NASA) 202
National Association for the Advancement of Colored People (NAACP) 38, 110, 111
National Bureau of Standards (U.S.) 115–116
National Center for Atmospheric Research (U.S.) 47–48, 207
National Commission on Product Safety (U.S.) 95
National Committee for Scientific Research (France) 106

National Defense Research Council (U.S.) 33
National Institute for Medical Research (U.K.) 137, 159
National Institutes of Health 129, 163
    guidelines governing rDNA research 18
    Human Genome Project Scientific Advisory Committee 18
    National Cancer Institute 204
    National Heart Institute 4
    National Institute for Arthritis, Metabolic and Digestive Diseases 4, 203
National Inventor's Hall of Fame 62
National Medal of Science (U.S.) 18, 34, 62, 70, 127, 163, 200, 204, 238
National Oceanic and Atmospheric Administration (U.S.) 48, 207
National Research Council (Argentina) 129–130
National Research Council (Canada) 92
National Research Council (U.S.) 111, 147, 208
National Science Foundation (U.S.) 43, 79, 138, 144, 163 *See also* National Medal of Science (U.S.)
National Taiwan University 126
National Tsinghua University 126
National Women's Hall of Fame (U.S.) 62
Natta, Giulio **149–150**
"The Nature of the Chemical Bond" (Pauling) 167–168
The Nature of the Chemical Bond, and the Structure of Molecules and Crystals (Pauling) 168
Nebraska, University of 130
neoprene 35
neptunium 139, 140
Nernst, Walther Hermann 84, **150–151**
Neuberg Medal in Biochemistry 160
The New Heat Theorem (Nernst) 151
New Ideas in Inorganic Chemistry (Werner) 229
Newlands, John Alexander Reina **151–153**

"A New Method for Determination of the Mobility of Proteins" (Tiselius) 218–219
New South Wales University of Technology 157
New System of Chemical Philosophy (Dalton) 52
A New View of the Origin of Dalton's Atomic Theory (Dalton, Harden) 87
New York University 61, 147, 160, 162, 218
Nichols, J. Burton 213–214
Nichols Medal 12, 57, 123, 209
nitrogen fixation 225–226
nitroglycerine 153–154
nitrous oxide 54
Nobel, Alfred **153–154**
Nobel Foundation 154, 219
Nobel Peace Prize 166–168
Nobel Prize in chemistry 1–2, 4–5, 5–7, 7–9, 13–14, 14–15, 17–18, 19–20, 23–24, 31, 33–34, 43–45, 47–48, 55–57, 57–58, 59–60, 65–66, 66–68, 68–69, 69–70, 73–74, 76–78, 81, 85–87, 87–88, 88–89, 89–91, 91–92, 92–94, 95–97, 97–98, 105–108, 113–114, 114–115, 117, 118–119, 121–123, 126–127, 127–128, 128–130, 132–133, 136–137, 139–140, 142–143, 143–145, 145–146, 146–148, 149–150, 150–151, 154–155, 155–157, 160–162, 163–164, 166–168, 170–171, 172–174, 174–175, 175–176, 177–179, 182–183, 185–186, 186–188, 188–189, 189–191, 192, 195–196, 196–197, 198–199, 200–201, 204–206, 207–209, 209–210, 211–213, 213–214, 214–215, 218–219, 219–221, 223–224, 225–226, 227–228, 228–229, 229–230, 230–232, 233–234, 234–235, 237–238, 241–242, 242–243
Nobel Prize in medicine or physiology 43, 45–47, 60–62, 159–160
Nobel Prize in physics 15–17, 48–50
Norges Tekniski Høgskole 161
Norris, James Flack 95

Norrish, Ronald G. W. **154–155,** 172–173
North Atlantic Research Corporation 95
North Carolina, University of, at Chapel Hill 108, 109–110
North Carolina Central University 138
Northrup, John Howard **155–157,** 212
Norwegian Chemical Society Guldberg-Waage Medal 89
Notable Service Medal (U.S.) 169
Notes on the Mathematical Theory of Electrodynamic Phenomena (Ampère) 3
nuclear accelerators 139, 140
nuclear disintegration theory 204–205
nuclear fission 85, 86
nuclear physics 189–191
nuclear power 105, 108
nuclear testing 132, 133, 166, 168, 197, 198, 199–200
nuclear winter 48
nucleotides 159, 160, 197, 221
Nuffield Foundation 157
nutrition 98–100 *See also* food chemistry; vitamins
Nutrition and Chemical Growth in Childhood (Hoobler) 100
Nyholm, Ronald Sydney **157–158**
nylon 34, 36

O

Ochoa, Severo **159–160,** *160*
Oesper Prize 34
Of Molecules and Men (Crick) 46
Ohio State University 36, 69–70
Ohio Wesleyan University 188
"On Depletion of Antarctic Ozone" (Garcia, Rowland, Solomon, Wuebbles) 189, 207
Onsager, Lars **160–162**
"On the Law of the Diffusion of Gases" (Graham) 80
"On the Photochemistry of Ozone in the Stratosphere and Troposphere..." (Crutzen) 47
optically active isomer complexes 228–229

orbital symmetry, conservation of 238
orbital theory of reactivity 73–74
Order of Culture (Japan) 74
Order of the Czechoslovak Republic 94
Order of Lenin 201
Order of Merit (U.K.) 98, 118, 191, 197, 221, 230
Order of the Red Banner of Labor (USSR) 201
Ordre National du Mérite (France) 128
Oregon State University 167
Organic Chemistry (Mendeleyev) 141
organic synthesis reactions 234
Ørsted Medal 25
Osborn, Mary Jane **162–163**
Oslo, University of 88, 89, 97
Ostwald, Wilhelm 6, 23, **163–164**
Otto Hahn Award 230, 235
Outline of General Chemistry (Ostwald) 164
Outline of Technology Chemistry on a Theoretical Basis (Haber) 84
Outlines of Physical Chemistry (Brønsted) 26
Oxford University 98, 148, 159, 187, 205–206, 220, 224, 238–239
    Balliol College 96, 238
    Exhibition Scholarships 44
    Merton College 172, 204
    St. Cross College 47
    St. John's College 115
    Somerville College 97–98
    Trinity College 96
oxidation, biological 230
oxygen, discovery of 177
ozone layer 47–48, 143–144, 188, 189, 206–207

P

Pan-American Association of Biochemical Sciences 130
Paracelsus Prize 128
Paris, University of (Sorbonne) 3, 49, 50, 72, 75, 94, 108, 142, 143, 181–182, 238
partition chromatography 136–137, 214, 215
Pasteur, Louis 72, **165–166**
pasteurization 165, 166

Pasteur Medal 234
Pauling, Linus 46, **166–168,** *167*
Pavia, University of 149
peanuts 37, 38
Pedagogical Institute 141
Pedersen, Charles John 127–128
penicillin 44, 97, 98
Pennington, Mary Engle **168–170**
Pennsylvania, University of 4, 119, 168, 177, 185
Pennsylvania State College 17, 125
pepsin 156
perfume industry 192, 227–228, 241
periodic table of the elements 140, 141–142, 152, 153, 183, 185, 186, 199
Perkin, William H., Jr. 187
Personal Knowledge (Polanyi) 172
Perutz, Max Ferdinand 114, 115, **170–171**
Peter Debye Award 127, 148
Pew Charitable Trusts Scholars Program 145
phase stability 139, 140
Philadelphia health department 169
Phillips Academy 130
photochemistry 155
photographic film developing 11
photosynthesis 33–34, 232, 233
pigments 68–69, 118, 119, 136, 225, 232–233, 237
Pisiello, George 126
The Planets: Their Origin and Development (Urey) 224
plant pigments 118, 119, 136, 225, 232–233, 237
plutonium 139, 140
Plutonium Project 148
Plymouth Marine Biological Laboratory 159
pneumatic transfer tube system 40–41
Polanyi, Michael 126, **171–172**
polarography 92, 93–94
polyethylene 149, 150
polymers 34, 35, 69–70, 94–95, 149, 150, 209–210, 241–242
polypropylene 149, 150
Polytechnic Institute (Copenhagen) 26

Polytechnic University (Brooklyn Polytechnic Institute) 61
Popják, George 44
Porter, George 154, 155, **172–174**
Practical Organic Chemistry (Harden, Garett) 87
Prague, University of 42
Pregl, Fritz **174–175**
Prelog, Vladimir 104, **175–176**
Priestley, Joseph **176–177**
Priestley Award/Medal 15, 34, 57, 70, 79
Prigogine, Ilya **177–179**, *178*
Princeton University 132, 139, 188, 219
Principles of Chemistry (Mendeleyev) 141
Principles of Polymer Chemistry (Flory) 70
"The Principle of the Specific Interaction of Ions" (Brønsted) 26
probability 2
progesterone 30
Project Sunshine 133
proteins 4, 65, 66, 114, 115, 145, 146, 196–197, 218–219, 238, 239
See also amino acids
Proust, Joseph-Louis **179–180**
purine 65, 66
pyrrole chemistry 68–69

Q

Quantitative Laws of Biological Chemistry (Arrenius) 6
quantum mechanics 167–168
Queens University 37

R

rabies 165, 166
radicals 21, 91, 92, 155, 241
radioactive decay 131–132, 190, 204–205
radioactive fallout 133, 197–198, 202
radioactivity 15, 17, 48–50, 105–108, 132, 189–191
Radio-Activity (Rutherford) 191
radiocarbon dating 132–133
Radiocarbon Dating (Libby) 133
radiochemistry 85–87
radioisotopes 86

radium 48, 50
Radium Institute 105, 106, 107–108
Ramart-Lucas, Pauline **181–182**
Ramsay, William **182–183**
Randall, Merle 130
rDNA (recombinant DNA) 17–18, 202, 203–204
reaction kinetics *See* kinetics
reaction mechanism theory 103–104
reaction rates, measuring 164
reagents, Grignard 81–82
Real Seminario Patriótico Vascongado 179
"Recherches sur le bleu de Prusse" (Proust) 180
reciprocal relations, Onsager's law of 161
"Reclaiming Agents for Rubber: Solvent Naphtha I" (Le Beau) 125
"Reclaiming of Elastomers" (Le Beau) 125
recombinant DNA (rDNA) 17–18, 202, 203–204
Reflexions sur le phlogistique (Lavoisier) 124
refrigeration 168–169
relaxation technique 59–60
"The Renaissance of Inorganic Chemistry" (Nyholm) 157
"Report on the Mode of Detecting Vegetable Substances…" (Graham) 80
Research Corporation Scientific Award 140
Research Institute of Molecular Pathology 171
resistance, biological 119
reversible processes 160, 161
ribonuclease 4, 146, 210, 211
Richards, Ellen Swallow **183–185**, *184*
Richards, Theodore William **185–186**
Riga Polytechnic University 164
Riverside Research Laboratory 95
RNA (ribonucleic acid) 159, 160, 197, 202, 203–204
Robertson Prize 173
Robinson, Robert **186–188**
Roche Institute for Molecular Biology 18, 160, 163

Rochester, University of (New York) 68
Rockefeller Foundation 4, 30, 88, 94, 98, 100–101, 129, 170, 192–193, 202, 239
Rockefeller University (Rockefeller Institute for Medical Research) 46, 145–146, 156, 171, 208, 211
Roger Adams Award 235
Rome, University of 149
Rosalind Franklin and DNA (Sayre) 72
Roscoe, Henry 29
Roswell Park Memorial Institute 42
Rowett Research Institute 215
Rowland, F. Sherwood 143–144, **188–189**
Royal Academy 74
Royal Academy of Sciences (Belgium) 114
Royal Academy of Sciences (Berlin) Helmholtz Medal 17
Royal Agricultural Society (U.K.) 152
Royal Artillery College (Spain) 179
Royal Australian Chemical Institute 158
Royal College of Chemistry (U.K.) 152
Royal Collège of Franche-Comté 165
Royal College of Science and Technology (Scotland) 80, 182
Royal Institution (London) 6, 54, 63–64, 173
Royal Medal 8, 45, 52, 80, 81, 91, 92, 97, 171, 187
Royal Military Academy (U.K.) 64
Royal Netherlands Academy of Science 45, 57, 114
Royal Norwegian Academy of Sciences 89
Royal Order of the Crown 228
Royal Society (London) 40, 45, 52, 53, 80, 88, 91, 96, 115, 118, 143, 171, 173, 177, 183, 191, 197, 221, 242
Albert Medal 29
Copley Medal 29, 45, 54, 55, 81, 97, 142, 171, 187
Davy Medal 7, 29, 34, 45, 50, 69, 88, 91, 97, 142, 143, 153, 155, 173, 183, 187, 196, 224, 228, 238

Royal Society (London) *(continued)*
  Faraday Medal 187
  Hughes Medal 8
  Longstaff Medal 187
  presidents 55, 96, 118, 187
  Royal Medal 8, 45, 52, 80, 81,
    91, 92, 97, 171, 187
  Rumford Medal 17, 54, 55, 57,
    123, 173, 179, 191
  Wolfson Research Professor 98
Royal Society of Arts (London) 38
Royal Society of Chemistry (U.K.)
  104, 197
Royal Society of Edinburgh 80
Royal Society of New South Wales
  158
Royal Swedish Academy of Sciences 6,
  7, 21, 179 *See also* Nobel Prizes
Royal Swedish Scientific Society 219
rubber 2, 34, 35, 124–125
Rumford Medal 17, 54, 55, 57, 123,
  173, 179, 191
Rutherford, Ernest **189–191**, *191,*
  204–205
Ružička, Leopold 31, **191–193**

S

Sabatier, Paul **195–196**
St. Andrews, University of 90, 187
St. Petersburg, University of 141, 200
Salk Institute 18, 46
sandwich compounds 66–68,
  230–231, 237
Sanger, Frederick 76, 78, **196–197**
Saruhashi, Katsuko **197–198**
Saruhashi Prize 198
Saskatchewan, University of 92
Sayre, Anne 72
Schott und Genossen 243
Science Advisory Council (Sweden)
  219
Science Council of Japan 198
Science Research Council (U.K.) 157
Scripps Institution of Oceanography
  48
Seaborg, Glenn T. 139, 140,
  198–200
Self Organization in Non-Equilibrium
  Systems (Nicolis, Prigogine) 179
Semenov, Nikolai **200–201**

Senderens, Abbé Jean-Baptiste 196
sex hormones 30–31, 186, 187, 192
Sheffield University 173
Shell Research, Ltd. 44
Shikata, Masuzo 94
Siedentopf, H. F. W. 243
Simon, Dorothy Martin **201–202**
Simpson College 38
Singer, Maxine 18–19, **202–204**, *203*
6-mercaptopurine (6MP) 62
Smith College 239
Société de Chimie Biologique 160
Société Chimique de France 97, 210,
  229, 238, 242
Society for Biochemical Research
  (Argentina) 130
Society of the Chemical Industry 187
Society of Sea Water Sciences 198
Society of Women Engineers 125,
  201, 202
Socony Vacuum Oil Company 22
Soddy, Frederick 190–191, **204–206**,
  205
solar energy 217–218
Solomon, Susan 48, 189, *206,*
  206–207
"Solutions to the Mathieu equation of
  period 4[pi] and certain related
  functions" (Onsager) 161
Some Problems of Chemical Kinetics and
  Reactivity (Semenov) 201
Sommerfeld, Arnold 56
Sorbonne *See* Paris, University of
Southampton, University of (Hartley
  University College) 103
South Dakota, University of 35
Southern California, University of 109
Southwest Missouri State College 201
Space Studies Board 163
Spector, Leonard 108, 109
spectroscopy 8, 28–29, 36–37, 91–92
Squibb Award 43
Stalin Prize 201
Standard Oil Development Company
  22, 70
Stanford University 18, 70, 168
Stanley, Wendell Meredith **207–209**
Stas, Jean-Servais 186
Statistical Mechanics of Chain Molecules
  (Flory) 70

Staudinger, Hermann 192, **209–210**
Stein, William H. 4, 145–146,
  210–211
stereochemistry 43, 44, 165, 175, 176
steroids 14–15, 30–31, 44, 186, 187,
  192, 237
sterols 233–234
Stevens Institute of Technology 121
Stockholm University 47
Strasbourg, University of 13, 65, 127,
  166
Strassman, Fritz 86–87
Strathclyde, University of 81
"A Structure of Deoxyribonucleic Acid"
  (Crick, Watson) 46
"The Structure of Iron
  Biscyclopentadienyl" (Wilkinson)
  231
Structure and Mechanism in Organic
  Chemistry (Ingold) 104
Structure of Physical Chemistry
  (Hinshelwood) 96
submarine detection 122
sugars 65, 66, 128, 129 *See also*
  carbohydrates
Sumner, James Batcheller **211–213**
supramolecular chemistry 127–128
surface chemistry 121, 122
"Sur les oxidations métalliques"
  (Proust) 180
Sussex, University of 44, 137
Svante Arrhenius Golden Medal 179
Svedberg, Theodor **213–214**
Swarthmore College 4, 138, 203
Swedish Medical Society 137
Swedish Natural Science Research
  Council 219
Sydney, University of 44, 157, 187
Sydney Technical College 157
Synge, Richard 136, **214–215**

T

Tamm, Konrad 59
Tarkio College 34
Tartu, University of (Dorpat
  University) 163–164
Technical Institute (St. Petersburg)
  141
Technical University (Stockholm) 6
Technische Hochschule (Aachen) 56

Technische Hochschule (Berlin) 23
Technische Hochschule
    (Braunschweig) 235
Technische Hochschule (Danzig) 30
Technische Hochschule (Graz) 243
Technische Hochschule (Hanover) 19
Technische Hochschule (Karlsruhe)
    171, 192, 209, 228
Technische Hochschule (Stuttgart) 101
Technische Hochschule (Vienna) 242
Technische Hochschule (Zurich) 100,
    232
Technische Universität (Darmstadt)
    91–92
Technische Universität (Munich)
    (Technische Hochschule, Technical
    University) 67, 69, 230
telegraph industry 40, 53
Telkes, Maria **217–218**
terpenes 90, 227–228
Terpene und Campher (Wallach) 228
Texas, University of, at Austin 179
Texas A&M University 15
Textbook of Chemistry (Berzelius) 21
Text-Book of Electrochemistry
    (Arrhenius) 6
Textbook of General Chemistry
    (Ostwald) 164
Textbook of Organic Chemistry (Karrer)
    114
Textbook of Stereochemistry (Werner)
    229
Thalén, Tobias 6
Theodore William Richards Medal
    211
Theorellin, Hugo 4
*Theoretical Chemistry from the
    Standpoint of Avogadro's Rule and
    Thermodynamics* (Nernst) 151
theoretical chemists 73–74, 177–179
Theories of Chemistry (Arrhenius) 6
Theories of Solutions (Arrhenius) 6
thermochemistry 150, 151
thermodynamics 26, 131, 150, 151,
    171, 172, 177–178
*Thermodynamics and the Free Energy of
    Chemical Substances* (Lewis, Randall)
    130
*The Thermodynamics of Technical Gas
    Reactions* (Haber) 84

Thiele, Johannes 209
Thomson, Joseph John 8
"Three Papers Containing Experiments
    on Factitious Airs" (Cavendish) 39
A Time to Remember (Todd) 221
Tiselius, Arne Wilhelm Kaurin
    218–219
tobacco mosaic virus 117, 207–209
Todd, Alexander **219–221**
Toho University 197
Tokyo, University of 198
Tomsk, University of 200
Tougaloo College 25
Toulouse University 196
Traité élémentaire de chimie (Lavoisier)
    123, 124
Transition Metal Carbene Complexes
    (Fischer, Dotz et al.) 68
Treatise on inorganic chemistry
    (Moissan) 143
Trinity College 8, 96, 114, 118, 190,
    214
trypsin 156
Tscherning, Kurt 30
Tübingen, University of 28, 182, 234,
    235
Turin, University of 10
Turin Polytechnic Institute 149
Tuskegee University (Institute) 24, 25,
    38

U

Über das Kolloide Gold (Zsigmondy)
    243
ultracentrifuge 213–214
ultramicroscope 242, 243
Union Carbide and Carbon Company
    12
United Nations Environmental
    Programme Global 500 Award 145
Université Descartes 179
Université Libre de Bruxelles 178
University College London 45, 80,
    85, 93, 103, 104, 157, 182–183,
    238
University College of Wales 204
Uppsala, University of 5–6, 21, 213,
    214, 215, 218, 219
urea synthesis 236
Urey, Harold **223–224**, *224*

U.S. Department of Agriculture 169
U.S. Department of Commerce 79,
    207
U.S. Department of Energy 127, 139
U.S. Department of the Interior 147
U.S. National Committee for
    Biochemistry 211
U.S. Naval Academy 137, 138
U.S. Navy Radio and Sound
    Laboratory 140
U.S. Office of Scientific Research and
    Development 4, 145, 156
Utrecht, University of 48, 56, 192

V

vacuum tubes 121
*Valence and the Structure of Atoms and
    Molecules* (Lewis) 130–131
Vanderbilt University 145, 146
Vassar College 184
Veksler, V. I. 139, 140
Velox 11
Vercelli, College of 9
Vermeil Medal of Paris 128
Vienna, University of 42, 118, 125,
    170, 238, 242
Vienna Academy of Sciences 175
Virtanen, Artturi Ilmari **225–226**
viruses 117, 207–209
Viruses and the Nature of Life (Stanley,
    Valens) 209
*The Viruses: Biochemical, Biological and
    Biophysical Properties* (ed. Burnet,
    Stanley) 209
vitalist theory 235, 236
vitamins 89–90, 97, 98, 113–114,
    119, 136, 168, 220, 221, 233–234,
    238 *See also* nutrition
voltaic cells 41, 52, 53
von Humboldt Prize 128

W

Wallach, Otto **227–228**
Warrington Academy 176
Warwick, University of 44
Washburn, E. W. 224
Washington, University of 162
Washington University (St. Louis) 18,
    43, 129, 160
Watson, James 46, 71, 72, 77

Weizmann Institute of Science 5, 115, 203

Wellcome Research Laboratories 61–62, 137

Werner, Alfred **228–229**

Werner, Gustaf 214

Werner Medal 176

Western Pennsylvania Hospital 99

Western Reserve University 17

Westford Academy 184

Westinghouse Electric 217

Westphal, Ulrich 30

What Mad Pursuit (Crick) 46

Wheeler, Schuyler Skaats 41

White House Initiative Lifetime Achievement Award 138

Wieland, Heinrich Otto **229–230**

Wilkins, Maurice 46, 71

Wilkinson, Geoffrey 67, 68, **230–232**

Willard Gibbs Medal 7, 57, 92, 148, 209, 224

Willstätter, Richard Martin 212, **232–233**

Windaus, Adolf Otto Reinhold **233–234**

Wisconsin, University of 68, 145, 213

Wittig, Georg **234–235**

Witwatersrand, University of 117

Wohl, Arthur C. 199

Wöhler, Friedrich **235–237**

Women's Medical College of Pennsylvania 169

Women's National Press Award 43

Woodward, Robert Burns 73–74, 231, **237–238**

Woodward-Hoffman rules 238

Woodward Institute 176

Woodward Research Institute 237

Wool Industries Research Association 136, 214–215

Wooster, Nora 136

Worlds in the Making (Arrhenius) 7

Wrinch, Dorothy Maud **238–239**, *239*

Würzburg, University of 28, 66, 151

X

X-ray crystallography 70–72, 88–89, 97–98, 117, 118, 167

X-ray diffraction 56

X-ray machines (diagnostic) 50, 107

X rays 16–17

Y

Yale University 6, 26, 99, 109, 110, 161, 168, 203

yeast *See* fermentation

Yerkes Observatory 92

ylides 235

Young, William 87

Z

Zagreb, University of 175

Zeitschrift für Physikalische Chemie 164

Ziegler, Karl 149, 150, **241–242**

Zonta International 79

Zoological Station (Italy) 111

Zosimos 135

Zsigmondy, Richard Adolf **242–243**

Zurich, University of 56, 113, 114, 219, 228, 229

Zurich Polytechnic 228

The Zymase: Fermentation (Buchner, Buchner, Hahn) 28